Partial Difference Equations

Advances in Discrete Mathematics and Applications

A series edited by ***Saber Elaydi***, *Trinity University, San Antonio, Texas, USA*, and ***Gerry Ladas***, *University of Rhode Island, Kingston, USA*

Volume 1
Analysis and Modelling of Discrete Dynamical Systems
Edited by Daniel Benest and Claude Froeschlé

Volume 2
Stability and Stable Oscillations in Discrete Time Systems
Aristide Halanay and Valdimir Răsvan

Volume 3
Partial Difference Equations
Sui Sun Cheng

This book is part of a series. The publisher will accept continuation orders which may be cancelled at any time and which provide for automatic billing and shipping of each title in the series upon publication. Please write for details.

Partial Difference Equations

Sui Sun Cheng
Tsing Hua University,
Republic of China

CRC Press is an imprint of the
Taylor & Francis Group, an **informa** business

CRC Press
Taylor & Francis Group
6000 Broken Sound Parkway NW, Suite 300
Boca Raton, FL 33487-2742

First issued in paperback 2019

ISBN-13: 978-0-415-29884-1 (hbk)
ISBN-13: 978-0-367-39547-6 (pbk)

Publisher's note:
This book was prepared from camera-ready-copy supplied by the authors

British Library Cataloguing in Publication Data
A catalogue record for this book is available from the British Library

Library of Congress Cataloging in Publication Data
A catalog record for this title has been requested

Visit the Taylor & Francis Web site at
http://www.taylorandfrancis.com

and the CRC Press Web site at
http://www.crcpress.com

Contents

Series Editors' Preface

This is the third volume in the book series *Advances in Discrete Mathematics and Applications*. The series will be a forum for all aspects of discrete mathematics and will act as a unifying force in the field, presenting books in areas such as numerical analysis, discrete dynamical systems, chaos theory, fractals, game theory, stability, control theory, complex dynamics, computational linear algebra, boundary value problems, oscillation theory, asymptotic theory, orthogonal polynomials, special functions, conbinatorics and functional equations. Volumes on applications of difference equations in science and engineering will also be considered for publication.

Advances in Discrete Mathematics and Applications will publish textbooks for both the upper undergraduate and graduate levels. In addition, it will publish advanced works at the research level.

We hope to meet the growing needs of the mathematical community for books in discrete mathematics.

Saber Elaydi
Gerry Ladas

Preface

Functional relations involving functions with two or more discrete variables appeared long before the introduction of analytical concepts such as a derivative or an integral. Yet renewed interest has only picked up momentum in the last fifty years. Such an active interest is undoubtedly due to the advance of modern digital computing devices. Indeed, since these relations can be simulated in a relatively easy manner by means of these devices and since such simulations often reveal important information about the behavior of complex systems, a large number of recent investigations related to image processing, population models, neural networks, social behaviors, digital control systems, etc. are described in terms of such functional relations.

In this book, we will treat a major class of such functional relations called partial difference equations. Roughly, they are functional relations with recursive structures so that the usual concepts of increments are important. Basic tools that can be used to deal with the qualitative behavior of such equations will be given. The main focus of this book, however, is to show by introductory examples how these tools are actually used in deriving qualitative results. Thus this book will be useful to anyone who has mastered the usual sophomore mathematical concepts and who wants a concise introduction to the tools and techniques that have been proven successful in obtaining results in partial difference equations.

A synopsis of the contents of the various chapters follows.

- The book begins with introductory examples of partial difference equations. Most of these examples stem from the familiar concepts of heat diffusion, heat control, temperature distribution, population growth and gambling, etc. They are elementary, but illustrative of more involved problems.

- In Chapter 2 basic definitions are introduced. Classifications of partial difference equations are also discussed. Since solutions of partial difference equations are sequences, basic information related to various types of sequences is also given.

- An important issue in partial difference equations is finding solutions. For this purpose, we introduce in Chapter 3 the concepts of operators and their manipulations. Roughly, operators are formal power series or generating functions. Although other methods exist such as the Z-transform, that are useful in finding solutions, our approach is distinguished by its algebraic nature and no analytic conditions are needed.

- In Chapter 4 we will be concerned with univariate sequences or bivariate sequences that satisfy "monotone" or "convex" functional relations. Maximum principles and discrete Wirtinger's inequalities will be derived. These principles and inequalities will be useful in yielding various qualitative properties of partial difference equations.

- In Chapter 5 we employ the method of operators as well as a number of other methods for constructing the explicit solutions of a number of basic partial difference equations.

- There are many concepts that are related to the "sizes" of solutions of partial difference equations in terms of the various parameters as well as auxiliary conditions. In Chapter 6 a number of methods that are useful for deriving stability criteria are introduced.
- A great variety of solutions of partial differential equations is possible and observable in simulations. In case general solutions cannot be explicitly constructed, we have to rely on different methods for proving the existence of solutions that exhibit specific behaviors. In Chapter 7 we introduce a number of techniques for obtaining existence criteria for partial difference equations.
- Besides the sufficient conditions for the existence of solutions with desired qualitative properties, it is also of interest to know when these solutions cease to exist. In the final chapter, we derive a number of conditions mainly for the nonexistence of solutions that are "eventually positive". In case our partial difference equations are linear, such nonexistence criteria become oscillation theorems.

Most of the material in this book is based on research work that has been carried out by myself and a number of friends and graduate students during the last twenty years. My special thanks go first to Professor Y. Z. Lin for his hard work and assistance. Next, thanks go to Professors B. G. Zhang, X. L. Xie, S. T. Liu, G. Zhang and W. T. Li.

Extra special thanks to my wife, Amelia, for her patience and support throughout my entire academic career. Finally, I would like to remark that without the indirect help of many other people, especially my mother and my father, this book would never have appeared. My last thanks go to them.

I have tried my best to eliminate any errors. If there are any that have escaped my attention, your comments will be much appreciated.

Chapter 1

Modelling

1.1 Introduction

A well known relation involving functions of two discrete variables is the following

$$C_m^{(n)} = C_{m-1}^{(n-1)} + C_m^{(n-1)}, \; 1 \le m < n, \tag{1.1}$$

and a solution to it is the celebrated binomial coefficient function $\left\{C_m^{(n)}\right\}$ defined by

$$C_m^{(n)} = \frac{n!}{m!(n-m)!}, \; 0 \le m \le n,$$

which appears in the early history of mathematics. The functional relation (1.1) is an example of a partial difference equation. Although partial difference equations such as (1.1) appear well before partial differential equations, the former equations have not drawn as much attention as their continuous counterparts. Renewed interest has, however, been picking up momentum during the last fifty years among mathematicians, physicists, engineers and computer scientists. Such active interest is undoubtedly due to the advance of modern computing devices, as well as the ease of modelling complex dynamical systems with such equations. Indeed, the unknown functions in our equations have discrete independent variables and thus computer simulations can be carried out in a relatively easy manner. Such simulations often reveal important quantitative as well as qualitative information which can help us understand the complex behavior of the systems represented by these equations.

In this book, introductory examples of partial difference equations will be presented. Most of these examples stem from the familiar concepts of heat diffusion, heat control, temperature distribution, population growth, and gambling, etc. They are elementary, but illustrative of more involved problems that are of interest in various branches of science. We will also present a number of techniques for obtaining quantitative as well as qualitative results for these equations. Since we are still at the early stage of development, our results are quite elementary compared with those that have been obtained for the continuous models. Sophisticated results, however, are expected in the near future.

There are some standard conventions and notations which we will adopt throughout the rest of this book. First of all, an empty sum will be taken to be zero, while an empty product will be taken as one. The symbols 0! and 0^0 will be taken as 1. The symbol $\lceil x \rceil$ stands for the least integer greater than or equal to the real number x, while $\lfloor x \rfloor$ is the greatest integral part of x. The set of real numbers will be denoted by R, the set of all complex numbers by C, the set of integers by Z, the set of positive integers by Z^+, and the set of nonnegative integers by N. A point (i, j) in the plane with integer coordinates is called a *lattice point* and the set of all lattice points is called the lattice plane and denoted by $Z \times Z$ or Z^2. In the sequel, $C_y^{(x)}$, where $x, y \in R$, will denote the extended binomial coefficient, i.e. $C_0^{(x)} = 1$ for $x \in R$,

$$C_m^{(x)} = \frac{x(x-1)...(x-m+1)}{m!}, \; m \in Z^+,$$

and $C_y^{(x)} = 0$ if $y \notin N$. The number of elements in a set Ω will be denoted by $|\Omega|$. The union of two sets A and B will be denoted by $A \cup B$ or $A + B$, while their intersection by $A \cap B$ or $A \cdot B$, and their difference by $A \backslash B$. A (univariate) sequence is a function defined over a set S of (usually) consecutive integers, and can be denoted by $\{u_k\}_{k \in S}$ or $\{u(k)\}_{k \in S}$. When S is finite and, say, equals $\{1, 2, ..., n\}$, a sequence is also denoted by $\{u_1, ..., u_n\}$. Bivariate or multivariate sequences are functions defined on subsets Ω of Z^2 or $Z^n = Z \times Z \times ... \times Z$ respectively. The set of all complex sequences defined on Ω will be denoted by l^{Ω}. There are many different ways to denote bivariate or double sequences. One way is to denote a bivariate sequence by $\{u_{i,j}\}$. However, we may also denote it by $\{u_{ij}\}$ if no confusion is caused. The other way is by $\{u_i^{(j)}\}$. In general, when the independent variables have different interpretations, the latter notation is employed. For instance, $u_i^{(t)}$ may represent the temperature of mass placed at the integral position i and in time period t. Finally, a real function (including a real sequence, a real matrix, etc.) f is said to be nonnegative if $f(x) \geq 0$ for each x in its domain of definition. In such a case, we write $f \geq 0$. Similarly, given two real functions with a common domain of definition Ω, we say that f is less than or equal to g if $f(x) \leq g(x)$ for each $x \in \Omega$. The corresponding notation is $f \leq g$. Other monotonicity concepts for real functions (such as $f < g$, $f > 0$, etc.) are similarly defined.

1.2 Examples

1.2.1 Discrete Heat Equations

As our first example of modelling realistic problems by partial difference equations, consider the temperature distribution of a "very long" rod. Assume that the rod is so long that it can be laid on top of the set Z of integers. Let $u_m^{(n)}$ be the temperature at the integral time n and integral position m of the rod. At time n, if the temperature $u_{m-1}^{(n)}$ is higher than $u_m^{(n)}$, heat will flow from the point $m - 1$ to m. The amount of increase is $u_m^{(n+1)} - u_m^{(n)}$, and it is reasonable to postulate that the increase is proportional to the difference $u_{m-1}^{(n)} - u_m^{(n)}$, say, $r\left(u_{m-1}^{(n)} - u_m^{(n)}\right)$ where r is a positive diffusion rate constant, that is

$$u_m^{(n+1)} - u_m^{(n)} = r\left(u_{m-1}^{(n)} - u_m^{(n)}\right), \; r > 0.$$

Similarly, heat will flow from the point $m+1$ to m if $u_{m+1}^{(n)} > u_m^{(n)}$. Thus, it is reasonable that the total effect is

$$u_m^{(n+1)} - u_m^{(n)} = r\left(u_{m-1}^{(n)} - u_m^{(n)}\right) + r\left(u_{m+1}^{(n)} - u_m^{(n)}\right), \quad m \in Z, n \in N. \tag{1.2}$$

Such a postulate can be regarded as a discrete Newton law of cooling.

We have assumed that the rod can be laid on top of the set of integers. Thus the domain of definition of equation (1.2) is naturally taken to be $Z \times N$. In case we assume that the rod is "semi-infinite" or finite, we are then defining the equation (1.2) over the domains

$$\{(m,n) |\ m \in Z^+;\ n \in N\},$$

and

$$\{(m,n) |\ m = 1,2,...,M;\ n \in N\}$$

respectively.

In the above example, we have employed the familiar concept of heat conservation for derivation of partial difference equations. More general equations can be derived by means of similar conservation principles. Recall that a conservation principle states that the change in the total amount of material density within some fixed sub-region in a diffusion medium is equal to the rate of flux of that material across the boundary of the region plus the rate of generation within this region. By means of these principles, we can build up various generalizations and variations of the above equation. For instance, if the rod is made of nonhomogeneous material, the corresponding equation may look like

$$u_m^{(n+1)} = au_{m-1}^{(n)} + bu_m^{(n)} + cu_{m+1}^{(n)}, \tag{1.3}$$

where $n \in N$ and $m \in Z$. Note that when the domain of definition is ignored, this equation includes (1.1) as a special case!

If a reaction term is introduced, equation (1.1) may be generalized to yield the following nonlinear reaction diffusion equation

$$u_m^{(n+1)} = u_m^{(n)} + f\left(u_m^{(n)}\right) + r\left(u_{m-1}^{(n)} - 2u_m^{(n)} + u_{m+1}^{(n)}\right), \tag{1.4}$$

or to

$$u_m^{(n+1)} = (1-\varepsilon)F\left(u_m^{(n)}\right) + \frac{\varepsilon}{2}\left[F\left(u_{m+1}^{(n)}\right) + F\left(u_{m-1}^{(n)}\right)\right],$$

where $n \in N$ and $m \in Z$, and f, F are real functions defined on R. When F is the *logistic map* $F(x) = 1 - ax^2$, computer simulations have been done (see e.g. Crutchfeld and Kaneko [53], Kaneko [96]) for various values of the parameter a and the "coupling constant" ε. Another function f which is popular in the theory of cellular neural networks is

$$f(u) = \frac{1}{2}\left(|u+\varepsilon| - |u-\varepsilon|\right), \ \varepsilon > 0.$$

The same postulate that works for the distribution of heat through a rod also works for that of a very large thin plate. Assume the plate is so large that it can be laid on top of the set Z^2 of lattice points. Let $u_{ij}^{(n)}$ be the temperature of the plate at position (i,j) and integral time n, then a corresponding heat equation may be given by

$$\begin{aligned} u_{ij}^{(n+1)} - u_{ij}^{(n)} &= r\left(u_{i-1,j}^{(n)} - 2u_{ij}^{(n)} + u_{i+1,j}^{(n)}\right) + r\left(u_{i,j-1}^{(n)} - 2u_{ij}^{(n)} + u_{i,j+1}^{(n)}\right) \\ &= r\left(u_{i-1,j}^{(n)} + u_{i+1,j}^{(n)} + u_{i,j-1}^{(n)} + u_{i,j+1}^{(n)} - 4u_{ij}^{(n)}\right), \end{aligned}$$

where $(i,j) \in Z^2$ and $n \in N$. For a thin plate made of composite materials, it is then reasonable to discuss an equation of the form

$$u_{ij}^{(n+1)} - u_{ij}^{(n)} = \alpha_{ij}u_{i-1,j}^{(n)} + \beta_{ij}u_{i+1,j}^{(n)} + \gamma_{ij}u_{i,j-1}^{(n)} + \delta_{ij}u_{i,j+1}^{(n)} - \sigma_{ij}u_{ij}^{(n)}. \quad (1.5)$$

1.2.2 Two-Level Equations

The features that are common among the above examples are as follows. We are given a discrete graph of interconnected sites or states, with each site having a value in the set of real numbers. These values evolve in discrete time steps according to a definite rule. The rule specifies the new value of a particular site in terms of its own old value, and the old values of some neighboring sites. The neighbors are typically taken to include sites within a finite distance from the one that is in concern. Partial difference equations reflecting such features are sometimes called (discrete time) *lattice dynamical systems*, or *coupled map lattices* (see e.g. Crutchfeld and Kaneko [53], Kaneko [96], Chow and Mallet-Paret [45], Chow and Shen [46]), or (discrete time) *cellular neural networks* (see e.g. the article by Harrer and Nossek in [83]).

These two-time-level equations may arise in many other ways. We give several examples.

Example 1 *By observing the first few algebraic equalities* $(a+b)^1 = a+b$, $(a+b)^2 = a^2 + 2ab + b^2$, $(a+b)^3 = a^3 + 3a^2b + 3ab^2 + b^3$, *it is clear that*

$$(a+b)^{n-1} = C_0^{(n-1)}a^{n-1} + C_1^{(n-1)}a^{n-2}b + ... + C_{n-2}^{(n-1)}ab^{n-2} + C_{n-1}^{(n-1)}b^{n-1},$$

for $n \geq 2$, *where* $C_0^{(1)} = 1 = C_1^{(1)}$. *Furthermore, from* $(a+b)^n = (a+b)(a+b)^{n-1}$, *we see that*

$$\begin{aligned} &C_0^{(n)}a^n + C_1^{(n)}a^{n-1}b + ... + C_{n-1}^{(n)}ab^{n-1} + C_n^{(n)}b^n \\ &= C_0^{(n-1)}a^n + C_1^{(n-1)}a^{n-1}b + ... + C_{n-2}^{(n-1)}a^2b^{n-2} + C_{n-1}^{(n-1)}ab^{n-1} \\ &+ C_0^{(n-1)}a^{n-1}b + C_1^{(n-1)}a^{n-2}b^2 + ... + C_{n-2}^{(n-1)}ab^{n-1} + C_{n-1}^{(n-1)}b^n, \end{aligned}$$

thus

$$C_0^{(n-1)} = C_0^{(n)} = 1,$$

$$C_{n-1}^{(n-1)} = C_n^{(n)} = 1,$$

and

$$C_m^{(n)} = C_m^{(n-1)} + C_{m-1}^{(n-1)}, \ 1 \leq m < n.$$

Example 2 *Suppose initially, the probability of finding a particle at one of the integral coordinates i of the x-axis is $P_i^{(0)}$. At the end of each time interval, the particle makes a decision to stay at its present position or move one unit in the positive direction along the x-axis. Assume that the probability that the particle does not move in a given unit of time is p, and the probability that the particle moves in a given unit of time is q. Let $P_i^{(t)}$ be the probability that the particle is at the point $x = i$ at the end of the t-th interval of time. Then by Bayes' formula, it is easy to see that the following partial difference equation holds:*

$$P_i^{(t)} = pP_i^{(t-1)} + qP_{i-1}^{(t-1)}. \tag{1.6}$$

Note that this equation has the same form as the previous one for the binomial coefficients!

Example 3 *Two persons A and B play a game. At each time period, suppose the probability of A getting one point is $p > 0$ and for B is $q > 0$, where $p + q = 1$. Let $P_{ij}^{(n)}$ denote the probability at time period n that A has i points and B has j points. Then by Bayes' formula,*

$$P_{ij}^{(n+1)} = pP_{i-1,j}^{(n)} + qP_{i,j-1}^{(n)}. \tag{1.7}$$

Note that this equation is a special case of (1.5).

Example 4 *As another example, recall that the factorial polynomials $x^{[n]}$ are defined as $x^{[0]} = 1$, and*

$$x^{[n]} = x(x-1)(x-2)\cdots(x-n+1),\ n \in Z^+.$$

Since $x^{[1]} = x$, $x^{[2]} = x^2 - x$, $x^{[3]} = x^3 - 3x^2 + 2x$, etc., we see that

$$x^{[n]} = \sum_{k=1}^{n} s_k^{(n)} x^k,$$

where the coefficients $s_k^{(n)}$ are called the Stirling numbers of the first kind. As in Example 1, we may infer from the relation $x^{[n+1]} = x^{[n]}(x - n)$ the following functional relationship

$$s_k^{(n+1)} = s_{k-1}^{(n)} - ns_k^{(n)},\ 1 \le k \le n, \tag{1.8}$$

where we define $s_n^{(n)} = 1$ for $n \in Z^+$, and $s_k^{(0)} = 0$ for $k \in N$. Similarly, we have

$$x^n = \sum_{k=1}^{n} S_k^{(n)} x^{[k]},$$

where the coefficients $S_k^{(n)}$ are called Stirling numbers of the second kind, and the partial difference equation for these numbers is

$$S_k^{(n+1)} = S_{k-1}^{(n)} + kS_k^{(n)},\ 1 \le k \le n \tag{1.9}$$

where we take $S_n^{(n)} = 1$ for $n \in Z^+$, and $S_k^{(0)} = 0$ for $k \in N$.

We remark that the factorial polynomial $x^{[n]}$ is only defined for a nonnegative integer n. For negative integer n, we may, however, define

$$x^{[-n]} = \frac{1}{(x+1)(x+2)...(x+n)},\ n \in Z^+, x \neq -1, -2, ..., -n.$$

We may further define a generalized factorial function corresponding to any function $f(x)$ by $f^{[0]}(x) = 1$,

$$f^{[n]}(x) = f(x)f(x-1)...f(x-n+1),\ n \in Z^+,$$

and

$$f^{[-m]}(x) = \frac{1}{f(x+1)f(x+2)...f(x+m)},\ m \in Z^+,$$

provided that none of the values in the denominator is zero. Note that $\{u_{mn}\} = \{m^{[n]}\}$ is a doubly indexed sequence defined over $Z^2 \setminus \{(m,n) \in Z^2 | n \leq m \leq -1\}$.

Example 5 *Consider an initial value problem involving the heat equation:*

$$u_t = u_{xx},\ t > 0 \tag{1.10}$$

$$u(x,0) = f(x),\ -\infty < x < +\infty. \tag{1.11}$$

By means of standard finite difference methods (see e.g. Forsythe and Wasow [67], Godunov and Ryabenkii [72], Strikwerda [151], Mitchell and Griffiths [125], Zhou [176]), we set up a grid in the x,t plane with grid spacings Δx and Δt, then we replace the second derivative u_{xx} with a central difference, and replace u_t with a forward difference. By writing $x_m = m\Delta x$, $t_n = n\Delta t$, $f_m = f(x_m)$, then under the assumption that $u_m^{(n)} \approx u(x_m, t_n)$, a finite difference scheme for the heat equation is obtained:

$$\frac{u_m^{(n+1)} - u_m^{(n)}}{\Delta t} = \frac{1}{(\Delta x)^2}\left(u_{m-1}^{(n)} - 2u_m^{(n)} + u_{m+1}^{(n)}\right)$$

or

$$u_m^{(n+1)} = ru_{m-1}^{(n)} + (1-2r)u_m^{(n)} + ru_{m+1}^{(n)},\ m \in Z, n \in N, \tag{1.12}$$

where $r = \Delta t/(\Delta x)^2 > 0$, with the initial condition (1.11) replaced by

$$u_m^{(0)} = f_m.$$

Example 6 *Next, we consider an initial value problem involving a hyperbolic equation*

$$u_t + au_x = 0,\ t > 0, a > 0.$$

By means of the forward-time forward-space finite difference scheme [59], we obtain

$$u_m^{(n+1)} = (1 + \lambda a)u_m^{(n)} - a\lambda u_{m+1}^{(n)},\ m \in Z, n \in N \tag{1.13}$$

where $\lambda = \Delta t/\Delta x$. If we now consider the same equation over a different domain

$$u_t + au_x = 0,\ t > 0, x > 0, a > 0,$$

and applying the Wendroff implicit finite difference method [59], we are led to the following difference scheme

$$(1+sa)u_{m+1}^{(n+1)} + (1-sa)u_m^{(n+1)} = (1-sa)u_{m+1}^{(n)} + (1+sa)u_m^{(n)},\ m,n \in N.$$

Example 7 *It is well known that the coefficients, c_i, of a polynomial,*

$$f(x) = x^n + c_1 x^{n-1} + ... + c_{n-1}x + c_n,$$

may be determined by the elementary symmetric functions of its zeros, r_i, $1 \le i \le n$, i.e.

$$\begin{aligned} r_1 + r_2 + ... + r_n &= -c_1, \\ r_1r_2 + r_1r_3 + ... + r_{n-1}r_n &= c_2, \\ ... &= ... \\ r_1r_2...r_n &= (-1)^n c_n. \end{aligned}$$

Given a set of numbers $\Phi = \{r_i\}_{i=1}^n$, denote the sum of the products, taken j at a time, of the first k elements of the set by $Q_j^{(k)}$. We observe that a simple recursion formula can be established by induction,

$$Q_j^{(k)} = Q_j^{(k-1)} + r_k Q_{j-1}^{(k-1)}, \ 1 < j \le k,$$

where we define $Q_0^{(k)} = 1$ for $k > 0$, and $Q_j^{(k)} = 0$ for $k < j$. Verification for $k = 2$ is immediate, for we have

$$Q_1^{(2)} = r_1 + r_2 = Q_1^{(1)} + r_2 Q_0^{(1)},$$

and

$$Q_2^{(2)} = r_1 r_2 = 0 + r_1 r_2 = Q_2^{(1)} + r_2 Q_1^{(1)}.$$

We assume validity for k and deduce it for $k+1$. First note that

$$Q_1^{(k+1)} = (r_1 + ... + r_k) + r_{k+1} = Q_1^{(k)} + r_{k+1} Q_0^{(k)},$$

and

$$Q_{k+1}^{(k+1)} = r_{k+1} Q_k^{(k)}.$$

For $1 < j < k+1$, we consider the expression

$$Q_j^{(k)} + r_{k+1} Q_{j-1}^{(k)}.$$

This is the sum of products of the elements $r_1, r_2, ..., r_{k+1}$ taken j at a time. $Q_j^{(k)}$ clearly contains all the terms not containing r_{k+1}. Those terms containing r_{k+1} are the products of r_{k+1} by the product of $j-1$ elements taken from $r_1, ..., r_k$, hence the sum of all the terms is $r_{k+1} Q_{j-1}^{(k)}$. Thus the above sum is the sum of all products which can be formed by taking j factors at a time from $r_1, r_2, ..., r_{k+1}$, i.e. $Q_j^{(k+1)}$.

Example 8 *Let $U^{(n)}$ be the number of individuals of a population in time period n. Assuming that the number of births in time period n is proportional to the size of the population in that time period, then the increase in size is given by $\alpha U^{(n)}$ for some positive proportional constant α. Likewise, if the number of deaths in time period n is also proportional to $U^{(n)}$, then the decrease is $\beta U^{(n)}$ for some $\beta > 0$. Combining these assumptions gives*

$$U^{(n+1)} - U^{(n)} = \alpha U^{(n)} - \beta U^{(n)},$$

or

$$U^{(n+1)} = (1+r)U^{(n)},$$

where $r = \alpha - \beta$ is the net growth rate for the population. This is the Malthusian model of population growth in one location. Suppose now we assume that the population is spatially distributed, say at sites $m \in Z$, and suppose the population size at the site m and at time period $n+1$ is accounted for by a constant multiple of the previous population size at the same site as well as another constant multiple of migrated offspring from a neighboring site. Then letting $U_m^{(n)}$ be the size at time period n and site m, we see that

$$U_m^{(n+1)} = (1+r)U_m^{(n)} + \rho U_{m-\sigma}^{(n)}, \ m \in Z, n \in N,$$

where σ is a nonzero integer.

We remark that more complicated population models can be built based on similar ideas. A well known model is the following. Consider a population in one location again, but this time we also assume that the environment can only support a certain number (called the carrying capacity), say L, of the species. Thus, if $U^{(n)} > L$, there will not be enough food available. It is therefore reasonable that

$$U^{(n+1)} - U^{(n)} = r\left(1 - \frac{U^{(n)}}{L}\right)U^{(n)},$$

since (1) if $U^{(n)}$ is small, then $1 - U^{(n)}/L$ is roughly 1, (2) if $U^{(n)} < L$, then $r(1 - U^{(n)}/L) > 0$, and (3) if $U^{(n)} > L$, then $r(1 - U^{(n)}/L) < 0$. If we divide through the above equation by L, we obtain

$$y^{(n+1)} - y^{(n)} = ry^{(n)}(1 - y^{(n)}), \ y^{(n)} = \frac{U^{(n)}}{L},$$

which can also be written as

$$x^{(n+1)} = \lambda x^{(n)}\left(1 - x^{(n)}\right), \tag{1.14}$$

since

$$x^{(n+1)} - x^{(n)} = \lambda x^{(n)}\left(1 - x^{(n)}\right) - x^{(n)} = (\lambda - 1)x^{(n)}\left(1 - \frac{\lambda}{\lambda - 1}x^{(n)}\right),$$

or

$$\frac{\lambda}{\lambda - 1}x^{(n+1)} - \frac{\lambda}{\lambda - 1}x^{(n)} = (\lambda - 1)\left(\frac{\lambda}{\lambda - 1}x^{(n)}\right)\left(1 - \frac{\lambda}{\lambda - 1}x^{(n)}\right).$$

The equation (1.14) is called the *logistic equation* and has been investigated to a great extent (though not completely). In particular, the investigations of the existence and qualitative properties of its periodic solutions lead to the concept of chaos in 1975 [106]. Now if we consider logistic models with spatial migrations, we may end up with partial difference equations of the form

$$y_m^{(n+1)} - y_m^{(n)} = ry_m^{(n)}\left(1 - y_m^{(n)}\right) + \xi y_{m-\sigma}^{(n)}.$$

The two-level partial difference equations described above all take on real state values. In case the state values are from a finite and discrete set, we are then talking about *cellular automata*. These partial difference equations have long been recognized and a well known example is the "Game of Life" invented by Conway in the 1970s. Conway imagined a two-dimensional square lattice of cells where each cell is connected to its eight nearest neighbors. Each node takes on either the value 0 or the value 1. Conway chose a particular rule inspired by interactions of living organisms with one another, then given an arbitrary initial distributions of cell values, as time evolves, various forms of fixed points, cycles, and transients are possible. There are a number of introductory books (see e.g. Wolfram [162]) on the subject of cellular automaton.

Example 9 *Consider an interesting mathematical game called "The four number game" proposed by Ducci [82]. Take a sequence* $\{x_1, x_2, x_3, x_4\}$ *of four nonnegative integers. Compute the sequence*

$$\{|x_1 - x_2|, |x_2 - x_3|, |x_3 - x_4|, |x_4 - x_1|\}.$$

Repeat this procedure again and we will find that a sequence $\{0,0,0,0\}$ *is obtained no matter what the initial sequence has been chosen. As a specific example, we may begin with* $\{5,19,2,70\}$. *Then the subsequent sequences are* $\{14,17,68,65\}$, $\{3,51,3,51\}$, $\{48,48,48,48\}$, $\{0,0,0,0\}$. *Note that this game can be described by a two-level equation as follows. Let* $x_m^{(n)}$ *be the m-th nonnegative integer during the n-th stage of the game, then*

$$x_m^{(n+1)} = \left|x_m^{(n)} - x_{m+1}^{(n)}\right|, \; m = 1,2,3,4; n \in N,$$

where

$$x_1^{(n)} = x_5^{(n)}, \; n \in N.$$

We remark that it is also possible that the state values take on abstract entities such as sets or functions or operators. Indeed, the theory of *fractals* and the theory of *time series* are often associated with difference equations involving unknown functions that take on set values or random variables.

Example 10 *Let* $f(x) = x/3$ *and* $g = x/3 + 2/3$. *Let* $F = f \cup g$ *be defined by*

$$F(\Omega) = f(\Omega) \cup g(\Omega)$$

for any subset Ω *of R. Now suppose* $\Omega^{(1)} = [0,1]$, *and*

$$\Omega^{(n+1)} = F(\Omega^{(n)}), \; n \in Z^+.$$

Then it is easily seen that

$$\Omega^{(2)} = \left[0, \frac{1}{3}\right] \cup \left[\frac{2}{3}, 1\right],$$

$$\Omega^{(3)} = \left[0, \frac{1}{9}\right] \cup \left[\frac{2}{9}, \frac{1}{3}\right] \cup \left[\frac{2}{3}, \frac{7}{9}\right] \cup \left[\frac{8}{9}, 1\right], \; \ldots,$$

and that the sequence $\{\Omega^{(n)}\}$ *of sets "tends" to the Cantor set. Partial difference equations for functions taking on set values can easily be generated by similar principles. An example is*

$$\Omega_m^{(n+1)} = f\left(\Omega_{m-1}^{(n)}\right) \cup g\left(\Omega_m^{(n)}\right),$$

where each $\Omega_m^{(n)}$ *is a subset of* R.

1.2.3 Multi-Level Equations

Even though lattice dynamical systems can be used to model many real systems, they are by no means exhaustive. For example, let us consider the heat equation (1.2) again. We have assumed that heat flow is instantaneous in the above model. However, in reality, it takes time for heat to flow from one point m to its neighboring points $m-1$ and $m+1$. Thus a corresponding model is the following discrete heat equation with delay

$$\begin{aligned} u_m^{(n+1)} - u_m^{(n)} &= r\left(u_{m-1}^{(n-\sigma)} - u_m^{(n-\sigma)}\right) + r\left(u_{m+1}^{(n-\sigma)} - u_m^{(n-\sigma)}\right) \\ &= ru_{m-1}^{(n-\sigma)} - 2ru_m^{(n-\sigma)} + ru_{m+1}^{(n-\sigma)}. \end{aligned} \tag{1.15}$$

On the other hand, even if we assume that heat flow is instantaneous, a "delayed" control mechanism of the form $du_m^{(n-1)}$ may be introduced, and the corresponding equation may look like

$$u_m^{(n+1)} = au_{m-1}^{(n)} + bu_m^{(n)} + cu_{m+1}^{(n)} + du_m^{(n-1)} + p_m^{(n)}, \tag{1.16}$$

where the term $p_m^{(n)}$ reflects additional heat sources or sinks.

Example 11 *An early model of population growth was proposed by Leonardo of Pisa. Assume that in time period* 1, *a pair of young rabbits of opposite sex is introduced. Assume that it takes one period of time for the young rabbits to grow up and then give birth to a pair of young rabbits (of opposite sex). Assume further that rabbits never die. Let* $f^{(k)}$ *be the number of rabbit pairs present in the time period* k. *Then since* $f^{(k)}$ *is the sum of the number of rabbit pairs in the previous period plus the number of new offspring, we see that*

$$f^{(t+1)} = f^{(t)} + f^{(t-1)}, \ t \in Z^+.$$

When $f^{(0)} = 0$ *and* $f^{(1)} = 1$, *this recurrence relation shows that* $f^{(2)} = 1, f^{(3)} = 2, f^{(4)} = 3, f^{(5)} = 5, \ldots$. *The numbers* $0, 1, 1, 2, 3, 5, \ldots$ *are known as the Fibonacci numbers and have remarkable properties. A more realistic population model assumes that* $f^{(t+1)}$ *is accounted for by a constant multiple of* $f^{(t)}$ *plus another constant multiple of* $f^{(t-1)}$. *The corresponding equation is then*

$$f^{(t+1)} = \alpha f^{(t)} + \beta f^{(t-1)}, \ t \in Z^+.$$

Yet another population model assumes that total population is spatially distributed, say at sites $i \in Z$, *and that the population size at time* $t+1$ *and site* i *is accounted* *for by a constant multiple of the previous population size at the same site as well as another constant multiple of migrated offspring from a neighboring site. More precisely, let $f_i^{(t)}$ be the number of entities of the population at time t and site i. Suppose it takes one period of time for the young to grow up and give birth to offspring, and suppose a portion of the new offspring migrate from site $i-\sigma$ to i, then our model can be described by the following equation*

$$f_i^{(t+1)} = bf_i^{(t)} + cf_{i-\sigma}^{(t-1)},\ i \in Z, t \in N.$$

Under properly posed initial population distributions, we are naturally interested in the corresponding dynamics governed by this equation.

1.2.4 Implicit Reaction Diffusion Equations

If we interpret $u_m^{(n)}$ as the price of a commodity at time n and at location m, then it is possible that speculation will cause the change $u_m^{(n+1)} - u_m^{(n)}$ to be proportional to $u_{m-1}^{(n+1)} - u_m^{(n+1)}$ as well as $u_{m+1}^{(n+1)} - u_m^{(n+1)}$. In this manner, we obtain a two-level implicit diffusion equation of the form

$$u_m^{(n+1)} - u_m^{(n)} = r\left(u_{m-1}^{(n+1)} - 2u_m^{(n+1)} + u_{m+1}^{(n+1)}\right),$$

which may also be called an equation with advancement. Again, precautionary control may be imposed, then we are faced with equations such as

$$u_m^{(n+1)} - u_m^{(n)} = r\left(u_{m-1}^{(n+1)} - 2u_m^{(n+1)} + u_{m+1}^{(n+1)}\right) + f\left(u_m^{(n+2)}\right),$$

or

$$\alpha u_{m-1}^{(n+1)} + \beta u_m^{(n+1)} + \gamma u_{m+1}^{(n+1)} = au_{m-1}^{(n)} + bu_m^{(n)} + cu_{m+1}^{(n)},$$

or others. In Example 6, we have already seen a similar implicit equation. Two more examples can be given as follows.

Example 12 *Consider the heat equation*

$$u_t = u_{xx},\ t > 0,$$

which has been considered in Example 5. If we set up a grid in the x, t plane with grid spacings Δx and Δt, replace the second derivative u_{xx} with a central difference, and replace u_t with a "backward" difference (instead of the forward difference), a finite difference scheme for the heat equation is obtained:

$$\frac{u_m^{(n)} - u_m^{(n-1)}}{\Delta t} = \frac{1}{(\Delta x)^2}\left(u_{m-1}^{(n)} - 2u_m^{(n)} + u_{m+1}^{(n)}\right),$$

which is a discrete implicit reaction diffusion equation.

Example 13 *In the previous example, if we employ the averaged difference or the Crank-Nicolson method, then we obtain*

$$\frac{u_m^{(n+1)} - u_m^{(n)}}{\Delta t} = \frac{1}{(\Delta x)^2}\left(u_{m-1}^{(n)} - 2u_m^{(n)} + u_{m+1}^{(n)} + u_{m-1}^{(n+1)} - 2u_m^{(n+1)} + u_{m+1}^{(n+1)}\right).$$

This can be considered as an equation with advancement.

1.2.5 Discrete Time Independent Equations

When the temperature at each integral point in our heat conducting rod (considered in Section 1.2.1) is stabilized after a long period of time, the temperature distribution in the rod is said to be in *equilibrium*. In this case, the temperature function satisfies $u_m^{(n+1)} = u_m^{(n)}$, so that the equation (1.4), for example, yields

$$u_{m+1} - 2u_m + u_{m-1} + f(u_m) = 0.$$

Note that we have suppressed the superscripts for the sake of convenience. Similarly, if the heat conducting plate described previously by (1.5) has an initial temperature distribution at time $n = 0$, then under suitable conditions, it is expected that after a long period of time, the temperature inside the plate will stabilize, and the subsequent temperature distribution $\left\{u_{ij}^{(n)}\right\} \equiv \{u_{ij}\}$ will satisfy the steady state equation

$$\alpha_{ij}u_{i-1,j} + \beta_{ij}u_{i+1,j} + \gamma_{ij}u_{i,j-1} + \delta_{ij}u_{i,j+1} - \sigma_{ij}u_{ij} = 0, \tag{1.17}$$

which states that the temperature at each interior lattice point is the weighted average of the temperature of the four neighboring lattice points. As a particular and important case, we have the discrete or finite Laplace equation

$$u_{i-1,j} + u_{i+1,j} + u_{i,j-1} + u_{i,j+1} - 4u_{ij} = 0, \tag{1.18}$$

which has been studied by Courant *et al.* [52], Heilbronn [80], Duffin [60], Duffin and Shelly [64] and others.

Discrete time independent equations may arise in other manners. The following two examples show that the technique of "separation of variables" for the loaded vibrating string and loaded vibrating net may lead to such equations.

Example 14 *A string of negligible mass is stretched between two points $x = 0$ and $x = n + 1$ and is loaded at $x = 1, 2, ..., n$ by n particles of equal mass m. The particles are then set into motion so that they vibrate transversely. Let y_k denote the displacement of the k-th particle from its original position. Consider the case where $y_k > y_{k-1}$ and $y_k > y_{k+1}$. Assuming that the tension τ in the string is constant, the vertical force acting on the k-th particle due to the $(k-1)$-th particle is then*

$$-\tau \sin\theta = -\frac{\tau(y_k - y_{k-1})}{\sqrt{(y_k - y_{k-1})^2 + 1}}.$$

Assuming that the displacements $y_1, ..., y_n$ are small, we can replace the above force by $-\tau(y_k - y_{k-1})$ to a high degree of approximation. Similarly, the vertical force on the k-th particle due to the $(k+1)$-th particle is approximately given by $-\tau(y_k - y_{k+1})$. The total force is then $\tau(y_{k+1} - 2y_k + y_{k-1})$. Assuming there are no other forces acting on the k-th particle, then by Newton's law, the equation of motion is

$$my_k''(t) = \tau(y_{k+1} - 2y_k + y_{k-1}).$$

Letting $y_k(t) = e^{\sqrt{-1}\omega t}v_k$ and substituting into the above equation and then dividing by $e^{\sqrt{-1}\omega t}$, we obtain the time independent equation

$$v_{k+1} - 2v_k + v_{k-1} = \frac{m}{\tau}\omega^2 v_k.$$

Example 15 *Let us be given a rectangular net of weightless cords loaded at each point of intersection (x_i, y_j), $1 \leq i \leq m$, $1 \leq j \leq n$, with a particle of mass m. We shall further assume vibrations small and perpendicular to the plane of the net when the net is at rest. We thus assume the tension in each string constant throughout its length. Denote the tension in the string over (x_i, y_{j-1}), (x_i, y_j), (x_i, y_{j+1}) by T_i and in the string over (x_{i-1}, y_j), (x_i, y_j), (x_{i+1}, y_j) by $\bar{T}_j$. We assume T_i and $\bar{T}_j$ are independent of t. We also assume vibrations so small that the same kind of approximations in the previous example can be applied. Denote the vertical coordinate of the intersection (x_i, y_j) by u_{ij}. Under the circumstances, the motion of the intersection is determined by the following differential equation*

$$m\frac{d^2u_{ij}}{dt^2} = \bar{T}_j\frac{u_{i+1,j} - u_{ij}}{x_{i+1} - x_i} - \bar{T}_j\frac{u_{ij} - u_{i-1,j}}{x_i - x_{i-1}}$$
$$+T_i\frac{u_{i,j+1} - u_{ij}}{y_{j+1} - y_j} - T_i\frac{u_{ij} - u_{i,j-1}}{y_j - y_{j-1}}.$$

for $1 \leq i \leq m$ and $1 \leq j \leq n$. For convenience, we let

$$b_{ij} = \frac{\bar{T}_j}{m(x_{i+1} - x_i)}, \quad c_{ij} = \frac{T_i}{m(y_{j+1} - y_j)}.$$

Then the above equation can be written as

$$\frac{d^2u_{ij}}{dt^2} = \Delta_1\left(b_{i-1,j}\Delta_1 u_{i-1,j}\right) + \Delta_2\left(c_{i,j-1}\Delta_2 u_{i,j-1}\right),$$

where we have introduced the convenient notations $\Delta_1 w_{ij} = w_{i+1,j} - w_{ij}$ and $\Delta_2 w_{ij} = w_{i,j+1} - w_{ij}$. We proceed by letting $u_{ij} = e^{\sqrt{-1}\omega t} v_{ij}$. Substituting into the above equation and dividing by $e^{\sqrt{-1}\omega t}$, we obtain the time independent partial difference equation

$$\Delta_1\left(b_{i-1,j}\Delta_1 v_{i-1,j}\right) + \Delta_2\left(c_{i,j-1}\Delta_2 v_{i,j-1}\right) - \omega^2 v_{ij} = 0, \tag{1.19}$$

where $1 \leq i \leq m$ and $1 \leq j \leq n$. We note that $b_{ij} > 0$ and $c_{ij} > 0$. We have here mn linear algebraic equations in $v_{11}, ..., v_{mn}$ which must be satisfied.

Example 16 *Two persons play a game. At the beginning, A has i points and B j points. At the end of each step, suppose the probability of A getting one point is $p > 0$ and for B is $q > 0$, where $p + q = 1$. The game is to terminate when either of the two has K points. What is the probability of A winning the game? Let P_{ij} denote the probability that A will win the game. Now A winning the game may be resolved into two alternatives, viz. first, his winning the first step, and afterwards wining the game. Secondly, his losing the first step, and afterwards winning the game. The probability of the first alternative is $pP_{i+1,j-1}$, while the probability of the second is $qP_{i-1,j+1}$. Then by Bayes' formula,*

$$P_{ij} = pP_{i+1,j-1} + qP_{i-1,j+1}, \quad i, j \in Z^+.$$

Example 17 *Another illustrative example [37] is as follows. Let Ω be the set of all lattice points in the plane. For each pair of lattice points $w = (m, n)$ and $z = (i, j)$*

in Ω, we associate a real value $f(w,z)$. Thus we have a function f, which we call a flow. In the theory of network flow, we are interested in steady state conversion equations such as the following

$$f(w,w) = f(w,w+e)) + f(w,w-e) + p(w)f(w,w-\delta),$$

where $e = (0,1)$ and $p(w)$ is a function defined on Ω. The value $p(w)$ may be regarded as an amplification factor at the point w. Roughly speaking then, the equation requires that at each point w, the residue flow $f(w,w)$ is balanced with the total amount of flows which leave w and enter the points $w-e$, $w+e$ and $w-\delta$, subject to the condition that the flow towards $w-\delta$ can be adjusted by means of a scaling factor $p(w)$.

1.3 Auxiliary Conditions

Physical models impose natural auxiliary conditions to be satisfied by solutions of our partial difference equations. For instance, the initial temperature distribution of our rod may be known. In such a case, we are then given

$$u_m^{(0)} = f_m,\ m \in Z, \tag{1.20}$$

for the infinite rod. One of our problems is then to find an explicit formula for all the solutions of (1.3),(1.20). There are other types of subsidiary conditions which depend on the physical mechanism surrounding our diffusion medium. For example, when we are dealing with a finite rod so that the domain of definition of our reaction diffusion equation is of the form

$$\Omega = \{(m,n)|\ m = 1,2,...,M;\ n \in N\},$$

the *Dirichlet type condition* of the form

$$u_0^{(n)} = \alpha,\ n \in N,$$

may be imposed. Physically, such a condition says that one end of our heat conducting rod is maintained at a certain fixed temperature. Or, the *mixed boundary condition* such as

$$u_0^{(n)} + \beta_n u_1^{(n)} = \gamma_n,\ n \in N,$$

$$G\left(u_0^{(n)}, u_1^{(n)}, ...u_k^{(n)}\right) = 0,\ n \in N,$$

may be imposed. Note that the *Neumann type condition*

$$u_1^{(n)} - u_0^{(n)} = \gamma,\ n \in N,$$

is a particular case. Boundary conditions may involve both ends. For instance, the *periodic boundary condition*

$$u_0^{(n)} = u_M^{(n)},\ u_1^{(n)} = u_{M+1}^{(n)},\ n \in N, \tag{1.21}$$

is a well known example. To see how these type of conditions may arise, let us consider a circular tube divided into M (≥ 2) identical cells. These cells are labeled with the integers $1, 2, ..., M$ in a counterclockwise manner. Suppose each cell contains a ceratin solute dissolved in a unit volume of solvent. These cells are separated from each other by semipermeable membranes through which the solute may flow but not the solvent. Let us denote by $u_i^{(n)}$ the amount of solute dissolved in the i-th cell and at different time periods $n = 0, 1, 2, ...$. Since each cell contains a unit volume of solvent, $u_i^{(n)}$ also represents the concentration of solute in the i-th cell. During the time period n, if the concentration $u_2^{(n)}$ is higher than $u_1^{(n)}$, solute will flow from the first and the third cells to the second cell. The amount of increase is $u_2^{(n+1)} - u_2^{(n)}$, and it is reasonable to postulate that the increase is proportional to the differences $u_1^{(n)} - u_2^{(n)}$ and $u_3^{(n)} - u_2^{(n)}$, that is,

$$u_2^{(n+1)} - u_2^{(n)} = \gamma\left(u_1^{(n)} - u_2^{(n)}\right) + \gamma\left(u_3^{(n)} - u_2^{(n)}\right),$$

where $\gamma \in [0, 1]$ is a proportionality constant which reflects the physical characteristic of the membranes. By similar considerations, we may then obtain the system of equations

$$\begin{aligned} u_1^{(n+1)} - u_1^{(n)} &= \gamma\left(u_M^{(n)} - 2u_1^{(n)} + u_2^{(n)}\right), \\ u_2^{(n+1)} - u_2^{(n)} &= \gamma\left(u_1^{(n)} - 2u_2^{(n)} + u_3^{(n)}\right), \\ \cdots &= \cdots \\ u_{M-1}^{(n+1)} - u_{M-1}^{(n)} &= \gamma\left(u_{M-2}^{(n)} - 2u_{M-1}^{(n)} + u_M^{(n)}\right), \\ u_M^{(n+1)} - u_M^{(n)} &= \gamma\left(u_{M-1}^{(n)} - 2u_M^{(n)} + u_1^{(n)}\right), \end{aligned} \tag{1.22}$$

which describes the evolution process of diffusion. This system can be written as

$$u_i^{(n+1)} - u_i^{(n)} = \gamma\left(u_{i-1}^{(n)} - 2u_i^{(n)} + u_{i+1}^{(n)}\right),\ i = 1, 2, ..., M,$$

subject to (1.21).

There are other types of boundary conditions, for example,

$$\begin{aligned} u_0^{(0)} &= \mu, \\ u_0^{(n+1)} &= \lambda u_0^{(n)}\left(1 - u_0^{(n)}\right),\ n \in N, \end{aligned}$$

in which case, the boundary conditions take the form of a (possibly "chaotic") recurrence relation!

In case one end (say, the right end) of our conducting rod is infinite, we may also impose conditions such as

$$\lim_{m\to\infty} u_m^{(n)} = 0,\ n \in N.$$

There are many possibilities. We remark that problems involving nonlinear auxiliary conditions are quite difficult to deal with in general. Not much has been done at this point in time.

1.4 Notes and Remarks

The binomial coefficients and the functional equation (1.1) were known to Chu Shih-chien in China in 1303, while partial differential equations were developed only after the 18th century. The population model of rabbit pairs was proposed by Leonardo of Pisa in 1202. The Fibonacci numbers are well known and have remarkable properties. Generalizations of these numbers and the corresponding functional equations are the subjects of numerous studies and a journal called The Fibonacci Quarterly is dedicated to publications of these studies.

Stirling numbers and their properties are well known. A generalization of these numbers can be found in [10].

Example 7 is due to Miller [122], while Example 15 is due to Fort [69].

Some of the examples in gambling can be found in the classic books of Jordan [95] and Feller [65].

We will not emphasize cellular automata in this book. However, for applications of such partial difference equations, see e.g. Khan *et al.* [99]. Since the inception of its basic architecture in 1987 [47, 48], cellular neural networks have received much attention. For a sample of results and applications, [83] may be consulted.

The four number game described in Example 9 has been extended to n number games. Complete analysis of these games, however, is still open. See Freedman [70], Lotan [115], Carlitz and Scoville [9], Miller [123], and Ullman [155] for more information.

Discrete analogs of the equations of mathematical physics have always been of great interest to many scholars. There are many examples other than those that are presented in this Chapter. The interested readers may consult [3], [4], [54], [55], [56], [61], [62], [63], [64], [68], [73], [71], [81], [85], [104], [117], [119], [140], [139], [145], [150].

In the literature, the term partial difference equation is also used for a functional relation involving functions of two or more *continuous* variables. As an example, the equation

$$\frac{f(x+t,y)-2f(x,y)+f(x-t,y)}{t^2}=\frac{f(x,y+s)-2f(x,y)+f(x,y-s)}{s^2}$$

has been called a nonsymmetric partial difference functional equation by Haruki [79]. We will, however, only deal with functional equations involving functions of two or more integral variables.

There are partial difference equations whose domains are abstract graphs instead of subsets of Z^n. See for examples the works of Shogenji and Yamasaki [147], Maeda *et al.* [117], Yamasaki [166, 167], Murakami *et al.* [130], as well as Murakami and Yamasaki [129].

Chapter 2

Basic Tools

2.1 Subsets of the Lattice Plane

We will deal mostly with functional partial difference equations which involve bivariate sequences or functions with two independent integer variables over subsets of the lattice plane. To facilitate discussions, it is necessary to discuss the geometry and topology of the lattice plane. The set of integers has been denoted by Z. A *lattice point* $z = (i, j)$ in the plane has also been defined as a point with integer coordinates. The set of lattice points is denoted by Z^2. Two lattice points are said to be 4-neighbors or *neighbors* if their Euclidean distance is one. Similarly, two lattice points are 8-neighbors if their Euclidean distance is less than or equal to $\sqrt{2}$. The concept of 8-neighbors is important in digital image processing (see e.g. Rosenfeld [146]) and in cellular neural networks (see e.g. [83]), but we will not use it in this book. A pair of neighbors is said to form an *edge*. The four neighbors of $z = (i, j)$, namely, $(i-1, j)$, $(i+1, j)$, $(i, j-1)$ and $(i, j+1)$, will sometimes be denoted by z_L, z_R, z_D and z_T respectively. The lattice points $z_1, z_2, ..., z_n$ are said to form a *path* with terminals z_1 and z_n if z_1 is a neighbor of z_2, z_2 is a neighbor of z_3, etc. A set of lattice points is said to be *connected* if any two of its points are terminals of a path of points contained in the set. A *component* of a set Ω of lattice points is a nonempty maximal connected subset of Ω. A nonempty connected set of lattice points is called a *domain*. The *degree* of a point in a domain Ω is the number of its neighbors in Ω. Our definition of connectedness is an "arcwise" definition, rather than a definition in terms of open and closed sets. We can, however, define a topology on Z^2 in which the standard notion of connectedness reduces to our definition (see e.g. Rosenfeld [146])

Given a domain Ω, a lattice point is an *exterior boundary point* of Ω if it does not belong to Ω but has at least one neighbor in Ω. We will denote the set of exterior boundary points of Ω by $\partial\Omega$. A lattice point is an *interior boundary point* of Ω if it belongs to Ω and is a neighbor of an exterior boundary point of Ω. The set of all interior boundary points of Ω is denoted by $\partial'\Omega$. The set of all neighboring pairs of interior and exterior points is denoted by

$$\Upsilon(\Omega) = \{((i,j),(m,n)) \in \partial\Omega \times \partial'\Omega |\ (i,j) \text{ and } (m,n) \text{ are neighbors}\}.$$

Note that the exterior boundary of any nonempty finite domain has at least four elements, and the interior boundary has at least one. For example, let $\Omega = \{(1,2),(2,2),(1,1),(2,1)\}$. Then

$$\partial\Omega = \{(1,3),(2,3),(3,2),(3,1),(2,0),(1,0),(0,1),(0,2)\},$$

and $\partial'\Omega = \Omega$. It will be necessary to classify boundary points as follows: $\partial_L\Omega = \{z \in \partial\Omega : z_R \in \Omega\}$, $\partial_R\Omega = \{z \in \partial\Omega : z_L \in \Omega\}$, $\partial_D\Omega = \{z \in \partial\Omega : z_T \in \Omega\}$, $\partial_T\Omega = \{z \in \partial\Omega : z_D \in \Omega\}$, $\partial'_L\Omega = \{z \in \partial'\Omega : z_L \in \partial_L\Omega\}$, $\partial'_R\Omega = \{z \in \partial'\Omega : z_R \in \partial_R\Omega\}$, $\partial'_D\Omega = \{z \in \partial'\Omega : z_D \in \partial_D\Omega\}$, and $\partial'_T\Omega = \{z \in \partial'\Omega : z_T \in \partial_T\Omega\}$.

Qualitative properties of domains will be useful. First of all, note that any two components of a domain Ω cannot have a common exterior boundary point. Note further that for any $z \in \Omega$, the set $\Omega\backslash\{z\}$ contains at most four components. Indeed, z has at most four neighbors in Ω, and for each w of its neighbors in Ω, there can be only one component containing w.

Theorem 1 *Let Ω be a finite domain of Z^2. Let $z \in \Omega$, and let Λ be a component of $\Omega\backslash\{z\}$. Then $\partial\Lambda$ is contained in $\partial\Omega + \{z\}$.*

Indeed, let $B(z)$ be the set of neighbors of z contained in Ω, and let $\Lambda(w)$ be the component of $\Omega\backslash\{z\}$ containing w. Then clearly

$$\Omega = \{z\} + \sum_{w \in B(z)} \Lambda(w).$$

Consequently, if u is an exterior boundary point of some component $\Lambda(w)$, then u is either equal to z, or else u cannot be in another component, which implies u is an exterior boundary point of Ω.

There are several common subsets of the lattice plane which are needed as domains of definitions of partial difference equations. A domain Ω of lattice points is said to be a *chain* if there are exactly two points in Ω with degree 1 and the remaining points in Ω with degree 2. To rule out degenerate cases, we will assume that a chain has at least two points. It is easily seen that a chain can be regarded as a path which neither crosses nor "touches" itself, i.e., its points can be numbered as $z_1, z_2, .., z_n$, so that z_i is a neighbor of z_j if, and only if, $i = j \pm 1$. A nonempty domain Ω of lattice points is said to be a *cycle* if every one of its points has degree 2. It is not difficult to see that the points in a chain can be numbered as $z_1, z_2, ..., z_n$ such that z_1 has exactly two neighbors z_n and z_2, z_2 has exactly two neighbors z_1 and z_3, etc.

Example 18 *The set*

$$\{(2,2),(2,3),(2,4),(3,4),(4,4),(4,5)\}$$

is a chain, while the set

$$\{(1,1),(1,2),(1,3),(1,4),(2,4),(3,4),(3,3),(4,3),(4,2),(4,1),(3,1),(2,1)\}$$

is a cycle.

A domain Ω is said to be *rectangular* if Ω consists of n rows of m lattice points. Such a domain can be denoted by $\{a,...,b\} \times \{c,...,d\}$. A *sphere* is the set of lattice points within k units from the origin, i.e.

$$\{(i,j) \in Z^2 |\ |i| + |j| \leq k\}, \ k \in N.$$

A *cylinder* is a set of lattice points of the form

$$\{(i,j) \in Z^2 |\ a \leq i \leq b, j \geq c\}.$$

A *cone* is a set of lattice points which are inside two rays from the origin, e.g.

$$\{(i,j) \in Z^2 |\ 0 \leq i \leq j\}$$

is a cone. A *quadrant* is a set of lattice points of the form

$$\{(i,j) \in Z^2 |\ i \geq a, j \geq b\}.$$

In particular, the *nonnegative quadrant* is the set

$$N \times N = \{(i,j) \in Z^2 |\ i,j \geq 0\}.$$

A *half plane* is the set of all lattice points with y-coordinates larger than or equal to some fixed integer. Finally, an *exterior domain* is the complement of a bounded domain.

All subsets of Z^2 or Z^n are countable. In our later investigations, we will see that such a property can be useful for deriving qualitative properties of partial difference equations. In order to do so, we need to label the elements in these subsets in orderly manners. In particular, let us introduce an ordering $\preccurlyeq$ for the lattice points in the nonnegative quadrant $N \times N$. First of all, we can partition $N \times N$ into equivalence classes $Q_0, Q_1, Q_2, ...$ defined by

$$Q_k = \{(i,j) \in N \times N |\ i + j = k\}, \quad k \in N.$$

Thus $Q_0 = \{(0,0)\}$, $Q_1 = \{(0,1),(1,0)\}, ...,$ etc. Each Q_k can be linearly ordered as follows: for any $(i,j),(m,n) \in Q_k$, (i,j) precedes (m,n) if $i > m$. Thus, $N \times N$ can be linearly ordered: for any $(i,j),(m,n) \in N \times N$, $(i,j) \preccurlyeq (m,n)$ if either (i) $(i,j) \in Q_k$ and $(m,n) \in Q_l$ and $k < l$, or (ii) $(i,j),(m,n) \in Q_k$ and $i > m$. Clearly, under the ordering just defined, the lattice points can be arranged in the following manners: $(0,0)$; $(1,0),(0,1)$; $(2,0),(1,1),(0,2)$; ... . A number of other linear orderings are possible and will be introduced when needed.

We have concerned ourselves with the lattice plane Z^2. Clearly, most of the concepts defined above have natural extensions to the set Z^n. In particular, the concepts of a path and a domain in Z^n are similarly defined. We will not repeat these definitions since most of our later discussions are restricted to partial difference equations involving bivariate sequences.

2.2 Classifications of Partial Difference Equations

We have already seen a number of examples of partial difference equations in the first chapter. To facilitate discussions, it is necessary to classify these equations and others. Roughly speaking, a partial difference equation is a functional relation defined over domains in Z^n which involves unknown functions of multiple integral variables, and a function is called its solution if the equation becomes an identity in the independent variables when this function is substituted in the equation. Therefore, a natural classification can be carried out by means of the types of domains involved. For instance, we have equations over cylinders, over quadrants, over half planes, over the lattice plane, etc. Equation (1.1) is a partial difference equation over the cone $\{(k,n)|\ 0 \le k < n\}$, while (1.2) is a partial difference equation over the upper half plane.

We have seen that some partial difference equations arise from real models. Since these models can also be described by partial differential equations, it is natural to give names to partial difference equations which are analogous to those given to the partial differential equations. Indeed, we have already called equation (1.2) a discrete or finite heat equation. It is also natural to call the following equation

$$u_{i,j+1} + u_{i+1,j} + u_{i,j-1} + u_{i-1,j} - 4u_{ij} = 0$$

a discrete or finite Laplace equation, the equation

$$u_{i,j+1} + u_{i+1,j} + u_{i,j-1} + u_{i-1,j} - 4u_{ij} = g_{ij}$$

a discrete or finite Poisson equation, etc.

When our partial difference equations are used to model processes which evolve in steps (or stages or levels), it is natural to name these equations according to the number of steps involved. For example, equation (1.2) is a two-level equation, as are the other examples in Section 2.2 in Chapter 1.

When our partial difference equation can be expressed in the form

$$F(u_{ij}, u_{i+1,j}, ...) = 0, \tag{2.1}$$

it is sometimes convenient to classify such an equation by the number of variables of F. For example, (1.2) is a four-point equation. The following equation

$$u_{i+1,j+1} = u_{i+1,j} + u_{i,j+1} + u_{ij}$$

is also a four-point equation, while the equation

$$u_{i+1,j+1} = u_{i+1,j} + u_{i,j+1} + u_{ij} + u_{i-2,j-4}$$

is a five-point equation, etc.

An important class of partial difference equation is the class of *linear equations*. We say that the partial difference equation

$$F(u_{ij}, u_{i+1,j}, ...) = g_{ij} \tag{2.2}$$

is linear if $F(x_1, ..., x_n)$ is linear, i.e.

$$F(\alpha x_1 + \beta y_1, ..., \alpha x_n + \beta y_n) = \alpha F(x_1, ..., x_n) + \beta F(y_1, ..., y_n),$$

for all α, β in C. For example, the equation

$$u_{i,j+1} - u_{i,j} = \gamma(u_{i-1,j} - 2u_{ij} + u_{i+1,j})$$

is linear. Indeed, the corresponding F is given by

$$F(u_{i,j+1}, u_{i-1,j}, u_{i,j}, u_{i+1,j}) = u_{i,j+1} + (2\gamma - 1)u_{ij} - \gamma u_{i-1,j} - \gamma u_{i+1,j},$$

which clearly satisfies our requirement. The important property of linear equations is stated in the principle of superposition: Let $g^{(1)} = \{g_{ij}^{(1)}\}, ..., g^{(t)} = \{g_{ij}^{(t)}\}$ be any functions and $\alpha_1, ..., \alpha_n$ any constants. If F is linear, and if $u^{(1)}, ..., u^{(t)}$ are solutions of the equations

$$F(u_{ij}, u_{i+1,j}, ...) = g_{ij}^{(1)}, ..., F(u_{ij}, u_{i+1,j}, ...) = g_{ij}^{(t)},$$

then $\alpha_1 u^{(1)} + ... + \alpha_t u^{(t)}$ is a solution of the equation

$$F(u_{ij}, u_{i+1,j}, ...) = \alpha_1 g_{ij}^{(1)} + ... + \alpha_t g_{ij}^{(t)}.$$

Finally, we remark that if the term g_{ij} in equation (2.2) is identically zero, then it is called a homogeneous equation, otherwise, it is nonhomogeneous. The equation (2.1) is called the associated homogeneous equation of (2.2).

2.3 Finite Differences

There are several useful mathematical tools for handling relations between sequences. Here, we introduce the difference and summation operations. Symbolic manipulations of sequences will be introduced in the next Chapter. Given a complex sequence $\{u_k\}$, the difference $u_{k+1} - u_k$ is usually denoted by Δu_k, and the difference $u_k - u_{k-1}$ by ∇u_k. The symbols Δ and ∇ are usually called the forward difference and backward difference operators. We easily see that

$$\nabla u_k = \Delta u_{k-1},$$

$$\Delta^2 u_k = \Delta(\Delta u_k) = u_{k+2} - 2u_{k+1} + u_k,$$

$$\Delta\nabla u_k = u_{k+1} - 2u_k + u_{k-1} = \Delta^2 u_{k-1},$$

$$\nabla^2 u_k = u_k - 2u_{k-1} + u_{k-2},$$

$$\Delta^2\nabla u_k = u_{k+2} - 3u_{k+1} + 3u_k - u_{k-1} = \Delta^3 u_{k-1},$$

$$\Delta(u_k v_k) = v_{k+1}\Delta u_k + u_k \Delta v_k, \tag{2.3}$$

$$\Delta(u_k\Delta v_k - v_k \Delta u_k) = u_{k+1}\Delta^2 v_k - v_{k+1}\Delta^2 u_k, \tag{2.4}$$

$$f_k \Delta g_k \Delta h_k + h_k \Delta(f_{k-1}\Delta g_{k-1}) = \Delta(f_{k-1} g_k h_k) - \Delta(f_{k-1} g_{k-1} h_k), \tag{2.5}$$

$$\Delta\left(\frac{f_k^2}{g_k}\right)\Delta g_k = (\Delta f_k)^2 - g_k g_{k+1}\left(\Delta\left(\frac{f_k}{g_k}\right)\right)^2, \; g_k \neq 0, g_{k+1} \neq 0, \tag{2.6}$$

etc. The telescoping identity

$$\sum_{k=a}^{b} \Delta u_k = u_{b+1} - u_a, \tag{2.7}$$

and Abel's summation by parts (which follows from (2.3) and (2.7))

$$\sum_{k=a}^{b} v_{k+1}\Delta u_k = u_{b+1}v_{b+1} - u_a v_a - \sum_{k=a}^{b} u_k \Delta v_k \tag{2.8}$$

are also well known.

There are some common difference formulas which will be useful in later sections:

$$\Delta c = 0,\ k \in Z,$$
$$\Delta(-1)^k = 2(-1)^{k+1},\ k \in Z,$$
$$\Delta b^k = b^k(b-1),\ k \in Z, b \neq 0,$$
$$\Delta b^k = b^k(b-1),\ k \in Z^+,$$
$$\Delta \cos(ak+b) = -2\sin\frac{a}{2}\sin\left(ak+\frac{a}{2}+b\right),\ k \in Z,$$
$$\Delta \sin(ak+b) = 2\sin\frac{a}{2}\cos\left(ak+\frac{a}{2}+b\right),\ k \in Z,$$

where the last equality follows from the trigonometric identity

$$\sin\alpha - \sin\beta = 2\sin\left(\frac{\alpha-\beta}{2}\right)\cos\left(\frac{\alpha+\beta}{2}\right).$$

Before introducing the partial differences, we first recall that there are various ways to represent bivariate sequences. We will use the notation $\{u_{ij}\}$ for defining the partial differences. For the notation $\left\{u_i^{(j)}\right\}$, it will be understood that the subscript has precedence over the superscripts, so that the partial differences may be defined in similar manners. The partial differences of a bivariate sequence $\{u_{ij}\}$ are defined by

$$\Delta_1 u_{ij} = u_{i+1,j} - u_{ij},$$
$$\Delta_2 u_{ij} = u_{i,j+1} - u_{ij},$$
$$\Delta_1^2 u_{ij} = \Delta_1(\Delta_1 u_{ij}),\ \Delta_{21}u_{ij} = \Delta_2(\Delta_1 u_{ij}),$$

etc., while those of $\left\{u_i^{(j)}\right\}$ will be

$$\Delta_1 u_i^{(j)} = u_{i+1}^{(j)} - u_i^{(j)},$$

$$\Delta_{12}u_i^{(j)} = \Delta_2\left(\Delta_1 u_i^{(j)}\right) = u_{i+1}^{(j+1)} - u_{i+1}^{(j)} - u_i^{(j+1)} + u_i^{(j)},$$

etc.

Example 19 *Let* $u_{mn} = m^{[n]}$ *for* $(m,n) \in Z^2\backslash\{(i,j)|\ j \le i \le -1\}$. *Then*

$$\Delta_1 u_{mn} = nm^{[n-1]},\ (m,n) \in Z^2\backslash\{(i,j)|\ j-1 \le i \le -1, j \le -1\}.$$

Example 20 *Let* $u_{mn} = (pm+q)^{[n]}$ *for* $n \in Z$ *and* $m \in N$, *then*

$$\Delta_1 u_{mn} = np(pm+q)^{[n-1]},\ n \in Z,\ m \in N.$$

Example 21 *Let* $u_{mn} = C_m^{(n)}$ *for* $m, n \in Z^+$ *and* $1 \le m \le n$, *then*

$$\Delta_2 C_m^{(n)} = \Delta_2 u_{mn} = C_{m-1}^{(n)},\ 1 \le m \le n.$$

Sometimes, the notations $\Delta_x u_{ij}, \Delta_y u_{ij}, ...$ will also be used in later sections instead of $\Delta_1 u_{ij}, \Delta_2 u_{ij}$, etc. They are convenient when we are dealing with bivariate sequences. For multivariate sequences, however, things are more complicated. In order to express the higher order differences in a convenient manner, let us call a vector $\alpha = (\alpha_1, \alpha_2, ..., \alpha_n)$ a n-index if $\alpha_1, \alpha_2, ..., \alpha_2$ are nonnegative integers. The number $|\alpha| = \alpha_1 + ... + \alpha_n$ is called the order of α. The partial differences $\Delta_1^{\alpha_1}\Delta_2^{\alpha_2} u_{ij}$ will be denoted by $\Delta^\alpha u_{ij}$, where $\alpha = (\alpha_1, \alpha_2)$ is a 2-index.

Example 22 *The* discrete Laplacian *of* u_{ij}

$$\Xi u_{ij} \equiv u_{i,j+1} + u_{i+1,j} + u_{i,j-1} + u_{i-1,j} - 4u_{ij}$$

can be written as $\Delta^{(2,0)} u_{i-1,j} + \Delta^{(0,2)} u_{i,j-1}$, $\Delta_1^2 u_{i-1,j} + \Delta_2^2 u_{i,j-1}$ *or* $\Delta_1^2 u(z_L) + \Delta_2^2 u(z_D)$, *etc.*

Example 23 *The following sum of partial differences*

$$\Delta^{(2,0)} u_{ij} + \Delta^{(0,2)} u_{ij} + \Delta^{(1,1)} u_{ij}$$

can be written as $\sum_{|\alpha|=2} \Delta^\alpha u_{ij}$.

Example 24 *The partial difference equation (1.2) can be written as*

$$\Delta_2 u_m^{(n)} = r\Delta_1^2 u_{m-1}^{(n)},\ m \in Z, n \in N.$$

By means of the telescoping identity (2.7), we have

$$\sum_{i=a}^{b}\sum_{j=c}^{d} \Delta_1\Delta_2 u_{ij} = \sum_{j=c}^{d} \Delta_2 u_{b+1,j} - \sum_{j=c}^{d} \Delta_2 u_{aj} = u_{b+1,d+1} - u_{b+1,c} - u_{a,d+1} + u_{ac}.$$

More generally, discrete analog of the Green's Theorem in the plane [22] can be expressed as follows: Let Ω be a finite domain and suppose $\{u_{ij}\}$ is a bivariate sequence such that $\Delta_1 u_{i-1,j}$ and $\Delta_2 u_{i,j-1}$ are defined on Ω. Then

$$\sum_{(i,j)\in\Omega} \Delta_1 u_{i-1,j} = \sum_{(i,j)\in\partial_R\Omega} u_{i-1,j} - \sum_{(i,j)\in\partial_L\Omega} u_{ij}, \tag{2.9}$$

and

$$\sum_{(i,j)\in\Omega} \Delta_2 u_{i,j-1} = \sum_{(i,j)\in\partial_T\Omega} u_{i,j-1} - \sum_{(i,j)\in\partial_D\Omega} u_{ij}. \tag{2.10}$$

Indeed, let $H(j)$ denote the horizontal straight line crossing the point $(0, j)$. Then for each j, by the telescoping identity (2.7),

$$\sum_{(i,j)\in H(j)\cdot\Omega} \Delta_1 u_{i-1,j} = \sum_{(i,j)\in H(j)\cdot\partial_R\Omega} u_{i-1,j} - \sum_{(i,j)\in H(j)\cdot\partial_L\Omega} u_{ij}.$$

Hence,

$$\begin{aligned}\sum_{(i,j)\in\Omega}\Delta_1 u_{i-1,j} &= \sum_{j\in Z}\sum_{(i,j)\in H(j)\cdot\Omega}\Delta_1 u_{i-1,j}\\ &= \sum_{j\in Z}\sum_{(i,j)\in H(j)\cdot\partial_R\Omega} u_{i-1,j}-\sum_{j\in Z}\sum_{(i,j)\in H(j)\cdot\partial_L\Omega} u_{ij}\\ &= \sum_{(i,j)\in\partial_R\Omega} u_{i-1,j}-\sum_{(i,j)\in\partial_L\Omega} u_{ij}.\end{aligned}$$

As immediate consequences, we have the following discrete analogs of the Green's identities.

Theorem 2 *Let Ω be a finite domain and suppose u and v are bivariate sequences defined over $\Omega+\partial\Omega$. Then*

$$\sum_{(i,j)\in\Omega}\left\{v_{ij}\left[\Delta_1^2 u_{i-1,j}+\Delta_2^2 u_{i,j-1}\right]-u_{ij}\left[\Delta_1^2 v_{i-1,j}+\Delta_2^2 v_{i,j-1}\right]\right\}$$

$$\begin{aligned}&=\sum_{(i,j)\in\partial_R\Omega}\left[v_{i-1,j}\Delta_1 u_{i-1,j}-u_{i-1,j}\Delta_1 v_{i-1,j}\right]-\sum_{(i,j)\in\partial_L\Omega}\left[v_{ij}\Delta_1 u_{ij}-u_{ij}\Delta_1 v_{ij}\right]\\ &+\sum_{(i,j)\in\partial_T\Omega}\left[v_{i,j-1}\Delta_2 u_{i,j-1}-u_{i,j-1}\Delta_2 v_{i,j-1}\right]-\sum_{(i,j)\in\partial_D\Omega}\left[v_{ij}\Delta_2 u_{ij}-u_{ij}\Delta v_{ij}\right].\end{aligned}$$

Indeed, in view of (2.4), the left hand side of the above equality is equal to

$$\begin{aligned}&\sum_{(i,j)\in\Omega}\Delta_1\left[v_{i-1,j}\Delta_1 u_{i-1,j}-u_{i-1,j}\Delta_1 v_{i-1,j}\right]\\ &+\sum_{(i,j)\in\Omega}\Delta_2\left[v_{i,j-1}\Delta_2 u_{i,j-1}-u_{i,j-1}\Delta_2 v_{i,j-1}\right],\end{aligned}$$

which by means of (2.9) and (2.10), is equal to the right hand side.

Similarly, the following two results follow from (2.3) and (2.5) as well as (2.9) and (2.10).

Theorem 3 *Let Ω be a finite domain and suppose u and v are bivariate sequences defined over $\Omega+\partial\Omega$. Then*

$$\begin{aligned}\sum_{(i,j)\in\Omega}&\left\{u_{ij}\left[\Delta_1^2 v_{i-1,j}+\Delta_2^2 v_{i,j-1}\right]+\left[\Delta_1 u_{ij}\Delta_1 v_{ij}+\Delta_2 u_{ij}\Delta_2 v_{ij}\right]\right\}\\ &=\sum_{(i,j)\in\partial_R\Omega} u_{ij}\Delta_1 v_{i-1,j}-\sum_{(i,j)\in\partial_L\Omega} u_{i+1,j}\Delta_1 v_{ij}\\ &+\sum_{(i,j)\in\partial_T\Omega} u_{ij}\Delta_2 v_{i,j-1}-\sum_{(i,j)\in\partial_D\Omega} u_{i,j+1}\Delta_2 v_{ij}.\end{aligned}$$

Theorem 4 *Let Ω be a finite domain and suppose f, g and h are bivariate sequences defined over $\Omega+\partial\Omega$. Then*

$$\sum_{(i,j)\in\Omega}\left\{f_{ij}\Delta_1 g_{ij}\Delta_1 h_{ij}+h_{ij}\Delta_1(f_{i-1,j}\Delta_1 g_{i-1,j})\right\}$$

$$=\sum_{(i,j)\in\partial_R\Omega} f_{i-1,j}h_{ij}\Delta_1 g_{i-1,j}-\sum_{(i,j)\in\partial_L\Omega} f_{ij}h_{i+1,j}\Delta_1 g_{ij},$$

and

$$\sum_{(i,j)\in\Omega}\{f_{ij}\Delta_2 g_{ij}\Delta_2 h_{ij}+h_{ij}\Delta_2(f_{i,j-1}\Delta_2 g_{i,j-1})\}$$
$$=\sum_{(i,j)\in\partial_T\Omega} f_{i,j-1}h_{ij}\Delta_2 g_{i,j-1}-\sum_{(i,j)\in\partial_D\Omega} f_{ij}h_{i,j+1}\Delta_2 g_{ij}.$$

2.4 Summable Infinite Sequences

Let Ω be a (finite or infinite) subset of $Z^n = Z\times Z\times ...\times Z$. Since Ω is countable, we may re-label the members of Ω, and assume without loss of generality that $\Omega=\{1,...,m\}$, N or Z. Let l^Ω be the set of all complex sequences defined on Ω. Strictly speaking, each member in l^Ω is a multiply indexed sequence of the form

$$\{f_{i,...,j}|\ (i,...,j)\in\Omega\}.$$

For the sake of simplicity, we will, however, write the above sequence in the form

$$f=\{f_k\}_{k\in\Omega},$$

and come back to the original notation when needed. For the sake of convenience, we will write $\{f_k\}$ instead of $\{f_k\}_{k\in\Omega}$ if no confusion is caused. For any $\alpha\in C$ and $f=\{f_k\}, g=\{g_k\}$, we define $-f$, αf, $f\cdot g$, and $f+g$ by $\{-f_k\}$, $\{\alpha f_k\}$, $\{f_k g_k\}$ and $\{f_k+g_k\}$ as usual, while $|f|$ denotes the sequence $\{|f_k|\}$; the sequence $\{f_k^p\}$ will be denoted by f^p, where p is a positive number. The sequence $\{...,0,0,0,...\}$ will be denoted by 0. Let i be an integer, the sequence in l^Z whose i-th term is 1 and the other terms are 0 will be called the *Kronecker delta sequence* and denoted by $\hat{h}^{[i]}$. Similarly, the sequence in l^N whose i-th term is 1 and the other terms are 0 will be called the *Dirac delta sequence* and denoted by $\hbar^i$.

A sequence $f=\{f_k\}$ is said to be nonnegative if each of its terms is nonnegative. A real sequence $g=\{g_k\}$ is said to be less than or equal to the real sequence $h=\{h_k\}$ if $\{h_k-g_k\}$ is nonnegative and denoted by $g\le h$. For a given real sequence $f=\{f_k\}$, we can always write it in the form f^+-f^- for some nonnegative sequences f^+, f^-. Indeed, the positive part f^+ is given by $(|f|+f)/2$, and the negative part by $(|f|-f)/2$. A sequence $f=\{f_k\}$ is said to have finite support if the number of nonzero terms of f is finite. The set $\Phi(f)$ of integers k for which $f_k\ne 0$ will be called the support of f. When $\{f^{(n)}\}_{n\in N}$ is a sequence of sequences in l^Ω, we say that $\{f^{(n)}\}_{n\in N}$ converges (pointwise) to $f\in l^\Omega$ if

$$\lim_{n\to\infty} f_k^{(n)}=f_k,\ k\in\Omega.$$

Note that for any nonnegative sequence $f=\{f_k\}\in l^\Omega$, we can always find a sequence $\{u^{(n)}\}_{n\in N}$ of nonnegative sequences in l^Ω such that

$$0\le u^{(0)}\le u^{(1)}\le ...\le f$$

and $u^{(n)}\to f$ as $n\to\infty$. For instance, if $f\in l^N$, we may pick

$$u^{(n)}=\sum_{k=0}^{n} f_k\hbar^k,\ n\in N.$$

We will need the concepts of summable infinite sequences. First of all, for a sequence f with finite support, we define its sum by the number

$$\sum_{\Omega} f = \sum_{k\in\Phi(f)} f_k.$$

For a nonnegative sequence $f = \{f_k\}$ in l^{Ω}, we define its sum by

$$\sum_{\Omega} f = \sup \sum_{\Omega} g,$$

where the supremum is taken over all sequences g with finite support such that $0 \le g \le f$. If the supremum on the right hand side is finite, we say that f is summable. Occasionally, it is convenient to allow the right hand side to be infinite and in such a case, we write

$$\sum_{\Omega} f = \infty.$$

Note that it easily follows from the definition that a finite linear combination of summable nonnegative sequences is summable and its sum is equal to the corresponding linear combination of the separate sums, and that if $0 \le f \le g$, then

$$0 \le \sum_{\Omega} f \le \sum_{\Omega} g.$$

We remark that the above definition of the sum of a nonnegative sequence in l^N can be simplified to

$$\sum_{N} f = \sum_{k=0}^{\infty} f_k \equiv \lim_{n\to\infty} \sum_{k=0}^{n} f_k,$$

and in l^Z to

$$\sum_{Z} f = \sup_{m,n\in Z} \sum_{k=m}^{n} f_k = \lim_{n\to\infty} \sum_{k=-n}^{n} f_k. \tag{2.11}$$

Indeed, for $f \in l^N$, since

$$\sum_{k=0}^{n} f_k = \sum_{N} w,$$

where $w = \{f_0, ..., f_n, 0, 0, ...\}$ is a sequence with finite support,

$$\sum_{k=0}^{n} f_k = \sum_{N} w \le \sup \sum_{N} g = \sum_{N} f.$$

Conversely, for any g which satisfies $0 \le g \le f$ and $\Phi(g)$ is finite, since $\Phi(g) \subseteq \{0, ..., m\}$ for some m, thus $0 \le g \le u \le f$, where $u = \{f_0, ..., f_m, 0, 0, ...\}$, and

$$\sum_{N} g \le \sum_{N} u = \sum_{k=0}^{m} f_k \le \sum_{k=0}^{\infty} f_k.$$

For $f \in l^Z$, (2.11) is similarly proved.

We remark also that our definition of a sum of infinite sequence is a special case of the Lebesgue integral for measurable functions. Thus standard results from the theory of Lebesgue integrals can be applied. In particular, Lebesgue's monotone convergence theorem holds: Let $g \in l^\Omega$ and let $\{f^{(n)}\}_{n\in N}$ be a sequence of nonnegative sequences $f^{(n)} \in l^\Omega$ such that

$$0 \le f_k^{(0)} \le f_k^{(1)} \le \ldots < \infty, \ k \in \Omega,$$

and

$$\lim_{n\to\infty} f_k^{(n)} = g_k, \ k \in \Omega,$$

then

$$\lim_{n\to\infty} \sum_\Omega f^{(n)} = \sum_\Omega g.$$

We provide a proof even though it follows from the general theory of Lebesgue integral. Since

$$\sum_\Omega f^{(n)} \le \sum_\Omega g,$$

there exists an $\omega \in [0, \infty]$ such that

$$\lim_{n\to\infty} \sum_\Omega f^{(n)} = \omega.$$

Since $f^{(n)} \le f$, we have

$$\sum_\Omega f^{(n)} \le \sum_\Omega g$$

for every $n \in N$, and thus

$$\omega \le \sum_\Omega g.$$

To see the converse, let u be a sequence with finite support that satisfies $0 \le u \le g$. Let c be a constant in $(0, 1)$. Since $f^{(n)} \to g$, we have $f^{(n)} \ge cu$ for all large n. Hence

$$\sum_\Omega f^{(n)} \ge \sum_\Omega cu = c \sum_\Omega u$$

for all large n. Since c and u are arbitrary, we must have

$$\lim_{n\to\infty} \sum_\Omega f^{(n)} = \omega \ge \sup \sum_\Omega u = \sum_\Omega g.$$

The proof is complete.

As an immediate corollary, if $\{f^{(n)}\}_{n\in N}$ is a sequence of nonnegative sequences in l^Ω such that

$$\sum_{n=0}^{\infty} f_k^{(n)} < \infty$$

for each $k \in \Omega$, then

$$0 \le \sum_{n=0}^{0} f_k^{(n)} \le \sum_{n=0}^{1} f_k^{(n)} \le \sum_{n=0}^{2} f_k^{(n)} \le \ldots < \infty, \ k \in \Omega,$$

and

$$\lim_{m\to\infty}\sum_{n=0}^{m} f_k^{(n)} = \sum_{n=0}^{\infty} f_k^{(n)},\ k\in\Omega,$$

hence

$$\sum_{\Omega}\sum_{n=0}^{\infty} f^{(n)} = \sum_{n=0}^{\infty}\sum_{\Omega} f^{(n)}. \tag{2.12}$$

As another corollary, we have Fatou's lemma: If $\{f^{(n)}\}_{n\in N}$ is a sequence of nonnegative sequences in l^{Ω} such that

$$\liminf_{n\to\infty} f_k^{(n)} < \infty,\ k\in\Omega,$$

then

$$\sum_{\Omega}\liminf_{n\to\infty} f^{(n)} \le \liminf_{n\to\infty}\sum_{\Omega} f^{(n)}.$$

Indeed, let $g_k^{(m)} = \inf_{n\ge m} f_k^{(n)}$ for each $k\in\Omega$ and $m\ge 0$, then $0\le g_k^{(0)}\le g_k^{(1)}\le \dots\le g_k^{(m)}\le f_k^{(m)}$ for each $k\in\Omega$ and $m\ge 0$, and

$$\lim_{m\to\infty} g_k^{(m)} = \liminf_{n\to\infty} f_k^{(n)} < \infty, \quad k\in\Omega,$$

so that

$$\sum_{\Omega}\liminf_{n\to\infty} f^{(n)} = \sum_{\Omega}\lim_{m\to\infty} g^{(m)} = \lim_{m\to\infty}\sum_{\Omega} g^{(m)} \le \liminf_{m\to\infty}\sum_{\Omega} f^{(m)}.$$

We have mentioned that the discrete set Ω can be linearly ordered so that it "looks like" $\{1,\dots,m\}$, N or Z. Note however, that for each ordering, the corresponding sum of a sequence defined over Ω may be different from the one that arises from another linear ordering. The theorem of Fubini states, however, that such cannot be the case. We will state Fubini's theorem for $\Omega = N\times N$, the general case being similar: Suppose $\{g_k\}_{k\in N}$ is any enumeration of the nonnegative doubly indexed sequence $\{f_{ij}\}_{i,j\in N}$, then $\{g_k\}_{k\in N}$ is summable if, and only if,

$$\sum_{j=0}^{\infty} f_{ij} < \infty \text{ for } i\in N, \text{ and } \sum_{i=0}^{\infty}\left\{\sum_{j=0}^{\infty} f_{ij}\right\} < \infty; \tag{2.13}$$

furthermore, if $\{g_k\}_{k\in N}$ is summable, then

$$\sum_{N} g = \sum_{i=0}^{\infty}\left\{\sum_{j=0}^{\infty} f_{ij}\right\}. \tag{2.14}$$

For a proof, let us first assume that (2.13) holds. Let M be any integer and choose integers I and J so large that $g_1,\dots,g_M$ occur among $\{f_{ij}\mid 0\le i\le I, 0\le j\le J\}$. Then

$$\sum_{m=0}^{M} g_k \le \sum_{i=0}^{I}\sum_{j=0}^{J} f_{ij} \le \sum_{i=0}^{\infty}\left\{\sum_{j=0}^{\infty} f_{ij}\right\} < \infty.$$

This shows that g is summable. Next, assume that g is summable. Let J be an integer and, for a fixed $i \in N$, choose the integer M so large that $f_{i1}, ..., f_{iJ}$ occur among $g_1, ..., g_M$. Then

$$\sum_{j=0}^{J} f_{ij} \leq \sum_{k=0}^{M} g_k \leq \sum_{k=0}^{\infty} g_k,$$

which implies

$$\sum_{j=0}^{\infty} f_{ij} < \infty, \ i \in N.$$

Now let $\{u^{(n)}\}_{n \in N}$ be a sequence of nonnegative sequences in $l^{N \times N}$ each of which has finite support and $0 \leq u^{(0)} \leq u^{(1)} \leq ... \leq f$. For each $n \in N$, let $v^{(n)}$ be the corresponding enumeration of $u^{(n)}$. Then since

$$\sum_{N} v^{(n)} = \sum_{i=0}^{\infty}\sum_{j=0}^{\infty} u_{ij}^{(n)},$$

and since

$$\sum_{j=0}^{\infty} u_{ij}^{(n)} \leq \sum_{j=0}^{\infty} f_{ij} < \infty, \ i \in N,$$

we may apply Lebesgue's monotone convergence theorem as well as (2.12) to obtain

$$\sum_{N} g = \lim_{n \to \infty} \sum_{N} v^{(n)} = \lim_{n \to \infty} \sum_{i=0}^{\infty}\sum_{j=0}^{\infty} u_{ij}^{(n)} = \sum_{i=0}^{\infty} \lim_{n \to \infty} \sum_{j=0}^{\infty} u_{ij}^{(n)} = \sum_{i=0}^{\infty}\sum_{j=0}^{\infty} f_{ij},$$

which shows that (2.13) and (2.14) hold.

The subset of all (complex) sequences $f \in l^{\Omega}$ for which $|f|$ is summable is denoted by l_1^{Ω}. Similarly the set of all sequences in $f = \{f_k\} \in l^{\Omega}$ for which

$$\|f\|_p \equiv \left\{\sum_{\Omega} |f|^p\right\}^{1/p} < \infty, \ p \in (0, \infty),$$

is denoted by l_p^{Ω}. The number $\|f\|_p$ is called the l_p^{Ω}-norm of f. It is convenient to define the infinity norm of f by

$$\|f\|_{\infty} = \max_{k \in \Omega} \{|f_k|\}.$$

The set of all sequences $f \in l^{\Omega}$ for which $\|f\|_{\infty} < \infty$ will be denoted by l_{∞}^{Ω}.

If $f \in l_1^{\Omega}$, we define its sum by

$$\sum_{\Omega} f = \sum_{\Omega} u^{+} - \sum_{\Omega} u^{-} + i\sum_{\Omega} v^{+} - i\sum_{\Omega} v^{-},$$

where $f = u + iv$, and $u^{+}, v^{+}, u^{-}, v^{-}$ are the positive parts and negative parts defined above. Note that each of the four sums on the right hand side exists. Occasionally, it is desirable to define the sum of a real sequence $g = \{g_k\}$ to be

$$\sum_{\Omega} g = \sum_{\Omega} g^{+} - \sum_{\Omega} g^{-},$$

provided that at least one of the sums on the right hand side is finite. The left side is then a number in the extended real number system $[-\infty, \infty]$.

Note that it easily follows from the definition of l_1^Ω that the sum of a finite linear combination of summable sequences in l_1^Ω is equal to the corresponding linear combination of the separate sums, and that for any $f \in l_1^\Omega$,

$$\left|\sum_\Omega f\right| \le \sum_\Omega |f|. \tag{2.15}$$

Lebesgue's dominated convergence theorem holds: Suppose $\{f^{(n)}\}_{n\in N}$ is a sequence of complex sequences in l^Ω such that $f = \lim_{n\to\infty} f^{(n)} \in l^\Omega$. If there is a sequence $g \in l_1^\Omega$ such that $|f^{(n)}| \le g$ for $n \in N$, then $f \in l_1^\Omega$,

$$\lim_{n\to\infty} \sum_\Omega \left|f^{(n)} - f\right| = 0, \tag{2.16}$$

and

$$\lim_{n\to\infty} \sum_\Omega f^{(n)} = \sum_\Omega f. \tag{2.17}$$

The proof is relatively easy. Since $|f| \le g$, and since $|f_n - f| \le 2g$, by Fatou's lemma, we see that

$$\sum_\Omega 2g \le \liminf_{n\to\infty} \sum_\Omega \left(2g - \left|f^{(n)} - f\right|\right) = \sum_\Omega 2g - \limsup_{n\to\infty} \left(\sum_\Omega \left|f^{(n)} - f\right|\right).$$

Thus

$$\lim_{n\to\infty} \sum_\Omega \left|f^{(n)} - f\right| = \limsup_{n\to\infty} \sum_\Omega \left|f^{(n)} - f\right| = 0.$$

Finally, (2.17) follows from (2.16) in view of (2.15).

We conclude this section by Fubini's theorem for l_1^Ω-sequences: Suppose $\{g_k\}_{k\in N}$ is any enumeration of the doubly indexed sequence $\{f_{ij}\}_{i,j\in N}$, then $\{g_k\} \in l_1^N$ if, and only if,

$$\sum_{j=0}^{\infty} |f_{ij}| < \infty \text{ for } i \in N, \text{ and } \sum_{i=0}^{\infty} \left\{\sum_{j=0}^{\infty} |f_{ij}|\right\} < \infty;$$

furthermore, if $\{g_k\} \in l_1^\Omega$, then

$$\sum_N g = \sum_{i=0}^{\infty} \left\{\sum_{j=0}^{\infty} f_{ij}\right\}.$$

The proof follows from breaking f into real, complex, positive and negative parts and then applying Fubini's theorem for nonnegative sequences.

2.5 Convolution of Doubly Infinite Sequences

Let l^Z be the set of all complex sequences of the form

$$f = \{..., f_{-1}, f_0, f_1, ...\} \equiv \{f_k\}_{\in Z}.$$

For the sake of convenience, we will write $\{f_k\}$ instead of $\{f_k\}_{k\in Z}$ if no confusion is caused. For any $\alpha \in C$ and $f = \{f_k\}, g = \{g_k\} \in l^Z$, we define $-f, \alpha f$ and $f+g$ by $\{-f_k\}, \{\alpha f_k\}$ and $\{f_k + g_k\}$ as usual.

There are a number of common sequences in l^Z which deserve special notations. First of all, the sequence $\{..., 0, 0, 0, ...\}$ will be denoted by 0. Let i be an integer, the sequence whose i-th term is 1 and whose other terms are 0 has been called the *Kronecker delta sequence* and denoted by $\hat{h}^{[i]}$, while the jump (or Heaviside) sequence is defined by

$$\hat{H}_k^{(m)} = \begin{cases} 0 & k < m \\ 1 & k \geq m \end{cases}.$$

The sequence $\{\Delta f_k\}_{k\in Z}$ obtained by taking the difference of the consecutive components of the sequence $\{f_k\}$ will be denoted by Δf and is called the *first difference* of f. The higher differences $\Delta^m f$, $m = 2, 3, ...$, are defined recursively. The sequence $\{f_{k+m}\}$ will be denoted by $E^m f$. This definition is good for any integer m. The sequence $E^m f$ will be called a *translated* or *shifted sequence* of f.

Example 25 *It is easy to see that $E^m(f+g) = E^m f + E^m g$, $E(Ef) = E^2 f$, and $\Delta f = Ef - f$ holds for any $f \in l^Z$. Thus,*

$$\Delta^2 f = E(Ef - f) - (Ef - f) = E^2 f - 2Ef + f,$$

$$\Delta^3 f = E^3 f - 3E^2 f + 3Ef + f,$$

etc.

For any complex number $\lambda \neq 0$, the sequence $\{\lambda^k f_k\}$ obtained by scaling each term of the sequence $f = \{f_k\}$ with a growth factor is called an attenuated sequence of f and is denoted by $T^\lambda f$. It is easily seen that $T^1 f = f$, $T^\lambda(T^\mu f) = T^{\lambda\mu} f$ and

$$T^\lambda(f+g) = T^\lambda f + T^\lambda g, \quad f, g \in l^Z.$$

For any two sequences $f = \{f_m\}$ and $g = \{g_m\}$ in l^Z, let Λ be the set of integers such that for each $m \in \Lambda$, the sum

$$\sum_{i=-\infty}^{\infty} |f_i g_{m-i}|$$

is finite, then the function $f * g : \Lambda \to C$ defined by

$$(f * g)_m = \sum_{i=-\infty}^{\infty} f_i g_{m-i}, \quad m \in \Lambda,$$

is called the convolution product of f and g. For the sake of convenience, we also write fg, instead of $f*g$. Note that $f*f$, $f*f*f$, ..., will also be written as $f^2, f^3, ...$, respectively.

Example 26 $\hat{h}^{[\alpha]} * \hat{h}^{[\beta]} = \hat{h}^{[\alpha+\beta]}$ *for* $\alpha, \beta \in Z$.

Example 27 $\hat{h}^{[0]} * f = f$, $0 * f = 0$,*and* $\hat{h}^{[1]} * f = E^{-1}f$.

There are a number of cases in which convolution products are defined on all of Z. First of all, when one of the sequences f or g has finite supports, the convolution product $f * g$ is defined on Z and is called the convolution sequence of f and g. In particular, when f is a finite linear combination of the delta sequences (e.g. $f = a\hat{h}^{[1]} + b\hat{h}^{[0]} + c\hat{h}^{[-1]}$), the convolution sequence of f and any $g \in l^Z$ is well defined.

Next, when the supports of $f, g \in l^Z$ are subsets of N, then $f * g$ is also defined on Z, and it is easily verified that

$$(f * g)_k = \begin{cases} 0 & k < 0 \\ \sum_{j=0}^{k} f_k g_{k-j} & k \geq 0 \end{cases}.$$

There is an elementary fact related to the convolution product of sequences: for any two sequences $f = \{f_k\}, g = \{g_k\}$ in l^Z, $f * g = g * f$. To see this, first consider the special case in which f and g are nonnegative sequences, and let Λ be the domain of definition of $f * g$. For each $t \in \Lambda$,

$$\begin{aligned}(f * g)_t &= \sum_{k=-\infty}^{\infty} f_k g_{t-k} = \sup_{m,n \in N} \sum_{k=m}^{n} f_k g_{t-k} \\ &= \sup_{m,n \in N} \sum_{k=t-m}^{t-n} f_{t-k} g_k = (g * f)_t,\end{aligned}$$

which shows that t is also in the domain of definition of $g * f$, and $(f * g)_t = (g * f)_t$. The general cases can now be proved by breaking f and g into their real parts, imaginary parts and/or positive, negative parts.

There are other important facts. Let l_c^Z be the subspace of l^Z consisting of sequences with finite supports. Then the following facts are true:

$$\|f * g\|_1 \leq \|f\|_1 \|g\|_1, \quad f, g \in l_c^Z,$$

$$\sum_Z f * g = \left(\sum_Z f\right)\left(\sum_Z g\right), \quad f, g \in l_c^Z,$$

$$\|f * g\|_\infty \leq \|f\|_p \|g\|_q, \quad f, g \in l_c^Z, p^{-1} + q^{-1} = 1,$$

$$\|f * g\|_p \leq \|f\|_1 \|g\|_p, \quad f, g \in l_c^Z, p \geq 1,$$

$$(f * g) * h = f * (g * h), \quad f, g, h \in l_c^Z$$

$$f * (g + h) = f * g + f * h, \quad f, g, h \in l_c^Z,$$

and

$$(\alpha f) * (\beta g) = (\alpha\beta)(f * g), \quad \alpha, \beta \in C; f, g \in l_c^Z.$$

The proofs of these statement are rather easy (since we are basically playing with finite sequences) and will be skipped. For sequences which may have infinite supports, we have the following facts. When $f, g \in l_1^Z$, we can show that $f * g$ is defined over Z. Furthermore,

$$\|f * g\|_1 \leq \|f\|_1 \|g\|_1$$

and

$$\sum_Z f * g = \left(\sum_Z f\right)\left(\sum_Z g\right).$$

If $f \in l_p^Z, g \in l_q^Z$ where $p^{-1} + q^{-1} = 1$ (and the case where $p = 1$ and $q = \infty$ is allowed), then $f * g$ is defined over Z, and

$$\|f * g\|_\infty \leq \|f\|_p \|g\|_q ;$$

while if $f \in l_1^Z, g \in l_p^Z$ where $p \geq 1$, then $f * g$ is defined on Z, $f * g \in l_p^Z$ and

$$\|f * g\|_p \leq \|f\|_1 \|g\|_p .$$

When $f, g, h \in l_1^Z$, then

$$(f * g) * h = f * (g * h),$$

$$f * (g + h) = f * g + f * h,$$

and

$$(\alpha f) * (\beta g) = (\alpha\beta)(f * g), \quad \alpha, \beta \in C.$$

To summarize, the linear space l_1^Z equipped with the convolution product forms a commutative Banach algebra with unit element $\hat{h}^{[0]}$. The proofs of these statements can be found in Kecs [98]. We skip them since we will not use these facts in later discussions.

2.6 Frequency Measures

Lebesgue's measure for the set of real numbers does not distinguish countable subsets from finite subsets. However, in some situations, it is necessary to distinguish these two classes of subsets, and regard the latter class as "trivial". For example, a real sequence $\{x_k\}_{k \in N}$ is said to be oscillatory if it is neither eventually positive nor eventually negative, that is, for every positive integer M, there is an integer $n \geq M$ such that $x_n x_{n+1} \leq 0$. However, it is clear that such a definition does not catch all the fine details of an oscillatory sequence. As an example, the sequences $u = \{1, -1, 1, -1, ...\}$ and $v = \{1, 1, 1, 1, -1, 1, 1, 1, 1, -1, ...\}$ are both oscillatory, but the sequence v appears to be "more positive" than the sequence u. In this section, we will introduce several concepts which will help to measure the "positivity" and/or the "negativity" of a real sequence. By introducing such measures, the original concept of an oscillatory sequence can be strengthened further.

Let S be a set of integers. We will denote the set of all integers in S which are less than or equal to an integer n by $S^{(n)}$, that is,

$$S^{(n)} = S \cdot \{..., n-1, n\},$$

and we will denote the set of translate $\{x+m|\ x\in S\}$ by E^mS, where m is an integer. Let α and β be two integers such that $\alpha\leq\beta$. The union

$$\sigma_\alpha^\beta(S)=\sum_{i=\alpha}^{\beta}E^iS$$

will be called a derived set of S. Note that an integer $j\in E^mS$ if, and only if, $j-m\in S$. Thus

$$j\in Z\backslash\left(\sigma_\alpha^\beta(S)\right)\Leftrightarrow j\in\prod_{i=\alpha}^{\beta}Z\backslash(E^iS)\Leftrightarrow j-k\in Z\backslash S \text{ for } \alpha\leq k\leq\beta. \tag{2.18}$$

Example 28 *The derived set* σ_{-1}^1 *of* $\{1,4,7,10,...\}$ *is* $\{0,1,2,3,...\}$.

Let S be a set of integers. If

$$\limsup_{n\to\infty}\frac{|S^{(n)}|}{n}$$

exists, then this limit, denoted by $\mu^*(S)$, will be called the upper frequency measure of S. Similarly, if

$$\liminf_{n\to\infty}\frac{|S^{(n)}|}{n}$$

exists, then this limit, denoted by $\mu_*(S)$, will be called the lower frequency measure of S. If $\mu^*(S)=\mu_*(S)$, then the common limit, denoted by $\mu(S)$, will be called the frequency measure of S.

We illustrate the above definition by several obvious facts, the first of which shows that the upper and lower frequency measures always exist for subsets of N: (1) $\mu(N)=1$, $\mu(\emptyset)=0$ and $0\leq\mu_*(S)\leq\mu^*(S)\leq1$ for any subset S of N. Furthermore, if S is finite, then $\mu^*(S)=0$. (2) Two subsets of Z will have the same upper and lower frequency measures if they differ by a finite number of points. (3) The upper and lower frequency measures are invariant under translation, e.g., $\mu^*(E^mS)=\mu^*(S)$.

Example 29 *Let* $S=\{2,2^2,2^3,...\}$. *Then* $\mu^*(S)=\mu_*(S)=\mu(S)=0$.

Example 30 *Consider the oscillatory sequence* $\{x_k\}_{k\in N}=\{1,-1,1,-1,...\}$. *Let* $S=\{k\in N|\ x_k\geq1/2\}$ *and* $\Gamma=\{k\in N|\ x_k\leq-1/2\}$. *Then* $\mu(S)=\mu(\Gamma)=1/2$.

Example 31 *Consider the oscillatory sequence* $\{x_k\}_{k\in N}=\{1,1,1,1,-1,1,1,1,1,-1,...\}$. *Let* $S=\{k\in N|\ x_k\geq1/2\}$ *and* $\Gamma=\{k\in N|\ x_k\leq-1/2\}$. *Then* $\mu(S)=4/5$ *and* $\mu(\Gamma)=1/5$.

Additional properties of the frequency measures will be considered later. For now, we will make use of these measures to define several concepts which supplement the concepts of oscillation and nonoscillation. For the sake of convenience, we will adopt the usual notation for level sets of a function, that is, let $f:A\to R$ be a real function, then the set $\{t\in A|\ f(t)\leq c\}$ will be denoted by $(f\leq c)$. The notations $(f\geq c)$, $(f<c)$, etc. will have similar meanings.

Let $x = \{x_k\}_{k\in N}$ be a real sequence. If $\mu^*(x \le 0) = 0$, then the sequence x is said to be frequently positive. If $\mu^*(x \ge 0) = 0$, then x is said to be frequently negative. The sequence x is said to be frequently oscillatory if it is neither frequently positive nor frequently negative.

Note that if a sequence x is eventually positive, then it is frequently positive; and if x is eventually negative, then it is frequently negative. Thus, if it is frequently oscillatory, then it is oscillatory.

Let $x = \{x_k\}_{k\in N}$ be a real sequence. If $\mu^*(x \le 0) \le \omega$, then x is said to be frequently positive of upper degree ω. If $\mu^*(x \ge 0) \le \omega$, then x is said to be frequently negative of upper degree ω. The sequence x is said to be frequently oscillatory of upper degree ω if it is neither frequently positive nor frequently negative of the same upper degree ω.

The concepts of frequently positive of lower degree, etc. are similarly defined by means of μ_*. For the sake of completeness, we say that x is frequently positive of the lower degree ω if $\mu_*(x \le 0) \le \omega$, frequently negative of the lower degree ω if $\mu_*(x \ge 0) \le \omega$, and frequently oscillatory if it is neither frequently positive nor frequently negative of the same lower degree ω.

Note that if the sequence x is frequently oscillatory of the lower degree ω, then it is also frequently oscillatory of upper degree ω. Note further that if the sequence x is frequently oscillatory of any upper degree, then it is frequently oscillatory.

Example 32 *Consider the sequence* $x = \{x_k\}_{k\in N} = \{1, 1, 1, -1, 1, 1, 1, -1, ...\}$. *Since* $\mu_*(x \le 0) = 1/4$ *and* $\mu_*(x \ge 0) = 3/4$, *the sequence* x *is frequently positive of lower degree* $1/4$, *frequently negative of lower degree* $3/4$, *and frequently oscillatory of lower degree* $1/5$.

Example 33 *Let the sequence* $x = \{x_k\}_{k\in N}$ *be defined by*

$$x_k = \begin{cases} -1 & k = 2^n,\ n \in N \\ 1 & \text{otherwise} \end{cases}.$$

Then the sequence is oscillatory, and frequently positive. However, it is neither frequently oscillatory nor frequently oscillatory of any positive degree.

Example 34 *It is not difficult to construct a subset* S *of* N *such that* $\mu^*(S) \ne \mu_*(S)$. *Indeed, let*

$$a_m = 2 + 2^{2\cdot 2} + 3^{2\cdot 3} + ... + (m-1)^{2(m-1)}, \quad m = 3, 4, ...$$

Let $A_1 = \{1\}$, $A_2 = \{2, 3, ..., a_3 - 1\}$, $A_3 = \{a_3, ..., a_4 - 1\}$, *and*

$$A_n = \{a_n, ..., a_{n+1} - 1\}, \quad n = 3, 4, ...$$

Then $|A_j| = j^{2j}$ *for* $j \ge 1$. *Finally, let*

$$S = \sum_{|A_i| \text{ is odd}} A_i.$$

We assert that $\mu_*(S) = 0$ *and* $\mu^*(S) = 1$. *To see this, let*

$$b_k = 1 + 2^4 + ... + k^{2k}, \quad k = 2, 4, ...$$

Then

$$\left|S^{(b_k)}\right| = 1 + 3^6 + ... + (k-1)^{2(k-1)},$$

so that

$$\frac{\left|S^{(b_k)}\right|}{b_k} \leq \frac{k(k-1)^{2(k-1)}}{k^{2k}} \leq \frac{1}{k} \to 0.$$

Thus $\mu_*(S) = 0$. *Next, let*

$$c_k = 1 + 2^4 + 3^6 + ... + (k+1)^{2(k+1)}, \quad k = 2, 4, 6, ...$$

Then

$$\left|S^{(c_k)}\right| = 1 + 3^6 + 5^{10} + ... + (k+1)^{2(k+1)},$$

so that

$$\frac{\left|S^{(c_k)}\right|}{c_k} \geq \frac{(k+1)^{2(k+1)}}{k \cdot k^{2k} + (k+1)^{2(k+1)}} = \left(1 + \frac{1}{k}\left(\frac{k}{k+1}\right)^{2(k+1)}\right)^{-1} \to 1.$$

Thus $\mu^*(S) = 1$.

We now consider several additional properties of the frequency measures. Our first result here is concerned with the monotonicity of these measures.

Theorem 5 *For any subsets* Ω *and* Γ *of* N *which satisfy* $\Omega \subseteq \Gamma$, *we have* $\mu^*(\Omega) \leq \mu^*(\Gamma)$ *and* $\mu_*(\Omega) \leq \mu_*(\Gamma)$.

Proof. Since $\Omega \subseteq \Gamma$, thus $\Omega^{(n)} \subseteq \Gamma^{(n)}$ so that

$$\mu^*(\Omega) = \limsup_{n \to \infty} \frac{\left|\Omega^{(n)}\right|}{n} \leq \limsup_{n \to \infty} \frac{\left|\Gamma^{(n)}\right|}{n} = \mu^*(\Gamma),$$

and

$$\mu_*(\Omega) = \liminf_{n \to \infty} \frac{\left|\Omega^{(n)}\right|}{n} \leq \liminf_{n \to \infty} \frac{\left|\Gamma^{(n)}\right|}{n} = \mu_*(\Gamma).$$

■

Additivities of the frequency measures are derived next.

Theorem 6 *Let* Ω *and* Γ *be subsets of* N. *Then* $\mu^*(\Omega + \Gamma) \leq \mu^*(\Omega) + \mu^*(\Gamma)$. *Furthermore, if* Ω *and* Γ *are disjoint, then*

$$\mu_*(\Omega) + \mu_*(\Gamma) \leq \mu_*(\Omega + \Gamma) \leq \mu_*(\Omega) + \mu^*(\Gamma) \leq \mu^*(\Omega + \Gamma) \leq \mu^*(\Omega) + \mu^*(\Gamma).$$

Proof. If Ω and Γ are disjoint, then $(\Omega + \Gamma)^{(n)} = \Omega^{(n)} + \Gamma^{(n)}$ so that

$$\begin{aligned} \mu^*(\Omega + \Gamma) &= \limsup_{n \to \infty} \frac{\left|\Omega^{(n)} + \Gamma^{(n)}\right|}{n} \\ &\geq \limsup_{n \to \infty} \frac{\left|\Gamma^{(n)}\right|}{n} + \liminf_{n \to \infty} \frac{\left|\Omega^{(n)}\right|}{n} = \mu_*(\Omega) + \mu^*(\Gamma). \end{aligned}$$

The other cases are similarly proved. ■

As an immediate consequence, we see that for any subset S of N,

$$1 = \mu_*(N) \leq \mu_*(S) + \mu^*(N \backslash S) \leq \mu^*(N) = 1,$$

and

$$\mu_*(S) + \mu_*(N \backslash S) \leq \mu_*(N) = 1.$$

Several other consequences of the above theorem follow. For example, if $S_1, ..., S_j$ are subsets of N such that $\mu^*(S_i) = 0$ for $1 \leq i \leq j$, then

$$\mu^* \left(\sum_{i=1}^{j} S_i \right) = 0.$$

Next, we have the following results for the differences of subsets of N.

Theorem 7 *Let Ω and Γ be subsets of N such that $\Omega \subseteq \Gamma$. Then*

$$\mu^*(\Gamma) - \mu^*(\Omega) \leq \mu^*(\Gamma \backslash \Omega) \leq \mu^*(\Gamma) - \mu_*(\Omega),$$

and

$$\mu_*(\Gamma) - \mu^*(\Omega) \leq \mu_*(\Gamma \backslash \Omega) \leq \mu_*(\Gamma) - \mu_*(\Omega).$$

Proof. Since $\Gamma = \Omega + (\Gamma \backslash \Omega)$, thus by Theorem 2.6, $\mu^*(\Gamma \backslash \Omega) \geq \mu^*(\Gamma) - \mu^*(\Omega)$ and $\mu^*(\Gamma) \geq \mu_*(\Omega) + \mu^*(\Gamma \backslash \Omega)$. The other cases are similarly proved. ■

Since $S + \Gamma = S + \Gamma \backslash (S \cdot \Gamma)$, the same technique can be used again to deduce the following result.

Theorem 8 *Let S and Γ be subsets of N. Then*

$$\mu_*(S) + \mu^*(\Gamma) - \mu^*(S \cdot \Gamma) \leq \mu^*(S + \Gamma) \leq \mu^*(S) + \mu^*(\Gamma) - \mu_*(S \cdot \Gamma)$$

and

$$\mu_*(S) + \mu_*(\Gamma) - \mu^*(S \cdot \Gamma) \leq \mu_*(S + \Gamma) \leq \mu_*(S) + \mu^*(\Gamma) - \mu_*(\Gamma \cdot S).$$

As seen before, if the upper frequency measure of a subset S of N is strictly greater than 0, then it cannot be a finite set. Thus, if the lower frequency measure of S is strictly less than 1, then $N \backslash S$ cannot be a finite set. An additional basic result of similar nature can also be derived.

Theorem 9 *Let S and Γ be subsets of N such that $\mu^*(S) + \mu_*(\Gamma) > 1$. Then $S \cdot \Gamma$ cannot be a finite set.*

Indeed, if $S \cdot \Gamma$ is finite, then $\mu^*(S \cdot \Gamma) = 0$ and in view of $S \subseteq (N \backslash \Gamma) + S \cdot \Gamma$, we have

$$\mu^*(S) \leq \mu^*(N \backslash \Gamma) + \mu^*(S \cdot \Gamma) = \mu^*(N \backslash \Gamma).$$

Thus by Theorem 6,

$$1 < \mu^*(S) + \mu_*(\Gamma) \leq \mu^*(N \backslash \Gamma) + \mu_*(\Gamma) = 1,$$

which is a contradiction.

The final property of the frequency measures is concerned with the translates.

Theorem 10 *For any subset Ω of N, we have*

$$\mu^*\left(\sigma_\alpha^\beta(\Omega)\right) \le (\beta-\alpha+1)\mu^*(\Omega),$$

and

$$\mu_*\left(\sigma_\alpha^\beta(\Omega)\right) \le (\beta-\alpha+1)\mu_*(\Omega).$$

Proof. The first inequality follows from Theorem 2.6 and the invariance of the frequency measures. Next, note that

$$\left|\left(\sigma_\alpha^\beta(\Omega)\right)^{(n)}\right| \le (\beta-\alpha+1)\left|\Omega^{(n)}\right| + |\alpha|\,(\beta-\alpha+1).$$

Thus

$$\mu_*\left(\sigma_\alpha^\beta(\Omega)\right) \le \liminf_{n\to\infty} \frac{(\beta-\alpha+1)\left|\Omega^{(n)}\right| + |\alpha|\,(\beta-\alpha+1)}{n} = (\beta-\alpha+1)\mu_*(\Omega).$$

■

We remark that the previous properties are stated for subsets of N. However, since finite sets of integers have frequency measures zero, these subsets can also be regarded as subsets of a ray $\{a, a+1, a+2, ...\}$ of integers. Such an observation will be useful in later discussions.

The concepts introduced in the previous discussions can be easily extended to suit multiple sequences. For simplicity, we will confine ourselves to a formal presentation of frequency measures for bivariate sequences of the form $\{x_{ij}\}_{i,j\in N}$.

Let S be a set of lattice points. We write

$$S^{(m,n)} = \{(i,j) \in S \mid i \le m, j \le n\}.$$

Given integers m and n, the translation operators E_x^m and E_y^n are defined by

$$E_x^m S = \{(i+m,j) \in Z^2 \mid m(i,j) \in S\}$$

and

$$E_y^n S = \{(i,j+n) \in Z^2 \mid (i,j) \in S\}$$

respectively. A subset of the form $E_x^m E_y^n S$ is said to be a translate of S. Let α, β, γ and τ be integers such that $\alpha \le \beta$ and $\gamma \le \tau$. The union

$$\sum_{i=\alpha}^{\beta}\sum_{j=\gamma}^{\tau} E_x^i E_y^j S$$

is called a derived set of S. Note that the following important relation holds:

$$(i,j) \in Z^2 \backslash \sum_{i=\alpha}^{\beta}\sum_{j=\gamma}^{\tau} E_x^i E_y^j S \tag{2.19}$$

$$\Longleftrightarrow \quad (i-k, j-l) \in Z^2\backslash S \text{ for } \alpha \le k \le \beta, \gamma \le l \le \tau.$$

Let S be a set of lattice points. If

$$\limsup_{m,n\to\infty} \frac{\left|S^{(m,n)}\right|}{mn}$$

exists, then this limit, denoted by $\mu^*(S)$, will be called the upper frequency measure of S. Similarly, if

$$\liminf_{m,n\to\infty} \frac{|S^{(m,n)}|}{mn}$$

exists, then this limit, denoted by $\mu_*(S)$, will be called the lower frequency measure of S. If $\mu^*(S) = \mu_*(S)$, then the common limit, denoted by $\mu(S)$, will be called the frequency measure of S.

We say that a bivariate sequence $x = \{x_{ij}\}_{i,j\in N}$ is eventually positive if $x_{ij} > 0$ for all large i, j. An eventually negative bivariate sequence is similarly defined.

Let $x = \{x_{ij}\}_{i,j\in N}$ be a real bivariate sequence. If $\mu^*(x \le 0) = 0$, then the sequence x is said to be frequently positive. If $\mu^*(x \ge 0) = 0$, then x is said to be frequently negative. The sequence x is said to be frequently oscillatory if it is neither frequently positive nor frequently negative.

We remark that if a bivariate sequence is eventually positive, then it is frequently positive; and if it is eventually negative, then it is also frequently negative.

Let $x = \{x_{ij}\}_{i,j\in N}$ be a real bivariate sequence. If $\mu^*(x \le 0) \le \omega$, then x is said to be frequently positive of upper degree ω. If $\mu^*(x \ge 0) \le \omega$, then x is said to be frequently negative of upper degree ω. The sequence x is said to be frequently oscillatory of upper degree ω if it is neither frequently positive nor frequently negative of the same upper degree ω. The concepts of frequently positive of lower degree, etc. are similarly defined by means of μ_*.

Clearly, if a bivariate sequence is frequently oscillatory of lower degree ω, then it is also frequently oscillatory of upper degree ω; and if it is frequently oscillatory of any upper degree, then it is frequently oscillatory.

Most of the properties of the two dimensional frequency measures are similar to those of the one dimensional measures. We list the important ones as follows.

Theorem 11 *$\mu(\emptyset) = 0$, $\mu(N^2) = 1$ and $0 \le \mu_*(S) \le \mu^*(S) \le 1$ for any subset S of N^2. Furthermore, $\mu^*(S) = 0$ for any finite subset S of Z^2. Two subsets of lattice points will have the same upper and lower frequency measures if they differ by a finite number of points or if they are translates of each other.*

Theorem 12 *Let $S, \Gamma \subseteq N^2$. Then $\mu^*(S + \Gamma) \le \mu^*(S) + \mu^*(\Gamma)$,*

$$\mu_*(S) + \mu^*(\Gamma) - \mu^*(S \cdot \Gamma) \le \mu^*(S + \Gamma) \le \mu^*(S) + \mu^*(\Gamma) - \mu_*(S \cdot \Gamma),$$

and

$$\mu_*(S) + \mu_*(\Gamma) - \mu^*(S \cdot \Gamma) \le \mu_*(S + \Gamma) \le \mu_*(S) + \mu^*(\Gamma) - \mu_*(S \cdot \Gamma).$$

Furthermore, if S and Γ are disjoint, then

$$\mu_*(S) + \mu_*(\Gamma) \le \mu_*(S + \Gamma) \le \mu_*(S) + \mu^*(\Gamma) \le \mu^*(S + \Gamma) \le \mu^*(S) + \mu^*(\Gamma).$$

As an immediate corollary of Theorems 11 and 12,

$$\mu_*(S) + \mu^*(N^2 \backslash S) = 1$$

for any subset of N^2. Therefore, if S has positive upper frequency measure, then it cannot be a finite subset of N^2.

Theorem 13 *Let $S, \Gamma \subseteq N^2$. If $\mu^*(S) + \mu_*(\Gamma) > 1$, then $S \cdot \Gamma$ is an infinite set.*

Theorem 14 *Suppose $S \subseteq \Gamma \subseteq N^2$, then $\mu^*(S) \le \mu^*(\Gamma)$, $\mu_*(S) \le \mu_*(\Gamma)$,*

$$\mu^*(\Gamma) - \mu^*(S) \le \mu^*(\Gamma \backslash S) \le \mu^*(\Gamma) - \mu_*(S),$$

and

$$\mu_*(\Gamma) - \mu^*(S) \le \mu_*(\Gamma \backslash S) \le \mu_*(\Gamma) - \mu_*(S).$$

Theorem 15 *Let $S \subseteq N^2$ and let α, β, γ and τ be integers such that $\alpha \le \beta$ and $\gamma \le \tau$. Then*

$$\mu_* \left(\sum_{i=\alpha}^{\beta} \sum_{j=\gamma}^{\tau} E_x^i E_y^j S \right) \le (\beta - \alpha + 1)(\tau - \gamma + 1)\mu_*(S),$$

and

$$\mu^* \left(\sum_{i=\alpha}^{\beta} \sum_{j=\gamma}^{\tau} E_x^i E_y^j S \right) \le (\beta - \alpha + 1)(\tau - \gamma + 1)\mu^*(S).$$

2.7 Useful Results For Matrices

We first recall [77] that the eigenvalues of the symmetric matrix

$$\begin{bmatrix} \beta & \alpha & 0 & \dots & 0 \\ \alpha & \beta & \alpha & \dots & 0 \\ & \cdot & \cdot & \cdot & \\ 0 & \dots & \alpha & \beta & \alpha \\ 0 & \dots & 0 & \alpha & \beta \end{bmatrix}_{n \times n}, \quad \alpha, \beta \in C,$$

are

$$\lambda_i = \beta + 2\alpha \cos \frac{i\pi}{n+1}, \quad i = 1, 2, \dots, n.$$

Now, suppose $a, c \in R$ and $ac \ne 0$, then by defining

$$p_1 = 1, p_2 = \sqrt{\frac{c}{a}}, \dots, p_n = \left(\frac{c}{a}\right)^{(n-1)/2},$$

we may easily verify that

$$\begin{bmatrix} p_1 & 0 & 0 & \dots & 0 \\ 0 & p_2 & 0 & \dots & 0 \\ & \cdot & \cdot & \cdot & \\ 0 & \dots & 0 & p_{n-1} & 0 \\ 0 & \dots & 0 & 0 & p_n \end{bmatrix} \begin{bmatrix} b & c & 0 & \dots & 0 \\ a & b & c & \dots & 0 \\ & \cdot & \cdot & \cdot & \\ 0 & \dots & a & b & c \\ 0 & \dots & 0 & a & b \end{bmatrix} \begin{bmatrix} p_1 & 0 & 0 & \dots & 0 \\ 0 & p_2 & 0 & \dots & 0 \\ & \cdot & \cdot & \cdot & \\ 0 & \dots & 0 & p_{n-1} & 0 \\ 0 & \dots & 0 & 0 & p_n \end{bmatrix}^{-1}$$

$$= \begin{bmatrix} b & \sigma\sqrt{ac} & 0 & \dots & 0 \\ \sigma\sqrt{ac} & b & \sigma\sqrt{ac} & \dots & 0 \\ & \cdot & \cdot & \cdot & \\ 0 & \dots & \sigma\sqrt{ac} & b & \sigma\sqrt{ac} \\ 0 & \dots & 0 & \sigma\sqrt{ac} & b \end{bmatrix},$$

where σ is the sign of a. Therefore, we have the following result.

Theorem 16 *Suppose $a, c > 0$, the eigenvalues of*

$$M = \begin{bmatrix} b & c & 0 & \dots & 0 \\ a & b & c & \dots & 0 \\ & \cdot & \cdot & \cdot & \\ 0 & \dots & a & b & c \\ 0 & \dots & 0 & a & b \end{bmatrix}_{n\times n}$$

are given by

$$\lambda_k(M) = b + 2\sigma\sqrt{ac}\cos\frac{k\pi}{n+1}, \quad k = 1, ..., n,$$

and the corresponding eigenvectors are given by

$$\text{col}\left(\sin\frac{k\pi}{n+1}, \left(\frac{c}{a}\right)^{-1/2}\sin\frac{2k\pi}{n+1}, ..., \left(\frac{c}{a}\right)^{-(n-1)/2}\sin\frac{nk\pi}{n+1}\right), \quad k = 1, 2, ..., n.$$

In case $ac = 0$, then the eigenvalues of M are all equal to b.

Next, the well known *Jacobi matrix*

$$\mathbb{J} = \begin{bmatrix} 2 & -1 & 0 & \cdot & \cdot & 0 \\ -1 & 2 & -1 & 0 & \cdot & \cdot \\ 0 & \cdot & \cdot & \cdot & \cdot & \cdot \\ \cdot & \cdot & \cdot & \cdot & \cdot & 0 \\ \cdot & \cdot & 0 & -1 & 2 & -1 \\ 0 & \cdot & \cdot & 0 & -1 & 2 \end{bmatrix}_{n\times n} \tag{2.20}$$

has an inverse $G = (g_{ij})$, which is given by

$$g_{ij} = \begin{cases} (n-i+1)j/(n+1) & 1 \le j \le i \le n \\ (n-j+1)i/(n+1) & 1 \le i \le j \le n \end{cases}. \tag{2.21}$$

For example, when $n = 5, 6$, the corresponding inverses are given by

$$\frac{1}{6}\begin{bmatrix} 5 & 4 & 3 & 2 & 1 \\ 4 & 8 & 6 & 4 & 2 \\ 3 & 6 & 9 & 6 & 3 \\ 2 & 4 & 6 & 8 & 4 \\ 1 & 2 & 3 & 4 & 5 \end{bmatrix}, \quad \frac{1}{7}\begin{bmatrix} 6 & 5 & 4 & 3 & 2 & 1 \\ 5 & 10 & 8 & 6 & 4 & 2 \\ 4 & 8 & 12 & 9 & 6 & 3 \\ 3 & 6 & 9 & 12 & 8 & 4 \\ 2 & 4 & 6 & 8 & 10 & 5 \\ 1 & 2 & 3 & 4 & 5 & 6 \end{bmatrix}$$

respectively.

Example 35 *The matrix $G = (g_{ij})$ defined above is symmetric, positive for $1 \le i, j \le n$ and*

$$\sum_{j=1}^{n} g_{ij}\Delta^2 v_{j-1} = -v_i + \delta_i, \quad 1 \le i \le n \tag{2.22}$$

where $\delta_1 = nv_0/(n+1), \delta_n = nv_{n+1}/(n+1)$ and $\delta_i = 0$ for $2 \le i \le n-1$.

Example 36 *The eigenvalue problem*

$$\begin{aligned}\Delta^2 w_{k-1} + \lambda w_k = 0, \quad 1 \le k \le n,\\ w_0 = 0 = w_{n+1},\end{aligned} \tag{2.23}$$

is equivalent to the symmetric eigenvalue problem

$$\mathbf{J}w = \lambda w,$$

where $w = \text{col}(w_1, ..., w_n)$, *or to the symmetric eigenvalue problem*

$$w = \lambda G w. \tag{2.24}$$

It possesses the positive eigenvalue

$$\mu_n = 4\sin^2\left(\frac{\pi}{2n+2}\right). \tag{2.25}$$

This eigenvalue is smaller than all other eigenvalues and its corresponding eigensolution $\{w_k\}$ *is given by*

$$w_k = \begin{cases} 0 & k = 0 \text{ or } n+1 \\ \sin(k\pi/(n+1)) & 1 \le k \le n \end{cases}. \tag{2.26}$$

More generally, we have the following result, which can be verified directly.

Theorem 17 *Let*

$$A = \begin{bmatrix} 2+a & -1 & 0 & & & & \\ -1 & 2 & -1 & 0 & & & \\ 0 & \cdot & \cdot & \cdot & & & \\ & & \cdot & \cdot & \cdot & 0 & \\ & & 0 & -1 & 2 & -1 & \\ & & & 0 & -1 & 2+b \end{bmatrix}_{n\times n}.$$

Then the principal determinants of A are $2+a, 3+2a, ..., n+(n-1)a$ *and* $n+1+na+nb+(n-1)ab$, *respectively. If the determinant of A (which is equal to* $n+1+na+nb+(n-1)ab$*) is not zero, the inverse* $H = (h_{ij})$ *of A is given by*

$$h_{ij} = \begin{cases} (j+(j-1)a)((n+1-i)+(n-i)b)/\det A & 1 \le j \le i \le n \\ (i+(i-1)a)((n+1-j)+(n-j)b)/\det A & 1 \le i \le j \le n \end{cases}.$$

A result (see e.g. Varga [156]) for nonnegative matrix due to Perron will be useful to us later.

Theorem 18 *Let B be a nonnegative square matrix. Then B has a nonnegative real eigenvalue equal to its spectral radius* $\rho(B)$, *and there corresponds to it a nonnegative eigenvector.*

2.8 Discrete Gronwall Inequalities

We will need the following discrete Gronwall type inequality.

Theorem 19 *Let $\{B_i\}_{i\in N}$ be a nonnegative sequence and $\{D_i\}_{i\in N}$ be a real sequence. Suppose $\{v_i\}_{i\in N}$ is a solution of the functional inequality*

$$v_{i+1} \le B_i v_i + D_i, \quad i \in N. \tag{2.27}$$

Then

$$v_j \le \prod_{i=0}^{j-1} B_i v_0 + \sum_{i=0}^{j-1} D_i \prod_{n=i+1}^{j-1} B_n, \; j \in N. \tag{2.28}$$

The proof is elementary. If we multiply (2.27) by the "integrating factor"

$$\prod_{k=i}^{j-1} B_k,$$

then since

$$\Delta\left\{v_i \prod_{k=i}^{j-1} B_k\right\} = \left(\prod_{k=i+1}^{j-1} B_k\right)(v_{i+1} - B_i v_i) \le D_i \prod_{k=i+1}^{j-1} B_k,$$

we see that (2.28) holds by summing both sides of the above inequality from $i = 0$ to $i = j - 1$.

As a corollary, if $\{b_n\}_{n\in N}$ is a nonnegative sequence, and if $\{x_i\}_{i\in N}$ is a solution of the functional inequality

$$x_n \le a_n + \sum_{s=0}^{n-1} b_s x_s, \; n \in N,$$

then we will have

$$x_n \le a_n + \sum_{i=0}^{n-1} a_i b_i \prod_{n=i+1}^{n-1} (1 + b_n), \; n \in N.$$

Indeed, setting

$$v_n = \sum_{s=0}^{n-1} b_s x_s, \; n \in N,$$

we see that $v_0 = 0$, and

$$\Delta v_n = b_n x_n \le a_n b_n + b_n v_n, \; n \in N.$$

Hence

$$v_{n+1} \le (1 + b_n) v_n + a_n b_n, \; n \in N,$$

so that

$$x_n \le a_n + v_n \le a_n + \sum_{i=0}^{n-1} a_i b_i \prod_{s=i+1}^{n-1} (1 + b_s), \; n \in N.$$

In case $\{a_n\}$ satisfy the additional conditions $a_n \geq 0$ and $\Delta a_n \geq 0$ for $n \in N$, then since

$$1 + \sum_{i=0}^{n-1} b_i \prod_{s=i+1}^{n-1} (1 + b_s) \leq \prod_{s=0}^{n-1} (1 + b_s),$$

and

$$1 + b_s \leq b_s, \ s \in N,$$

we see further that (Lees [105])

$$x_n \leq a_n \prod_{s=0}^{n-1} (1 + b_s) \leq a_n \exp\left(\sum_{s=0}^{n-1} b_s\right), \ n \in N.$$

2.9 Miscellaneous

We will need two fixed point theorems in our later chapters. The first one is the Brouwer fixed point theorem.

Theorem 20 *Let Ω be a nonempty, closed, bounded and convex subset of R^n, and let $f : \Omega \to \Omega$ be a continuous mapping. Then f has a fixed point in Ω.*

The second is Banach's fixed point theorem.

Theorem 21 *Let Ω be a nonempty complete metric space and let $T : \Omega \to \Omega$ be a contraction mapping. Then T has a fixed point in Ω.*

Recall that Ω is a metric space if there is a metric $d : \Omega \times \Omega \to [0, \infty)$ which satisfies (i) for every pair of $x, y \in \Omega$, $d(x, y) = 0$ if, and only if, $x = y$, (ii) $d(x, y) = d(y, x)$ for $x, y \in \Omega$, and (iii) $d(x, z) \leq d(x, y) + d(y, x)$ for $x, y, z \in \Omega$. Ω is said to be complete if every Cauchy sequence in Ω converges to a point in Ω. $T : \Omega \to \Omega$ is a contraction if there is number λ in $[0, 1)$ such that $d(Tx, Ty) \leq \lambda d(x, y)$ for all $x, y \in \Omega$.

A large number of metric spaces are normed linear spaces, that is, linear spaces whose metrics are induced by norms. Recall that a norm $\|\cdot\|$ on a linear space Ω is a function that maps Ω into $[0, \infty)$ such that (i) for every $x \in \Omega$, $\|x\| = 0$ if, and only if, $x = 0$, (ii) $\|\alpha x\| = |\alpha| \, \|x\|$ for any scalar α and $x \in \Omega$, and (iii) $\|x + y\| \leq \|x\| + \|y\|$ for $x, y \in \Omega$. When a normed linear space is also a complete metric space, it is called a Banach space.

Let Ω be a convex subset of R^n, and let $f : \Omega \to R$. Recall that f is said to be convex if

$$f(\alpha x + (1 - \alpha) y) \leq \alpha f(x) + (1 - \alpha) f(y), \ x, y \in \Omega; \alpha \in (0, 1).$$

By induction, it is then not difficult to see that

$$f\left(\sum_{i=1}^{n} \alpha_i x_i\right) \leq \sum_{i=1}^{n} \alpha_i f(x_i)$$

holds for $x_1, ..., x_n \in \Omega$ and $\alpha_1, ..., \alpha_n \in (0, 1)$ satisfying $\alpha_1 + ... + \alpha_n = 1$. The above relation is called Jensen's inequality.

We will need the following result.

Theorem 22 *Consider the linear homogeneous difference equation*

$$x_{n+\sigma} + c_1 x_{n+\sigma-1} + c_2 x_{n+\sigma-2} + \ldots + c_\sigma x_n = 0,\ n \in N,$$

where $\sigma \in N$ and $c_1, \ldots, c_\sigma$ are real numbers. Then it has an eventually positive solution if, and only if, its characteristic equation

$$\lambda^\sigma + c_1\lambda^{\sigma-1} + \ldots + c_\sigma = 0$$

has a positive root.

Another elementary result is also needed [165].

Theorem 23 *Let $\{x_i\}_{i=m}^{\infty}$ be a real sequence such that*

$$|x_i| \le \alpha + \beta \sup_{m\le s\le i} |x_s|,\ i \ge T \ge m,$$

where $\alpha \ge 0$, $1 > \beta \ge 0$ and T is a fixed integer. Then

$$\sup_{m\le s\le i} |x_s| \le \frac{\alpha}{1-\beta} + \frac{1}{1-\beta} \sup_{m\le s\le T} |x_s|,\ i \ge T. \tag{2.29}$$

Indeed, since

$$\sup_{T\le s\le i} |x_s| \le \alpha + \beta \sup_{m\le s\le i} |x_s|$$

thus

$$\sup_{m\le s\le i} |x_s| \le \sup_{m\le s\le T} |x_s| + \sup_{T\le s\le i} |x_s| \le \sup_{m\le s\le T} |x_s| + \alpha + \beta \sup_{m\le s\le i} |x_s|$$

as required.

2.10 Notes and Remarks

The lattice plane and its subsets are simple but useful concepts and probably well known to many authors (see e.g. Cheng [14], Cheng and Lu [22]). Theorem 1 is due to Cheng *et al.* [20].

Discrete analogs of the Green's theorems are also simple and useful results. The results stated in this book are from Cheng and Lu [22].

Summable sequences are standard subjects of study in analysis and there are a great many results related to them. We have only presented some basic results which will be useful in later discussions.

Convolution products or Cauchy products are naturally related to the principle of superposition and hence they are important in any linear functional relations. Again, we have only presented some basic results. The interested reader may consult books in standard analysis for additional information.

Frequency measures appear in Tian *et al.* [153]. Since these and related concepts are new, we expect some significant follow up result in the future. Right now, there are only some results in oscillation theory of partial difference equations which make use of these concepts, and one such result will be presented in Chapter 8.

Theorem 16 is a correction of Lemma 3 in Cheng and Lin [32], and Theorem 17 is due to Cheng and Lu [24].

Discrete Gronwall type inequalities can be found in the book by Bainov and Simeonov [5]. The elementary proof of Theorem 19 is due to Zhang and Cheng [175].

Theorem 22 is known to a number of authors. A proof by means of the Z-transform can be found in Györi and Ladas [78]. A more general approach which does not use the Z-transform method can also be found in [44].

Chapter 3

Symbolic Calculus

3.1 Introduction

There are several methods that are helpful in manipulating sequences such as those implicitly defined by functional relations. One well known method is the method of generating functions [143, 144], while another one is the method of Z-transform (see e.g. Vich [158], Gregor [74]). The operational calculus of Mikusinski [121] is also helpful in this respect. However, his calculus is based on the Titchmarsh's theorem for convolution product which is technically complicated. To circumvent this problem, several authors introduced a number of elementary calculi by considering convolution rings of sequences. In this chapter, we intend to gather the essential features of these calculi into an elementary symbolic calculus and expound on its various diversifications and possible applications in partial difference equations.

3.2 Semi-Infinite Univariate Sequences

3.2.1 Ring of Sequences

In practice, an infinite univariate sequence can either be singly infinite or doubly infinite: $\{u_k\}_{k=a}^{\infty}$, or $\{u_k\}_{k=-\infty}^{\infty}$. It turns out the set of all singly infinite sequences form a ring under the usual addition and a "convolution" product.

Let C denote the set of all complex numbers. Let l^N be the set of all complex sequences of the form $f = \{f_k\}_{k\in N}$. We will call f_k the k-th term of the sequence f. Note that the first k terms of f are $f_0, ..., f_{k-1}$ respectively. For the sake of convenience, we will write $\{f_k\}$ instead of $\{f_k\}_{k\in N}$ if no confusion is caused. For any $\alpha \in C$ and $f = \{f_k\}, g = \{g_k\} \in l^N$, we define $-f, \alpha f$ and $f+g$ by $\{-f_k\}, \{\alpha f_k\}$ and $\{f_k + g_k\}$ as usual.

There are a number of common sequences in l^N which deserve special notations. First of all, let α be a complex number, the sequence $\{\alpha, 0, 0, ...\}$ will be denoted by $\overline{\alpha}$ and is called a *scalar sequence*. In particular, the sequences $\{0, 0, ...\}$ and $\{1, 0, ...\}$ will be denoted by $\overline{0}$ and $\overline{1}$ respectively. The sequence $\{1, 1, 1, ...\}$ will be denoted by σ.

Let m be a nonnegative integer. Recall that $\hbar^m$ denotes the (Dirac) sequence defined by

$$\hbar_k^m = \begin{cases} 1 & k = m \\ 0 & k \neq m \end{cases},$$

and $H^{(m)}$ denotes the jump (or Heaviside) sequence defined by

$$H_k^{(m)} = \begin{cases} 0 & 0 \leq k < m \\ 1 & k \geq m \end{cases}.$$

Note that we have also used $\hbar^0$ to denote $\bar{1}$ and $H^{(0)}$ for σ. It is also convenient to write $\hbar$ instead of $\hbar^1$ and this practice will be assumed for similar situations in the sequel.

We have adopted the convention that for any complex number z, $z^{[0]} = 1$ and that

$$z^{[m]} = z(z-1)(z-2)\cdots(z-m+1), \ m \in Z^+.$$

Note that $n^{[n]} = n!$, $0^{[0]} = 0! = 1$, and $0 = n^{[n+1]} = n^{[n+2]} = \dots$ for $n \in N$. Therefore, the sequence $\{1, -3, 3, -1, 0, \dots\}$ can be written as $\{(-1)^k 3^{[k]}/k!\}$, and the sequence $\{1, 2, 3, \dots\}$ as $\{(k+1)^{[1]}\}$.

The Gamma function can be used to extend the meaning of $z^{[m]}$. We first recall that the Gamma function $\Gamma(z)$ is meromorphic on C, has poles at $0, -1, -2, \dots$, and satisfies the properties $\Gamma(z+1) = z\Gamma(z)$ for $z \in C$, $\Gamma(n) = (n-1)!$ for $n \in Z^+$ as well as $\Gamma(1/2) = \sqrt{\pi}$. Thus $z^{[m]} = \Gamma(z+1)/\Gamma(z-m+1)$ for $m \in Z^+$ and $z \neq m-1, m-2, \dots$.

The sequence $\{\Delta f_0, \Delta f_1, \dots\}$ obtained by taking the difference of the consecutive components of the sequence $\{f_k\}$ will be denoted by Δf and is called the *first difference* of f. The higher differences $\Delta^m f$, $m = 2, 3, \dots$, are defined recursively. The sequence $\{f_m, f_{m+1}, \dots\}$ obtained by deleting the first m terms of the sequence $\{f_0, \dots, f_m, f_{m+1}, \dots\}$ will be denoted by $E^m f$, and the sequence $\{0, \dots, 0, f_0, f_1, \dots\}$ obtained by adding m zeros to the front of the terms of f by $E^{-m} f$. These definitions require $m \geq 1$. However, it is natural to define $\Delta^0 f = f$ and $E^0 f = f$. Note that we have

$$E^m \{f_k\}_{k \in N} = \{f_{m+k}\}_{k \in N}, \ m \in Z^+,$$

and

$$E^m E^{-m} f = f, \ m \in Z^+,$$

but for $m \in Z^+$, $E^{-m} E^m f$ is not equal to f in general. The sequence $E^m f$ will be called a *translated* or *shifted sequence* of f.

Example 37 *It is easy to see that* $E^m(f+g) = E^m f + E^m g$, $E(Ef) = E^2 f$, *and* $\Delta f = Ef - f$ *holds for any* $f \in l^N$. *Thus,*

$$\Delta^2 f = E(Ef - f) - (Ef - f) = E^2 f - 2Ef + f,$$

$$\Delta^3 f = E^3 f - 3E^2 f + 3Ef + f,$$

etc.

For any complex number λ, where $\lambda = 0$ is allowed, the sequence $\{\lambda^k f_k\}$ obtained by scaling each term of the sequence $f = \{f_k\}$ with a growth factor is called an *attenuated sequence* of f and is denoted by $T^\lambda f$. It is easily seen that $T^0 f = \overline{f_0}$, $T^1 f = f$, $T^\lambda(T^\mu f) = T^{\lambda\mu} f$ and

$$T^\lambda(f+g) = T^\lambda f + T^\lambda g, \quad f, g \in l^N.$$

For any $f = \{f_k\}$, $g = \{g_k\} \in l^N$, we define the *convolution product* $f * g$, by

$$(f * g)_k = \sum_{i=0}^{k} f_{k-i} g_i, \quad k \in N.$$

Example 38 $\overline{1} * f = f$, $\overline{0} * f = \overline{0}$, $\overline{\alpha} * \overline{\beta} = \overline{\alpha\beta}$, $\overline{\alpha} * f = (\alpha\overline{1}) * f = \alpha(\overline{1} * f) = \alpha f$ *and* $\hbar * f = E^{-1} f$.

For the sake of convenience, we write fg, instead of $f * g$. Note that $f * f$, $f * f * f, \ldots$, will also be written as $f^2, f^3, \ldots$, respectively.

Several elementary facts related to the convolution product of sequences will be useful later. First of all, we may easily show that for any two sequences $f = \{f_k\}, g = \{g_k\}$ in l^N, $f * g = g * f$. Furthermore, if $f * g = \overline{0}$, then $f = \overline{0}$ or $g = \overline{0}$. Indeed, suppose that $f_0 = \ldots = f_{m-1} = 0$, $f_m \neq 0$, $g_0 = \ldots = g_{n-1} = 0$ and $g_n \neq 0$. Then assume without loss of generality that $m \leq n$, we have

$$(fg)_{m+n} = f_0 g_{m+n} + \ldots + f_m g_n + \ldots f_{m+n} g_0 = f_m g_n \neq 0.$$

This shows that $f * g \neq \overline{0}$.

Theorem 24 *Let* $f = \{f_k\}$, $g = \{g_k\}$ *be sequences in* l^N. *If* $g_0 \neq 0$, *then there is a unique sequence* $x = \{x_k\} \in l^N$ *such that* $g * x = f$.

The proof is elementary. We simply note that the infinite linear system

$$g_0 x_0 = f_0,$$

$$g_0 x_1 + g_1 x_0 = f_1,$$

$$g_0 x_2 + g_1 x_1 + g_2 x_0 = f_2, \ \ldots,$$

can be solved successively in the following unique manner: $x_0 = f_0/g_0$, $x_1 = (f_1 - g_1 x_0)/g_0, \ \ldots$.

Theorem 25 *Let* $f = \{f_k\} \in l^N$. *If* $f_0 = 0$, *then the first* n *terms of the convolution product* f^n *are equal to zero.*

Indeed, it is not difficult to show that the two terms of the sequence $f * f = f^2$ are equal to zero, and then by induction that the first n terms of the sequence f^n are equal to zero.

Theorem 26 *Let* $f = \{f_k\}$, $g = \{g_k\} \in l^N$. *Then* $T^\lambda(f * g) = (T^\lambda f) * (T^\lambda g)$ *for* $\lambda \in C$.

Indeed,

$$T^\lambda(fg) = \left\{\sum_{j=0}^{k} \lambda^k f_j g_{k-j}\right\} = \left\{\sum_{j=0}^{k} \lambda^j f_j \lambda^{k-j} g_{k-j}\right\} = (T^\lambda f)\,(T^\lambda g).$$

3.2.2 Operators

It is easily verified that under the above addition and convolution product, l^N is a commutative ring with no zero divisor, i.e. $f * g = \bar{0}$ implies $f = \bar{0}$ or $g = \bar{0}$, and the additive and multiplicative identities are $\bar{0}$ and $\bar{1}$ respectively. Consequently, we can construct from l^N a field l^N/l^N of quotients in a routine manner (see e.g. Dubisch [58]). Briefly, on the set

$$\{(f,g) | \ f,g \in l^N, g \neq 0\},$$

we define a relation $\sim$ by

$$(f,g) \sim (p,q) \Longleftrightarrow fq = pg.$$

Then it is easily verified that $\sim$ is an equivalence relation. Each equivalence class can be represented by an ordered pair (f,g), which is written in the form of a quotient f/g in analogy with the rationals. The set of all such quotients is denoted by l^N/l^N. Hence, two quotients f/g and p/q are equal if, and only if, $fq = pg$. We define product and addition of quotients by

$$\frac{f}{g}\frac{p}{q} = \frac{fp}{gq},$$

and

$$\frac{f}{g} + \frac{p}{q} = \frac{fq + pg}{gq}$$

respectively. Under these definitions, l^N/l^N is a field with the unique additive identity $\bar{0}/\bar{1}$ and the unique multiplicative identity $\bar{1}/\bar{1}$. For convenience, we will call a quotient f/g an **operator**. As usual, the additive inverse of an operator f/g is $-f/g$, which is also denoted by $-(f/g)$. Operators can be divided by one another. More specifically, if $\phi = p/q$ and $\psi = u/v$ are operators where $p, q, u, v \in l^N$ with $u \neq 0$, then the quotient ϕ/ψ of the operators ϕ and ψ is pv/qu. In particular, when ψ is a nonzero operator, $\bar{1}/\psi$, which is also denoted by ψ^{-1}, is called the multiplicative inverse of the operator ψ. Note that a quotient of the form f/g may have different meanings. For definiteness, by a quotient operator, we will always mean an operator of the form f/g where $f, g \in l^N$.

Every sequence $f \in l^N$ can be regarded as an operator since it can be identified with $f/\bar{1}$, and in such a case, the corresponding operator is said to be *ordinary*. In particular, an ordinary operator of the form $\bar{\alpha}/\bar{1}$, where $\alpha \in C$, will be called a *scalar operator*. If no confusion arises, we will denote an ordinary operator $f/\bar{1}$ by f.

Not all operators are ordinary. For example, the operator $\bar{1}/(\sigma - \bar{1})$ cannot be ordinary. For otherwise there would exist a sequence $f \in l^N$ such that $\sigma * f - f = \bar{1}$. But then $f_0 - f_0 = 1$, which is a contradiction. However, any quotient operator f/g is ordinary when the first term of the sequence g is nontrivial. This is true in view of Theorem 24.

As is customary, for any operators ϕ and ψ, we denote $\phi + \phi$ by 2ϕ, $\phi + \psi + \psi$ by $\phi + 2\psi$, $\phi\phi$ by ϕ^2, etc. The notation 2ϕ, for example, is actually a definition. Fortunately, $\phi + \phi = \bar{2}\phi$ so that we may identify 2ϕ as $\bar{2}\phi$. In the sequel, for a complex number α, a positive integer n and an operator ϕ, the notation $\alpha\phi$ will be taken to mean $\overline{\alpha}\phi$, ϕ^0 to mean $\overline{1}$, and ϕ^n to mean $\phi\phi\cdots\phi$. In an expression involving operators, the scalar α will be taken to mean $\overline{\alpha}$. For instance, the following notations

$$\frac{\overline{1}}{\delta-\overline{\alpha}}, \frac{\overline{1}}{\delta-\alpha}, \frac{1}{\delta-\overline{\alpha}}, \frac{1}{\delta-\alpha}$$

are all equivalent but the last one is much simpler. Instead of sticking to one particular type of notation, we will use the full notations in the immediately following sections and gradually use the simplified notations in later sections. Additional notations will be defined as we go along.

3.2.3 Summation Operators

We have denoted the sequence $\{1,1,...\}$ by σ. Note that

$$\sigma\{f_k\} = \left\{\sum_{i=0}^{k} f_i\right\}.$$

Therefore, if σ is regarded as an operator, it is natural to call it the summation operator. There are several useful identities related to the summation operator. For instance,

$$\sigma * \sigma \equiv \sigma^2 = \left\{\sum_{i=0}^{k} 1\right\} = \left\{(k+1)^{[1]}\right\},$$

and in general,

$$\sigma^n = \left\{\frac{(k+n-1)^{[n-1]}}{(n-1)!}\right\}, \; n \in Z^+. \tag{3.1}$$

The inverse of the operator σ will be denoted by δ. The operator δ is ordinary. Indeed, $\delta = \{1,-1,0,0,...\} = \overline{1} - \hbar$. Similarly, $\delta^2 = \{1,-2,1,0,...\}$, and in general,

$$\delta^n = \{(-1)^k n^{[k]}/k!\}, \; n \in Z^+.$$

Note that

$$\delta * \{f_k\} = \{f_0, f_1 - f_0, f_2 - f_1, ...\} = \overline{f_0} + \hbar(\Delta f), \tag{3.2}$$

so that δf cannot be "identified" as Δf.

Theorem 27 *Let f be a sequence in l^N, Δf its first difference, and $\overline{f_0}$ the sequence $\{f_0, 0, 0, ...\}$. Then the corresponding ordinary operators satisfy the following relation*

$$\delta f = \Delta f + \overline{f_0} - \delta(\Delta f) = \overline{f_0} + \hbar(\Delta f). \tag{3.3}$$

The first equality in (3.3) is obtained by substituting $\hbar = \overline{1} - \delta$ into (3.2). We give an alternate proof as follows by making use of the fundamental fact

$$\sum_{i=0}^{k} \Delta f_i = f_{k+1} - f_0 = \Delta f_k + f_k - f(0).$$

Thus

$$\{f_k\} = \left\{\sum_{i=0}^{k} \Delta f_i\right\} - \{\Delta f_k\} + \{f_0, f_0, ...\} = \sigma(\Delta f) - \Delta f + \overline{f_0}\sigma.$$

By multiplying both sides of the above equality with δ, we obtain $\delta f = \Delta f + \overline{f_0} - \delta(\Delta f)$ as required.

If we multiply $\delta f = \Delta f + \overline{f_0} - \delta(\Delta f)$ by δ, then we see that

$$\begin{aligned}\delta^2 f &= \delta(\Delta f) + \delta\overline{f_0} - \delta^2(\Delta f)\\ &= \Delta^2 f + \overline{\Delta f_0} - \delta(\Delta^2 f) + \delta\overline{f_0} - \delta^2(\Delta f),\end{aligned}$$

or

$$\Delta^2 f = \delta^2 f - \delta\overline{f_0} - \overline{\Delta f_0} + \delta^2(\Delta f) + \delta(\Delta^2 f).$$

In general, we obtain

$$\begin{aligned}\Delta^n f &= \delta^n f - \delta^{n-1}\overline{f_0} - \delta^{n-2}\overline{\Delta f_0} - \dots\\ &\quad -\overline{\Delta^{n-1} f_0} + \delta^n(\Delta f) + \delta^{n-1}(\Delta^2 f) + \dots + \delta(\Delta^n f)\end{aligned}$$

for $n \geq 1$.

Since $\Delta f = Ef - f$,

$$\delta f = Ef - f + \overline{f_0} - \delta(Ef - f) = Ef - f + \overline{f_0} - \delta(Ef) + \delta f$$

which implies

$$(\overline{1} - \delta)(Ef) = f - \overline{f_0}.$$

As a consequence

$$(\overline{1} - \delta)(E^2 f) = (\overline{1} - \delta)(EEf) = Ef - \overline{(Ef)_0} = Ef - \overline{f_1}.$$

By induction, we then see that the following holds.

Theorem 28 *Let f be a sequence in l^N, $E^n f$ its translated sequence, and $\overline{f_n}$ the sequence $\{f_n, 0, 0, \dots\}$. Then the corresponding ordinary operators satisfy the following relation*

$$(\overline{1} - \delta)^n(E^n f) = f - \overline{f_0} - (\overline{1} - \delta)\overline{f_1} - \dots - (\overline{1} - \delta)^{n-1}\overline{f_{n-1}}, \tag{3.4}$$

for $n \in Z^+$.

As an immediate application of Theorem 27, let $f_k = 2^k$ for $k \geq 0$. Since $\Delta f_k = 2^k$,

$$f = \delta f - \overline{1} + \delta f,$$

from which we obtain

$$\{2^k\} = \frac{\overline{1}}{\overline{2}\delta - \overline{1}}.$$

Similarly, the same principle leads to

$$\{\alpha^k\} = \frac{\overline{1}}{\overline{\alpha}\delta - \overline{\alpha} + \overline{1}}. \tag{3.5}$$

Substituting $\alpha = 1/(1 - \beta)$ into the above formula, we obtain

$$\frac{\overline{1}}{\delta - \overline{\beta}} = \left\{\left(\frac{1}{1 - \beta}\right)^{k+1}\right\},\ \beta \neq 1. \tag{3.6}$$

Now in view of Theorem 26,

$$\{c^k\} * \{c^k\} = (T^c\sigma)(T^c\sigma) = T^c\sigma^2 = \left\{c^k(k+1)^{[1]}\right\},$$

we see that

$$\frac{\overline{1}}{(\delta-\overline{\beta})^2} = \left\{(1-\beta)^{-k-2}(k+1)^{[1]}\right\}. \tag{3.7}$$

By induction, it is not difficult to see that for any scalar $\beta \neq 1$, the following extension of formula (3.1) holds

$$\frac{\overline{1}}{(\delta-\overline{\beta})^n} = \left\{\frac{(k+n-1)^{[n-1]}}{(n-1)!}(1-\beta)^{-k-n}\right\},\ n \in Z^+. \tag{3.8}$$

Substituting $\alpha = e^\beta$ into (3.5), we obtain further that

$$\{e^{\beta k}\} = \frac{\overline{1}}{\overline{e^\beta}\delta - \overline{e^\beta} + \overline{1}} = \frac{\overline{e^{-\beta}}}{\delta - (\overline{1} - \overline{e^{-\beta}})}$$

As a further application of (3.3), let us rewrite it as

$$\delta(f + \Delta f) = \delta(Ef) = \Delta f + \overline{f_0}.$$

Consider the difference equation

$$\Delta f_k = \frac{1}{2} f_{k+1},\ k \in N,$$

subject to the initial condition $f_0 = 1$. Since this equation can be written as

$$\Delta f = \frac{1}{2} Ef,$$

and the initial condition as $\overline{f_0} = \overline{1}$, we see that

$$\frac{1}{2} Ef = \Delta f = \delta(Ef) - \overline{f_0} = \delta(Ef) - \overline{1},$$

and hence that

$$Ef = \frac{\overline{1}}{\delta - \overline{1/2}} = \{2^{1+k}\}.$$

This implies that $f = \{2^k\}$, which is the correct solution of the above initial value problem.

In terms of the Gamma function, we can write (3.8) in the form

$$(\delta - \overline{\alpha})^{-n} = \left\{\frac{\Gamma(k+n)}{\Gamma(n)\Gamma(k)}(1-\alpha)^{-k-n}\right\},\ n \in Z^+, \alpha \neq 1.$$

We will generalize this formula by defining the operator $(\delta - \overline{\alpha})^{-\lambda}$ as

$$(\delta - \overline{\alpha})^{-\lambda} = \left\{\frac{\Gamma(k+\lambda)}{\Gamma(\lambda)\Gamma(k)}(1-\alpha)^{-k-\lambda}\right\},\ \alpha \neq 1, \tag{3.9}$$

when $\lambda > 0$. It will also be defined as $\overline{1}$ when $\alpha \neq 1$ and $\lambda = 0$, and as

$$(\delta - \overline{\alpha})^{-\lambda} = \frac{1}{(\delta - \overline{\alpha})^{\lambda}}, \ \alpha \neq 1$$

when $\lambda < 0$. It is natural to call $(\delta - \overline{\alpha})^{-\lambda}$ the *fractional summation operator.*

Theorem 29 *Suppose the complex number α is not equal to 1. The following formula holds for any $\lambda, \mu \in R$:*

$$(\delta - \overline{\alpha})^{-\lambda}(\delta - \overline{\alpha})^{-\mu} = (\delta - \overline{\alpha})^{-\lambda-\mu}.$$

In view of the above theorem, the fractional summation operators can be treated as ordinary powers in algebraic manipulations.

3.2.4 Translation or Shift Operators

Recall that when m is a nonnegative integer, $\hbar^m = \{\hbar_k^m\}_{k\in N}$ denotes the (Dirac) sequence, which is defined by

$$\hbar_k^m = \begin{cases} 1 & k = m \\ 0 & k \neq m \end{cases},$$

and $H^{(m)} = \left\{H_k^{(m)}\right\}_{k\in N}$ the jump (or Heaviside) sequence

$$H_k^{(m)} = \begin{cases} 0 & 0 \leq k < m \\ 1 & k \geq m \end{cases}.$$

Recall that $\delta = \{1, -1, 0, ...\}$. Thus we have the fundamental relationship $\delta + \hbar = \overline{1}$. When the sequence $\hbar^m$ is identified as an operator, it is called a translation operator. This term is justified by the following simple observation.

Theorem 30 *Let m be a nonnegative integer. For any $f = \{f_k\} \in l^N$, the convolution $\hbar^m * f$ is given by*

$$g_k = \begin{cases} 0 & 0 \leq k < m \\ f_{k-m} & k \geq m \end{cases},$$

that is, $\hbar^m * f = E^{-m} f$.

This observation implies that

$$f = \frac{E^{-m} f}{\hbar^m}, \quad f \in l^N,$$

and that

$$\hbar^m \sigma = H^{(m)},$$

which can also be written as

$$\hbar^m = \frac{H^{(m)}}{\sigma} = \delta H^{(m)}, \quad m \in N. \tag{3.10}$$

Recall that $\delta + \hbar = \overline{1}$. By means of this simple relation, some of the previous formulas can also be expressed in terms of the translation operator $\hbar$. For instance, (3.5) can be written as

$$\{\alpha^k\} = \frac{\overline{1}}{\overline{1} - \overline{\alpha}\hbar}.$$

Similarly, (3.6) and (3.8) can be written in the form

$$\{\alpha^{-k-1}\} = \frac{\overline{1}}{\overline{\alpha} - \hbar},\ \alpha \neq 0, \tag{3.11}$$

and

$$\frac{\overline{1}}{(\overline{\alpha} - \hbar)^n} = \left\{\frac{(k+n-1)^{[n-1]}}{(n-1)!}\alpha^{-k-n}\right\},\ \alpha \neq 0,\ n \in Z^+, \tag{3.12}$$

or

$$\frac{\overline{1}}{(\overline{1} - \overline{\lambda}\hbar)^n} = \left\{\frac{(k+n-1)^{[n-1]}}{(n-1)!}\lambda^k\right\},\ n \in Z^+, \tag{3.13}$$

while (3.4) can be written as the following important formula

$$\hbar^n(E^n f) = f - \overline{f_0} - \hbar\overline{f_1} - \dots - \hbar^{n-1}\overline{f_{n-1}}, \quad n \in Z^+. \tag{3.14}$$

We can make use of (3.11) to derive operator forms of the sequences $\{\sin k\theta\}$ and $\{\cos k\theta\}$. Indeed, in view of the well known formulas of Euler,

$$\sin x = \frac{e^{ix} - e^{-ix}}{2i}, \quad \cos x = \frac{e^{ix} + e^{-ix}}{2},$$

we see that

$$\{\sin k\theta\} = \frac{1}{2i}\left\{e^{ik\theta}\right\} - \frac{1}{2i}\left\{e^{-ik\theta}\right\} = \frac{\hbar\overline{\sin\theta}}{\overline{1} - \overline{2}\hbar\overline{\cos\theta} + \hbar^2},$$

and

$$\{\cos k\theta\} = \frac{\overline{1} - \hbar\overline{\cos\theta}}{\overline{1} - \overline{2}\hbar\overline{\cos\theta} + \hbar^2}.$$

Similarly, we have

$$\{\sinh k\theta\} = \frac{\hbar\overline{\sinh\theta}}{\overline{1} - \overline{2}\hbar\overline{\cosh\theta} + \hbar^2},$$

and

$$\{\cosh k\theta\} = \frac{\overline{1} - \hbar\overline{\cosh\theta}}{\overline{1} - \overline{2}\hbar\overline{\cosh\theta} + \hbar^2}.$$

Theorem 30 also implies that

$$\hbar^m\hbar^n f = \hbar^{m+n} f,\ m, n \in N, \tag{3.15}$$

so that

$$\hbar^m\hbar^n = \hbar^{m+n},\ m, n \in N. \tag{3.16}$$

It is convenient to extend the definition of the translation operator by

$$\hbar^{-m} = \frac{\bar{1}}{\hbar^m}, \; m \in Z^+.$$

Then formula (3.16) is now valid for all integers m and n even though the same may not be true for formula (3.15). For example, $\hbar^{-1}f$ is not equal to $\{f_{k+1}\}$ in general since

$$\hbar\{f_{k+1}\} = \{0, f_1, f_2, ...\}.$$

However, note that $\hbar^{-1}f = \hbar^{-1}\overline{f_0} + Ef$. Note further that for each $m \geq 1$, $\hbar^{-m}$ is not an ordinary operator.

We may also generalize formula (3.12) by defining

$$\frac{\bar{1}}{(\bar{\alpha} - \hbar)^\lambda} = \left\{\frac{\Gamma(k+\lambda)}{\Gamma(\lambda)\Gamma(k)}\alpha^{-k-\lambda}\right\}, \; \alpha \neq 0, \lambda > 0,$$

$$\frac{\bar{1}}{(\bar{\alpha} - \hbar)^0} = \bar{1},$$

and

$$(\bar{\alpha} - \hbar)^{-\lambda} = \frac{\bar{1}}{(\bar{\alpha} - \hbar)^\lambda}, \; \alpha \neq 0, \lambda < 0.$$

The translation operators are useful for representing various common sequences. For instance, the following are valid:

$$\frac{\bar{1} - \hbar^m}{\delta} = \sigma\left(\bar{1} - \hbar^m\right) = \left\{\bar{1} - H_k^{(m)}\right\}, \; m \in N,$$

$$\frac{\hbar^m - \hbar^n}{\delta} = \sigma\left(\hbar^m - \hbar^n\right) = \left\{H_k^{(m)} - H_k^{(n)}\right\}, \; m, n \in N,$$

and

$$\frac{\hbar^m}{\delta^2} = \sigma^2\hbar^m = \{f(k)\}, \; m \in N,$$

where

$$f(k) = \begin{cases} 0 & 0 \leq k < m \\ k - m + 1 & k \geq m \end{cases}.$$

Finally, we remark that for any quotient operator f/g, since $g = \{g_k\} \neq 0$, there must be some nonnegative integer m such that $g_0 = g_1 = ... = g_{m-1} = 0$ and $g_m \neq 0$. In this case, we have $g = \hbar^m * p$, where $p = \{g_m, g_{m+1}, ...\}$. In other words,

$$f/g = f/(\hbar^m p) = (f/p)\,\hbar^{-m},$$

which says, in view of Theorem 24, that any quotient operator must be of the form $\hbar^{-m}q$ for some nonnegative integer m and some ordinary operator q.

3.2.5 Rational Operators

Given nonnegative integers m and n, and an arbitrary operator z, we may form polynomial operators of the form

$$\overline{\alpha_m}z^m + ... + \overline{\alpha_1}z + \overline{\alpha_0}, \tag{3.17}$$

and rational operators of the form

$$\frac{\overline{\alpha_m}z^m + ... + \overline{\alpha_1}z + \overline{\alpha_0}}{\overline{\beta_n}z^n + ... + \overline{\beta_1}z + \overline{\beta_0}}, \ 0 \leq m < n$$

where $\alpha_0, ..., \alpha_m, \beta_0, ..., \beta_n$ are complex numbers and $\beta_n \neq 0$. These operators are respectively called polynomial operators and rational operators in z, and can be manipulated as if they were ordinary rational numbers. In particular, if a polynomial operator in δ is equal to zero, then all its coefficients are zero. Indeed, multiplying the left hand side of

$$\overline{\alpha_m}\delta^m + ... + \overline{\alpha_1}\delta + \overline{\alpha_0} = \overline{0}$$

by σ^{m+1}, we see that

$$\overline{\alpha_m}\sigma + ... + \overline{\alpha_1}\sigma^m + \overline{\alpha_0}\sigma^{m+1} = \overline{0}.$$

This implies

$$\alpha_m + \sum_{j=1}^{m} \frac{(j+k-1)^{[j-1]}}{(j-1)!}\alpha_{m-j} = 0$$

for $k = 0, 1, ...$. Thus the coefficients $\alpha_0, ..., \alpha_m$ must all be zero.

It is well known that rational expressions with real coefficients such as the one given above can be decomposed into simple fractions of the following types:

$$\frac{\overline{1}}{(z-\overline{\alpha})^p}, \frac{\overline{1}}{((z-\overline{\alpha})^2+\overline{\beta}^2)^p}, \frac{z}{((z-\overline{\alpha})^2+\overline{\beta}^2)^p}, \tag{3.18}$$

where α and β are real numbers and p is a nonnegative integer. Some of these rational operators are ordinary operators. Indeed, the formula (3.8) shows that this is true for fractions of the first type when $\alpha \neq 1$. We may further show that some of the fractions of the second and third types are ordinary operators. To see this, note that when α, β are real numbers such that $\beta \neq 0$, we have

$$\begin{aligned}\frac{\overline{1}}{(\delta-\overline{\alpha})^2+\overline{\beta}^2} &= \frac{\overline{1}}{\overline{2i\beta}}\left(\frac{\overline{1}}{\delta-(\overline{\alpha}+\overline{i\beta})} - \frac{\overline{1}}{\delta-(\overline{\alpha}-\overline{i\beta})}\right) \\ &= \frac{1}{2i\beta}\left\{\frac{1}{(1-\alpha-i\beta)^{1+k}} - \frac{1}{(1-\alpha+i\beta)^{1+k}}\right\},\end{aligned} \tag{3.19}$$

and

$$\begin{aligned}\frac{\delta-\overline{\alpha}}{(\delta-\overline{\alpha})^2+\overline{\beta}^2} &= \frac{\overline{1}}{\overline{2}}\left(\frac{\overline{1}}{\delta-(\overline{\alpha}+\overline{i\beta})} + \frac{\overline{1}}{\delta-(\overline{\alpha}-\overline{i\beta})}\right) \\ &= \frac{1}{2}\left\{\frac{1}{(1-\alpha-i\beta)^{1+k}} + \frac{1}{(1-\alpha+i\beta)^{1+k}}\right\},\end{aligned} \tag{3.20}$$

as well as

$$\begin{aligned}\frac{\delta}{(\delta-\overline{\alpha})^2+\overline{\beta}^2} &= \frac{\delta-\overline{\alpha}}{(\delta-\overline{\alpha})^2+\overline{\beta}^2}+\frac{\overline{\alpha}}{(\delta-\overline{\alpha})^2+\overline{\beta}^2} \\ &= \left\{\frac{\alpha+i\beta}{2i\beta(1-\alpha-i\beta)^{1+k}}-\frac{\alpha-i\beta}{2i\beta(1-\alpha+i\beta)^{1+k}}\right\}. \quad (3.21)\end{aligned}$$

For rational operators of the form (3.18) with $p > 1$, we may then obtain the corresponding sequences by convolution. The explicit formulae of these sequences are rather complicated and will be given later.

We remark that in view of the formulae of Euler,

$$\sin x = \frac{e^{ix}-e^{-ix}}{2i}, \quad \cos x = \frac{e^{ix}+e^{-ix}}{2},$$

we may rewrite the sequences in (3.19), (3.20) and (3.21) by means of trigonometric and power sequences. For instance, when $\beta \neq 0$,

$$\frac{\overline{1}}{(\delta-\overline{\alpha})^2+\overline{\beta}^2} = \left\{\frac{\sin((1+k)\theta)}{\beta((1-\alpha)^2+\beta^2)^{(1+k)/2}}\right\},$$

and

$$\frac{\delta-\overline{\alpha}}{(\delta-\overline{\alpha})^2+\overline{\beta}^2} = \left\{\frac{\cos((1+k)\theta)}{((1-\alpha)^2+\beta^2)^{(1+k)/2}}\right\},$$

where θ is the argument of the complex number $1-\alpha+i\beta$.

For polynomial or rational operators in $\hbar$, we have similar results. For instance, if a polynomial operator in $\hbar$ is equal to zero, then all its coefficients are zero. Furthermore, when $\alpha, \beta \in R$ and $\beta \neq 0$, we have

$$\begin{aligned}\frac{\overline{1}}{(\hbar-\overline{\alpha})^2+\overline{\beta}^2} &= \frac{\overline{-1}}{\overline{2i\beta}}\left(\frac{\overline{1}}{(\overline{\alpha}+\overline{i\beta})-\hbar}-\frac{\overline{1}}{(\overline{\alpha}-\overline{i\beta})-\hbar}\right) \\ &= \frac{-1}{2i\beta}\left\{(\alpha+i\beta)^{-k-1}-(\alpha-i\beta)^{-k-1}\right\} \\ &= \left\{\frac{\sin((1+k)\phi)}{\beta(\alpha^2+\beta^2)^{(1+k)/2}}\right\} \quad (3.22)\end{aligned}$$

and

$$\begin{aligned}\frac{\hbar-\overline{\alpha}}{(\hbar-\overline{\alpha})^2+\overline{\beta}^2} &= \frac{\overline{-1}}{\overline{2}}\left(\frac{\overline{1}}{(\overline{\alpha}+\overline{i\beta})-\hbar}+\frac{\overline{1}}{(\overline{\alpha}-\overline{i\beta})-\hbar}\right) \\ &= \frac{-1}{2}\left\{(\alpha+i\beta)^{-k-1}+(\alpha-i\beta)^{-k-1}\right\} \\ &= \left\{\frac{-\cos((1+k)\phi)}{(\alpha^2+\beta^2)^{(1+k)/2}}\right\}, \quad (3.23)\end{aligned}$$

where ϕ is the argument of the number $\alpha+i\beta$.

3.2.6 Attenuation Operators

Recall that the attenuated sequence of $f \in l^N$ is $T^\lambda f = \{\lambda^k f_k\}$. Given a quotient operator f/g, we denote the operator

$$\frac{T^\lambda f}{T^\lambda g}, \; \lambda \in C, \lambda \neq 0,$$

by $T^\lambda (f/g)$, and we call it an attenuated operator of the quotient operator f/g. T^λ is well defined since

$$T^\lambda (f/1) = (T^\lambda f)/(T^\lambda 1) = T^\lambda f.$$

It is easily seen that $T^\lambda \overline{\alpha} = \overline{\alpha}$ for any complex number α, and for any quotient operator ϕ, $T^\lambda(T^\mu \phi) = T^{\lambda\mu}\phi$. Furthermore, by means of Theorem 26, for quotient operators $\phi = f/g$ and $\psi = p/q$,

$$T^\lambda(\phi + \psi) = \frac{T^\lambda(fq + pg)}{T^\lambda(gq)} = \frac{T^\lambda f T^\lambda q + T^\lambda p T^\lambda g}{T^\lambda g T^\lambda q} = T^\lambda \phi + T^\lambda \psi,$$

and

$$T^\lambda(\phi\psi) = T^\lambda \left(\frac{fp}{qg}\right) = \frac{T^\lambda f T^\lambda p}{T^\lambda g T^\lambda q} = (T^\lambda \phi)(T^\lambda \psi).$$

Consequently, we have

$$T^\lambda(\overline{\alpha}\phi) = T^\lambda \overline{\alpha} T^\lambda \phi = \overline{\alpha} T^\lambda \phi, \; \alpha \in C, \phi \in l^N/l^N,$$

$$T^\lambda (f/g) = \frac{\overline{1}}{T^\lambda(g/f)}, \; f, g \in C, f \neq 0, g \neq 0,$$

$$T^\lambda \delta^n = (\overline{1} - \overline{\lambda}\hbar)^n, \; n \in Z^+,$$

and

$$T^\lambda \hbar^n = \overline{\lambda}^n \hbar^n = \overline{\lambda}^n (\overline{1} - \delta)^n, \; n \in Z^+.$$

It is interesting to note that if $R(\phi)$ is an arbitrary rational expression of the operator ϕ,

$$R(\phi) = \frac{\overline{\alpha_m}\phi^m + ... + \overline{\alpha_0}}{\overline{\beta_n}\phi + ... + \overline{\beta_0}},$$

then

$$T^\lambda R(\hbar) = R(\overline{\lambda}\hbar) = R(\overline{\lambda}(\overline{1} - \delta)),$$

and

$$T^\lambda R(\delta) = R(\overline{1} - \overline{\lambda}\hbar).$$

In particular,

$$T^\lambda \frac{\overline{1}}{(\overline{1} - \hbar)^n} = \frac{\overline{1}}{(\overline{1} - \overline{\lambda}\hbar)^n}, \; n \in Z^+,$$

which we have already shown (see formula (3.13)).

3.2.7 Sequences and Series of Operators

Given a sequence $\{f^{(j)}\}$ of sequences in l^N, we say that $\{f^{(j)}\}$ converges (pointwise) to $f = \{f_k\}$ in l^N if

$$\lim_{j\to\infty} f_k^{(j)} = f_k, \quad k \in N.$$

A sequence of operators $\left\{\phi^{(j)}\right\}$ is said to converge to an operator f/g if there are two sequences $\{f^{(j)}\}$ and $\{g^{(j)}\}$ of sequences in l^N such that $\phi^{(j)} = f^{(j)}/g^{(j)}$ for each j, that $\{f^{(j)}\}$ and $\{g^{(j)}\}$ converge pointwise to f and g respectively, and that $g \neq \bar{0}$. Clearly, these two definitions are compatible in the sense that if a sequence $\{f^{(j)}\}$ of sequences converges to f, then the corresponding operator sequence $\{f^{(j)}/\bar{1}\}$ converges to the operator $f/\bar{1}$. Note that the limit of the sequence $\left\{\phi^{(n)}\right\}$ of operators is independent of the sequences $\{f^{(j)}\}$ and $\{g^{(j)}\}$. Indeed, suppose to the contrary that there are four sequences $\{f^{(n)}\}$, $\{g^{(n)}\}$, $\{p^{(n)}\}$ and $\{q^{(n)}\}$ converging pointwise to f, $g \neq 0$, p and $q \neq 0$ respectively where $f^{(n)}/g^{(n)} = p^{(n)}/q^{(n)}$ for each n. Since $\{f^{(n)}q^{(n)}\} = \{g^{(n)}p^{(n)}\}$ and converges pointwise, the limits fq and gp must be equal. This says that $f/g = p/q$.

Properties of limits of sequences of operators are similar to those of ordinary numerical sequences. If $\left\{\phi^{(n)}\right\} = \{\phi, \phi, ...\}$ where ϕ is an arbitrary but fixed operator, then $\lim_{n\to\infty} \phi^{(n)} = \phi$. Subsequences of a sequence of operators which converges to ϕ have the same limit.

Theorem 31 *Let $\left\{\phi^{(n)}\right\}$ and $\left\{\psi^{(n)}\right\}$ be two sequences of operators which converge to ϕ and ψ respectively. Then*

$$\lim_{n\to\infty} (\phi^{(n)} + \psi^{(n)}) = \phi + \psi,$$

and

$$\lim_{n\to\infty} \phi^{(n)}\psi^{(n)} = \phi\psi.$$

The proof is quite elementary and is thus omitted.

Theorem 32 *Let $\{f^{(n)}/g^{(n)}\}$ be a sequence of quotient operators which converges to ϕ. Suppose further that $\phi \neq 0$ and $f^{(n)} \neq 0$ for all n, then*

$$\lim_{n\to\infty} g^{(n)}/f^{(n)} = \bar{1}/\phi.$$

Let $\left\{\phi^{(n)}\right\}_{n=0}^{\infty}$ be a sequence of operators. The series $\phi^{(0)} + \phi^{(1)} + ...$ is said to converge to the sum A if the sequence of partial sums $A^{(n)} = \phi^{(0)} + \phi^{(1)} + ... + \phi^{(n)}$ converges to A. We will use the notation

$$\sum_{n=0}^{\infty} \phi^{(n)} \in l^N/l^N$$

to denote convergence. Note that if $\phi^{(n)}$ is ordinary, i.e. $\phi^{(n)} = \{\phi_k^{(n)}\}$, and if $\phi^{(0)} + \phi^{(1)} + \ldots$ converges, then we have

$$\sum_{n=0}^{\infty} \phi^{(n)} = \sum_{n=0}^{\infty} \{\phi_k^{(n)}\} = \left\{ \sum_{n=0}^{\infty} \phi_k^{(n)} \right\}, \tag{3.24}$$

that is, the k-th term of the series is obtained by adding all the k-th terms of the individual sequences.

Theorem 33 *Let $\{\alpha_n\}_{n=0}^{\infty}$ be a sequence of complex numbers and $\{\beta_n\}_{n=0}^{\infty}$ be a strictly increasing integral sequence which diverges to ∞. Then the series*

$$\sum_{n=0}^{\infty} \overline{\alpha_n} \hbar^{\beta_n}$$

is convergent.

Proof. Note that when $\beta_0 \geq 0$, the above series is a sequence whose component at $k = \beta_n$ is equal to α_n. In this case, our theorem is clear. If $\beta_0 < 0$, we can multiply and divide the series by $\hbar^{\beta_0}$ so that the numerator is the convergent series $\sum_{n=0}^{\infty} \overline{\alpha_n} \hbar^{\beta_n - \beta_0}$, and the denominator is the convergent sequence $\left\{\hbar^{\beta_0}\right\}$. ■

Series of translation operators enjoy several interesting properties. First of all, we have

$$\{f_0, f_1, f_2, \ldots\} = \overline{f_0} + \overline{f_1}\hbar + \overline{f_2}\hbar^2 + \ldots, \tag{3.25}$$

where the right hand side is a series of translation operators which is always convergent, and converges to the left hand side. We remark that, for historical reasons, the series on the right hand side is also called the "moment generating function" of the sequence $\{f_0, f_1, \ldots\}$. We will have more to say in a later section. Next, we recall that any operator must be of the form $\hbar^{-m} q$ where m is a nonnegative integer and $q \in l^N$. Thus any operator must be of the form

$$\overline{q_0}\hbar^{-m} + \overline{q_1}\hbar^{-m+1} + \overline{q_2}\hbar^{-m+2} + \ldots, \tag{3.26}$$

which appears in the form of a Laurent series.

Theorem 34 *Let $\{\alpha_n\}_{n=0}^{\infty}$ be a sequence of complex numbers and $\{\beta_n\}_{n=0}^{\infty}$ be a strictly increasing integral sequence which tends to ∞. If*

$$\sum_{n=0}^{\infty} \overline{\alpha_n} \hbar^{\beta_n} = 0,$$

then $\alpha_i = 0$ for $i \geq 0$.

Proof. If $\beta_0 \geq 0$, the above series is the sequence whose component at $k = \beta_n$ is equal to α_n. Thus $\alpha_n = 0$ for $n \geq 0$. If $\beta_0 < 0$, we can multiply and divide the series by $\hbar^{\beta_0}$ so that the numerator is now the series $\sum_{n=0}^{\infty} \overline{\alpha_n} \hbar^{\beta_n - \beta_0}$. ■

Operators of the form $\bar{1}/(\bar{1}-\bar{\beta}\hbar^m)^{1+n}$ can be expanded into series of translation operators as follows. Note that

$$\begin{aligned} & (\bar{1}-\bar{\beta}\hbar^m)(\bar{1}+\overline{2\beta}\hbar^m+\overline{3\beta}^2\hbar^{2m}+\ldots) \\ = & \ \bar{1}+\bar{\beta}\hbar^m+\bar{\beta}^2\hbar^{2m}+\bar{\beta}^3\hbar^{3m}+\ldots \\ = & \ \frac{\bar{1}}{\bar{1}-\bar{\beta}\hbar^m} \end{aligned}$$

for $m \geq 1$, thus

$$\frac{\bar{1}}{(\bar{1}-\bar{\beta}\hbar^m)^2} = \sum_{j=0}^{\infty} \overline{(j+1)\beta^j}\hbar^{jm}, \ m \in Z^+.$$

Similarly, we have

$$\frac{\bar{1}}{(\bar{1}-\bar{\beta}\hbar^m)^{1+n}} = \sum_{j=0}^{\infty} \overline{C_n^{(j+n)}\beta^j}\hbar^{jm}, \ n \in N, m \in Z^+.$$

We remark that periodicity of sequences is related to the series

$$\frac{\bar{1}}{\bar{1}-\bar{\beta}\hbar^m} = \bar{1}+\bar{\beta}\hbar^m+\bar{\beta}^2\hbar^{2m}+\ldots,$$

where β is a complex number and m is positive integer. Let $f = \{f_k\} \in l^N$ such that $f_k = 0$ for $k \geq m \geq 0$. Then multiplying f with the operator $\bar{1}/(\bar{1}-\hbar^m)$, we obtain a sequence whose graph is found by repeating infinitely often the fragment of the graph of f over the set of integers $\{0,1,2,\ldots,m\}$. Thus $f/(\bar{1}-\hbar^m)$ represents a periodic sequence provided $f_k = 0$ for $k \geq m$. On the other hand, if we are given a periodic sequence $g = \{g_k\}$ such that $g_{k+\tau} = g_k$ for all k, then $(\bar{1}-\hbar^\tau)g$ can be identified as a sequence $f = \{f_k\}$ which vanishes for $k \geq \tau$. We summarize these as follows.

Theorem 35 *A sequence $g = \{g_k\} \in l^N$ is periodic with periodic $\tau \in Z^+$ if, and only if,*

$$g = \frac{f}{(\bar{1}-\hbar^\tau)}, \quad \tau \in Z^+,$$

where $f = \{f_k\} \in l^N$ satisfies $f_k = 0$ for $k \geq \tau$.

We now deal with more general series of operators of the form

$$\overline{\alpha_0}+\overline{\alpha_1}f+\overline{\alpha_2}f^2+\ldots, \tag{3.27}$$

where $\{\alpha_n\}_{n=0}^{\infty} \in l^N$ and $f = \{f_k\}$ is an ordinary operator.

Theorem 36 *Let $\{\alpha_n\}_{n=0}^{\infty}$ be a sequence of complex numbers and $f = \{f_k\}$ an operator. If $f_0 = 0$ or if the power series $\alpha_0+\alpha_1 z+\alpha_2 z^2+\ldots$ in the complex variable z is entire, then (3.27) is also convergent.*

Proof. In view of Theorem 3.25, the first n terms of the sequence f^n are equal to zero. Thus

$$\alpha_0 + \alpha_1 f_k + \alpha_2 f_k^2 + ... = \alpha_0 + \alpha_1 f_k + ... + \alpha_k f_k^k < \infty.$$

The first case is proved. Next, let J be an arbitrary nonnegative integer. Let $M = \max_{0 \le i \le J} |f_i|$. Then for $0 \le k \le J$,

$$|f_k^2| = \left| \sum_{j=0}^{k} f_j f_{k-j} \right| \le \sum_{j=0}^{k} |f_j| \, |f_{k-j}| \le M^2 (J+1),$$

$$|f_k^3| = \left| \sum_{j=0}^{k} f_j^2 f_{k-j} \right| \le \sum_{j=0}^{k} |f_j^2 f_{k-j}| \le M^3 (J+1)^2,$$

and in general,

$$|f_k^n| \le M^n (J+1)^{n-1}.$$

Consequently,

$$\left| \alpha_0 + \sum_{n=1}^{\infty} \alpha_n f_k^n \right| \le |\alpha_0| + \frac{1}{J+1} \sum_{n=1}^{\infty} |\alpha_n| \, M^n (J+1)^n < \infty,$$

in view of the fact that the power series $\alpha_0 + \alpha_1 z + \alpha_2 z^2 + ...$ in the complex variable z is entire. The proof is complete. ■

We remark that if the power series $\alpha_0 + \alpha_1 z + \alpha_2 z^2 + ...$ is not entire, the above result may not hold. For instance, the geometric series $1 + z + z^2 + ...$ is analytic in the unit disk, nevertheless, the series $\overline{1} + \overline{1} + ...$ is not convergent.

By means of Theorem 36, we see that for any $f = \{f_k\} \in l^N$, the series

$$\sum_{n=0}^{\infty} \frac{f^n}{n!}, \quad f \in l^N$$

is convergent. We will denote this series by e^f or $\exp f$ and call it the *exponential* of the ordinary operator f. By means of (3.24), we see that e^f is ordinary and the k-th term of e^f is given by

$$f_k + \frac{1}{2} \sum_{j=0}^{k} f_j f_{k-j} + ..., \; k \in Z^+,$$

and that $e^f \neq 0$ for any $f \in l^N$ since the 0-th term is equal to e^{f_0}, and that e^f is a scalar operator when f is a scalar sequence. We may also show that

$$e^f e^g = e^{f+g}, \; f, g \in l^N.$$

Indeed,

$$e^f * e^g = \sum_{n=0}^{\infty} \frac{f^n}{n!} \sum_{n=0}^{\infty} \frac{g^n}{n!} = \sum_{n=0}^{\infty} \sum_{j=0}^{n} \frac{f^j}{j!} \frac{g^{n-j}}{(n-j)!} = \sum_{n=0}^{\infty} \frac{(f+g)^n}{n!} = e^{f+g}.$$

3.2.8 Algebraic Derivatives

Given a sequence $f = \{f_k\}_{k=0}^{\infty} \in l^N$, we denote the sequence $\{(k+1)f_{k+1})\}_{k=0}^{\infty}$ by Df and call it the *algebraic derivative* [66] of f. The higher algebraic derivatives $D^n f$ are defined recursively. Thus we have

$$D\{f_0, f_1, f_2, ...\} = \{f_1, 2f_2, 3f_3, ...\},$$

$$D^2 f = \{(k+1)(k+2)f_{k+2}\}, ..., D^n f = \{(k+1)\cdots(k+n)f_{k+n}\}.$$

For instance, for any complex number α, $D\overline{\alpha} = \overline{0}$, and we have

$$D^n \sigma = \{(k+1)(k+2)...(k+n)\}$$

for $n \in Z^+$.

It can easily be verified that for $\alpha, \beta \in C$ and $f, g \in l^N$,

$$D(\alpha f + \beta g) = \alpha Df + \beta Dg$$

and

$$D(f * g) = f * Dg + g * Df.$$

Given a quotient operator f/g, we denote the operator

$$\frac{gDf - fDg}{g^2}$$

by $D(f/g)$, and again we call it the algebraic derivative of the quotient operator f/g. Clearly, if f is an ordinary operator, then its algebraic derivative is ordinary. It is also easy to verify that for any complex numbers α, β and two operators ϕ and ψ, $D(\overline{\alpha}\phi + \overline{\beta}\psi) = \overline{\alpha}D\phi + \overline{\beta}D\psi$ and $D(\phi\psi) = \phi D\psi + \psi D\phi$.

Algebraic derivatives of some common operators can easily be found. More complicated derivatives can be obtained by employing the following list of useful formulae:

$$\hbar D\{f_k\} = \{kf_k\}, \tag{3.28}$$

$$D\hbar^n = n\hbar^{n-1},\ n \in Z, \tag{3.29}$$

$$D\delta^n = -n\delta^{n-1},\ n \in Z, \tag{3.30}$$

$$D\left(q_0\hbar^{-m} + q_1\hbar^{-m+1} + ...\right) = -mq_0\hbar^{-m-1} + (-m+1)q_1\hbar^{-m} + ..., \tag{3.31}$$

where m is a nonnegative integer and $\{q_k\} \in l^N$,

$$De^f = e^f Df,\ f \in l^N, \tag{3.32}$$

and

$$D\phi^n = D(\phi^{n-1}\phi) = \phi^{n-1}D\phi + \phi D\phi^{n-1} = ... = \overline{n}\phi^{n-1}D\phi,\ n \in Z^+, \tag{3.33}$$

where ϕ is an operator,

$$D^n\left(\hbar^m\{f_k\}\right) = \hbar^{m-n}\left\{(m+k)^{[n]}f_k\right\},\ m \geq n \geq 1, \tag{3.34}$$

and finally

$$\hbar^m D^n\{f_k\} = \left\{(k+n-m)^{[n]} f_{k+n-m}\right\},\ n \geq m \geq 0. \tag{3.35}$$

To see that (3.32) holds, simply observe that

$$De^f = \sum_{n=0}^{\infty} D\left(\frac{f^n}{n!}\right) = \sum_{n=1}^{\infty} \frac{nf^{n-1}Df}{n!} = e^f Df.$$

To see that (3.34) holds, let $f = \{f_k\}$, then

$$D(\hbar^m f) = \hbar^m Df + fD\hbar^m = \hbar^{m-1}(\hbar Df + mf) = \hbar^{m-1}\{(k+m)f_k\}.$$

Similarly,

$$\begin{aligned} D(D(\hbar^m f)) &= \hbar^{m-1} D\{(k+m)f_k\} + \{(k+m)f_k\}(m-1)\hbar^{m-2} \\ &= \hbar^{m-2}\{k(k+m)f_k\} + \hbar^{m-2}(m-1)\{(k+m)f_k\} \\ &= \hbar^{m-2}\{(k+m)(k+m-1)f_k\}. \end{aligned}$$

The general formula is then obtained by induction.

It is interesting to note that if $D\phi = 0$ where ϕ is an operator, then ϕ must be a scalar operator. Indeed, since any operator is of the form (3.26), thus in view of (3.31), we have

$$-mq_0\hbar^{-m-1} + (-m+1)q_1\hbar^{-m} + ... = 0,$$

so that $-mq_0 = (-m+1)q_1 = ... = 0$. If $m = 0$, then $q_1 = q_2 = ... = 0$ and $\phi = \{q_0, 0, ...\}$. If $m > 0$, then $q_0 = q_1 = ... = q_{m-1} = 0$ and $q_{m+1} = q_{m+2} = ... = 0$. Hence $\phi = \overline{q_m}$. As an application, we show that if e^f is a scalar operator, then f is a scalar sequence. Indeed, $De^f = e^f Df = 0$, so that (recall that $e^f \neq 0$) $Df = 0$.

The algebraic derivative may be used to derive the k-th term of an ordinary operator $f = \overline{f_0} + \overline{f_1}\hbar + \overline{f_2}\hbar^2 + \overline{f_3}\hbar^3 + ...$. Indeed, since

$$Df = \overline{f_1} + \overline{2f_2}\hbar + \overline{3f_3}\hbar^2 + \overline{4f_4}\hbar^3 + ...,$$

$$D^2 f = \overline{2f_2} + \overline{3\cdot 2f_3}\hbar + \overline{4\cdot 3f_4}\hbar^2 + ...,$$

etc., it is clear that f_0 can be obtained from f by formally substituting $\hbar$ by 0 in its infinite sum representation and then identifying the resulting scalar sequence as f_0, and f_k, $k \in Z^+$, by formally substituting $\hbar$ with 0 in $\overline{1/k!}D^k f$, and then identifying the resulting scalar sequence as f_k.

Example 39 *By replacing $\hbar$ with zero in $\overline{1}/(\overline{1} - \hbar)$, we obtain $\overline{1}$ and the corresponding scalar is 1. Next, since*

$$D\left(\frac{\overline{1}}{\overline{1}-\hbar}\right) = \frac{\overline{1}}{(\overline{1}-\hbar)^2},$$

formally substituting $\hbar = 0$ in the above equation, we obtain $\overline{1}$ again and the corresponding scalar is 1. Similarly,

$$\frac{\overline{1}}{\overline{2!}}D^2\left(\frac{\overline{1}}{\overline{1}-\hbar}\right) = \frac{\overline{1}}{\overline{2!}}\frac{\overline{2}}{(\overline{1}-\hbar)^3} = \frac{\overline{1}}{(\overline{1}-\hbar)^3},$$

formally substituting $\hbar = 0$ in the above equation, we obtain $\overline{1}$ again and the corresponding scalar is 1. These scalars are expected, since $\overline{1}/(\overline{1}-\hbar)$ is just the summation operator $\sigma = \{1, 1, 1, ...\}$.

The algebraic derivatives may also be used to derive operational forms for some common sequences in l^N. For instance, the equality

$$D\{\alpha^k\} = D\left(\frac{1}{1-\alpha\hbar}\right)$$

implies

$$\{(k+1)\alpha^{k+1}\} = \frac{\alpha}{(1-\alpha\hbar)^2}.$$

Similarly, the equalities $\hbar\sigma^2 = \{k\}$ and $D\{k\} = \{(k+1)^2\}$ imply

$$\{(k+1)^2\} = D(\hbar\sigma^2) = \hbar D\sigma^2 + \sigma^2 = 2\hbar\sigma^3 + \sigma^2,$$

The same principle leads to

$$\{(k+1)^3\} = D\left(\hbar D\left(\hbar\sigma^2\right)\right),$$

$$\{(k+1)^4\} = D\left(\hbar D\left(\hbar D(\hbar\sigma^2)\right)\right),$$

etc., and

$$\{(k+1)^3\} = 6\hbar^2\sigma^4 + 6\hbar\sigma^3 + \sigma^2,$$

$$\{(k+1)^4\} = 24\hbar^3\sigma^5 + 36\hbar^2\sigma^4 + 14\hbar\sigma^3 + \sigma^2,$$

etc. In general, assuming

$$\{(k+1)^{n+1}\} = \sum_{j=0}^{n} c_{nj}\hbar^j\sigma^{j+2}, \; n \in Z^+,$$

then $c_{10} = 1$, $c_{11} = 2$, and

$$\begin{aligned} &\{(k+1)^{n+2}\} \\ = &\sum_{j=0}^{n} c_{nj} D\left(\hbar^{j+1}\sigma^{j+2}\right) \\ = &c_{n0}\sigma^2 + \sum_{j=1}^{n}(j+1)\left[c_{n,j-1} + c_{nj}\right]\hbar^j\sigma^{j+2} + (n+2)c_{nn}\hbar^{n+1}\sigma^{n+3}, \end{aligned}$$

so that

$$c_{n+1,0} = c_{n0}, \; n \in Z^+,$$

$$c_{n+1,n+1} = (n+2)c_{nn}, \; n \in Z^+,$$

and

$$c_{n+1,j} = (j+1)\left[c_{n,j-1} + c_{nj}\right], \quad 1 \le j \le n, \; n \in Z^+.$$

It is easy to construct a table for the coefficients c_{nj} for $0 \le j \le n$ and $n \ge 1$, or we may write a simple algorithm to generate the coefficients c_{nj}.

Next we make use of the algebraic derivatives to find the powers of the operator $1/((\hbar-\alpha)^2+\beta^2)$ where $\beta \neq 0$. First of all, we rewrite (3.22) and (3.23) in the form

$$\frac{1}{(\hbar-\alpha)^2+\beta^2} = \frac{1}{\beta}\rho,$$

and

$$\frac{\hbar-\alpha}{(\hbar-\alpha)^2+\beta^2}=\omega$$

respectively, where

$$\rho=\{\rho_k\}=\left\{\frac{(\alpha-i\beta)^{-k-1}-(\alpha+i\beta)^{-k-1}}{2i}\right\}=\left\{\frac{\sin((1+k)\phi)}{(\alpha^2+\beta^2)^{(1+k)/2}}\right\},$$

$$\omega=\{\omega_k\}=\left\{\frac{(\alpha-i\beta)^{-k-1}+(\alpha+i\beta)^{-k-1}}{-2}\right\}=\left\{-\frac{\cos((1+k)\phi)}{(\alpha^2+\beta^2)^{(1+k)/2}}\right\},$$

and ϕ is the argument of the complex number $\alpha+i\beta$. Note that

$$D\left(\frac{\hbar-\alpha}{(\hbar-\alpha)^2+\beta^2}\right)=\frac{2\beta^2}{((\hbar-\alpha)^2+\beta^2)^2}-\frac{1}{(\hbar-\alpha)^2+\beta^2},$$

thus,

$$\begin{aligned}\frac{1}{((\hbar-\alpha)^2+\beta^2)^2} &= \frac{1}{2\beta^2}\left[\frac{1}{(\hbar-\alpha)^2+\beta^2}+D\left(\frac{\hbar-\alpha}{(\hbar-\alpha)^2+\beta^2}\right)\right]\\ &= \frac{1}{2\beta^3}\rho+\frac{1}{2\beta^2}D\omega.\end{aligned}$$

By means of

$$D\left(\frac{1}{((\hbar-\alpha)^2+\beta^2)^2}\right)=\frac{-4(\hbar-\alpha)}{((\hbar-\alpha)^2+\beta^2)^3},$$

We may now obtain

$$\frac{(\hbar-\alpha)}{((\hbar-\alpha)^2+\beta^2)^3}=-\frac{1}{8\beta^3}D\rho-\frac{1}{8\rho^2}D^2\omega.$$

The more general cases can be dealt with by means of the following two formulas: for $n=2,3,...,$

$$D^2\left(\frac{1}{((\hbar-\alpha)^2+\beta^2)^{n-1}}\right)=\frac{2(2n-1)(n-1)}{((\hbar-\alpha)^2+\beta^2)^n}-\frac{4\beta^2 n(n-1)}{((\hbar-\alpha)^2+\beta^2)^{n+1}}, \tag{3.36}$$

and

$$D\left(\frac{1}{((\hbar-\alpha)^2+\beta^2)^n}\right)=\frac{-2n(\hbar-\alpha)}{((\hbar-\alpha)^2+\beta^2)^{n+1}}. \tag{3.37}$$

Thus we see, by substituting $n=2$ into (3.36), that

$$\begin{aligned}\frac{1}{((\hbar-\alpha)^2+\beta^2)^3} &= \frac{1}{8\beta^2}\left[\frac{6}{((\hbar-\alpha)^2+\beta^2)^2}-D^2\left(\frac{1}{((\hbar-\alpha)^2+\beta^2)}\right)\right]\\ &= \frac{3}{8\beta^5}\rho+\frac{3}{8\beta^4}D\omega-\frac{1}{8\beta^3}D^2\rho.\end{aligned}$$

By a similar argument, we may derive

$$\frac{1}{((\hbar-\alpha)^2+\beta^2)^4}=\frac{5}{16\beta^7}\rho+\frac{5}{16\beta^6}D\omega-\frac{1}{\beta^5}D^2\rho-\frac{1}{48\beta^4}D^3\omega.$$

Assume by induction that

$$\frac{1}{((\hbar-\alpha)^2+\beta^2)^n}=\frac{c_{n1}}{\beta^{2n-1}}\tilde{D}^0+\frac{c_{n2}}{\beta^{2n-1}}\tilde{D}^1+\frac{c_{n3}}{\beta^{2n-3}}\tilde{D}^2+\ldots+\frac{c_{nn}}{\beta^n}\tilde{D}^n,$$

where $\tilde{D}^0\equiv\rho$, and $\tilde{D}^k$ denotes $D^k\rho$ when k is odd, or $D^k\omega$ when k is even. Then $c_{21}=1/2$, $c_{22}=1/2$, and in view of (3.36), we have

$$\begin{aligned}
&\frac{4\beta^2 n(n-1)}{((\hbar-\alpha)^2+\beta^2)^{n+1}}\\
=\;& 2(2n-1)(n-1)\left(\frac{c_{n1}}{\beta^{2n-1}}\tilde{D}^0+\frac{c_{n2}}{\beta^{2n-2}}\tilde{D}^1+\frac{c_{n3}}{\beta^{2n-3}}\tilde{D}^2+\ldots+\frac{c_{nn}}{\beta^n}\tilde{D}^n\right)\\
&-\left(\frac{c_{n-1,1}}{\beta^{2n-3}}\tilde{D}^2+\frac{c_{n-1,2}}{\beta^{2n-4}}\tilde{D}^3+\frac{c_{n-1,3}}{\beta^{2n-5}}\tilde{D}^4+\ldots+\frac{c_{n-1,n-1}}{\beta^{n-1}}\tilde{D}^{n+1}\right)\\
=\;& \frac{2(2n-1)(n-1)c_{n1}}{\beta^{2n-1}}\tilde{D}^0+\frac{2(2n-1)(n-1)}{\beta^{2n-2}}\tilde{D}^1\\
&+\frac{2(2n-1)(n-1)c_{n3}-c_{n-1,1}}{\beta^{2n-3}}\tilde{D}^2+\ldots+\frac{2(2n-1)(n-1)c_{nn}-c_{n-1,n-2}}{\beta^n}\\
&-\frac{c_{n-1,n-1}}{\beta^{n-1}}\tilde{D}^{n+1}.
\end{aligned}$$

Therefore,

$$c_{n+1,1}=\frac{2n-1}{2n\beta^{2n+1}},\quad c_{n+1,2}=\frac{2n-1}{2n\beta^{2n}},$$

$$c_{n+1,j}=\frac{2(2n-1)(n-1)c_{nj}-c_{n-1,j-2}}{4n(n-1)\beta^{2n-j+2}},\quad j=3,4,\ldots,n,$$

and

$$c_{n+1,n+1}=\frac{c_{n-1,n-1}}{4n(n-1)\beta^{n+1}}.$$

It is easy to construct a table for the coefficients c_{nj} for $1\le j\le n$ and $n\ge 2$, or we may write a simple algorithm to generate the coefficients c_{nj}.

Finally, we may employ (3.37) to deduce

$$\begin{aligned}
&\frac{\hbar-\alpha}{((\hbar-\alpha)^2+\beta^2)^{n+1}}\\
=\;&\frac{-1}{2n}D\left(\frac{1}{((\hbar-\alpha)^2+\beta^2)^n}\right)\\
=\;&\frac{-1}{2n}\left(\frac{c_{n1}}{\beta^{2n-1}}\tilde{D}^1+\frac{c_{n2}}{\beta^{2n-1}}\tilde{D}^2+\frac{c_{n3}}{\beta^{2n-3}}\tilde{D}^3+\ldots+\frac{c_{nn}}{\beta^n}\tilde{D}^{n+1}\right)
\end{aligned}$$

for $n=2,3,\ldots$.

Example 40 *Given a sequence $\{f_k\}$ in l^N, let us define the* exponential (moment) generating operator

$$\bar{f}=\sum_{k=0}^{\infty}\frac{\overline{f_k}}{\overline{k!}}\hbar^k. \tag{3.38}$$

Then its j-th algebraic derivative is

$$D^j \tilde{f} = \sum_{k=0}^{\infty} \frac{\overline{f_{k+j}}}{\overline{k!}} \hbar^k, \quad j \in Z^+.$$

Thus, f_0 can be obtained from $\tilde{f}$ by formally substituting $\hbar$ by 0 in $\tilde{f}$, and f_k, $k \in Z^+$, by formally substituting $\hbar$ by 0 in $D^k \tilde{f}$. Consider the difference equation

$$f_{k+2} = f_{k+1} + f_k, \quad k \in N.$$

If we write this equation as

$$\overline{f_{k+2}} = \overline{f_{k+1}} + \overline{f_k}, \; n \in N,$$

then convoluting both sides with $\hbar^k/\overline{k!}$, and then sum from $k = 0$ to ∞, we see that the following algebraic differential equation

$$D^2 \tilde{f} = D\tilde{f} + \tilde{f}$$

holds.

There are many types of "algebraic differential" equations such as the one above. For instance,

$$\hbar^n D^n f + \hbar^{n-1} D^{n-1} f + ... + \hbar D f + f = q,$$

$$v_n(\hbar) D^n f + ... + v_1(\hbar) D f + v_0(\hbar) f = q,$$

where $v_i(\hbar)$ stands for an algebraic expression in $\hbar$ and $v_n(\hbar) \neq 0$, are two examples. Standard questions concerning existence and uniqueness of solutions, qualitative behaviors of solutions, etc. of these equations can be developed [66], but we will not discuss them here.

3.2.9 Algebraic Integrals

As in calculus, we may define the concept of a primitive of an operator. Let ϕ be an operator, if there is an operator ψ such that $D\psi = \phi$, then ψ is called the primitive or *algebraic integral* [66] of ϕ and is denoted by

$$\psi = \int \phi.$$

In such a case, we also say that ϕ is integrable. Not all operators are integrable. For instance, the operator $1/\hbar$ is not integrable. To see this, we first note that no sequence $f = \{f_k\}$ can satisfy $\hbar D f = 1$ since the 0-th term of $\hbar D f$ is zero. Now let f/g be a quotient operator, then $\hbar D(f/g) = 1$ would imply

$$g^2 = \hbar g D f - \hbar f D g.$$

Since $g = \{g_k\} \neq 0$, we may assume that $g_0 = .. = g_{m-1} = 0$ and $g_m \neq 0$. Note that g^2 is then a sequence whose first $2m$ terms are zero and the $2m$-th term is g_m^2. Note further that the first $2m+1$ terms of $\hbar g Df - \hbar f Dg$ are given by 0, $g_0 f_1 - f_0 g_1, ...,$ and

$$\sum_{j=0}^{k}(k+1-j)g_j f_{k+1-j} - \sum_{t=0}^{k}(t+1)g_{t+1}f_{k-t}, \quad k = 1, ..., 2m,$$

respectively. Thus we have

$$0 = \sum_{j=0}^{m-1}(m-j)g_j f_{m-j} - \sum_{t=0}^{m-1}(t+1)g_{t+1}f_{m-1-t} = mg_m f_0,$$

which implies $f_0 = 0$. Assume by induction that $f_0 = ... = f_m = 0$, then

$$\begin{aligned} 0 &= \sum_{j=0}^{2m-1}(2m-j)g_j f_{2m-j} - \sum_{t=0}^{2m-1}(t+1)g_{t+1}f_{2m-1-t} \\ &= mg_m f_{m+1}, \end{aligned}$$

so that $f_{m+1} = 0$. But then

$$g_m^2 = \sum_{j=0}^{2m}(2m+1-j)g_j f_{2m+1-j} - \sum_{t=0}^{2m}(t+1)g_{t+1}f_{2m-t} = 0,$$

which is a contradiction.

It is easy to show that

$$\int(\alpha\phi + \beta\theta) = \alpha\int\phi + \beta\int\theta, \ \alpha, \beta \in C; \quad \phi, \theta \in l^N/l^N,$$

provided ϕ and θ are integrable. It is also easy to show that any two primitives of an operator ϕ must differ only by a scalar operator.

The important question arises as to how a primitive is obtained. In case ϕ is an operator represented by the infinite series

$$q_0\hbar^{-m} + q_1\hbar^{-m+1} + q_2\hbar^{-m+2} + ...,$$

where m is some nonnegative integer, and $q = \{q_k\} \in l^N$ such that $q_0 \neq 0$. Then we may integrate this series term by term and obtain the primitive

$$\int\phi = \frac{q_0\hbar^{-m+1}}{-m+1} + \frac{q_1\hbar^{-m+2}}{-m+2} + ... ,$$

provided that $q_{m-1} = 0$.

Other techniques for obtaining primitives are similar to those for primitives of ordinary functions in calculus. For instance, we have the following "integration by parts" formula:

$$\int\phi D\theta = \int D(\phi\theta) - \int\theta D\phi.$$

Example 41 *Consider the following discrete Volterra-type equation*

$$(k+1)f_{k+1} - \sum_{j=0}^{k} f_j g_{k-j} = 0, \quad k \in N,$$

where $g = \{g_k\}$ is a given sequence. This equation can be written as

$$Df = gf,$$

where $f = \{f_k\}$. Since $De^{\phi} = e^{\phi}D\phi$, if ϕ is a primitive of g, then $f = e^{\phi}$ is a solution of the above equation. For instance, if $g = \{(k+1)2^{k+1}\}$, then its primitive ϕ is given by $\{2^k\}$ and the corresponding solution is $e^{\{2^k\}}$. To find other solutions of the equation $Df = gf$, let us assume that p is also a solution. Then $Dp = gp$ so that $pDf - fDp = pgf - fgp = 0$. But then $D(p/f) = 0$, which implies $p = \overline{\alpha} f$ for some scalar operator $\overline{\alpha}$. In other words, all solutions are of the form $\overline{\alpha} f$ when f is a nontrivial solution.

3.2.10 Ordinary Difference Equations

We consider in this section several methods for solving difference equations of the form

$$\alpha_0 f_{k+n} + \alpha_1 f_{k+n-1} + \ldots + \alpha_n f_k = g_k, \quad k \in N, \tag{3.39}$$

where n is a positive integer, and $\alpha_0, \alpha_1, \ldots, \alpha_n$ are complex numbers such that $\alpha_0 \neq 0$.

Before we turn to these methods, let us first illustrate the standard method of generating functions for solving ordinary difference equations. For this purpose, we recall the population model of Leonardo de Pisa:

$$f_{n+2} = f_{n+1} + f_n, \; n = 0, 1, 2, \ldots . \tag{3.40}$$

We first multiply both sides of (3.40) by t^n, and then sum from $n = 0$ to ∞. The resulting equation is

$$\sum_{n=0}^{\infty} f_{n+2}t^n = \sum_{n=0}^{\infty} f_{n+1}t^n + \sum_{n=0}^{\infty} f_n t^n. \tag{3.41}$$

We will say that the power series $f_0 + f_1 t + f_2 t^2 + \ldots$ is the *generating function* of the sequence $f_0, f_1, f_2, \ldots$, and denote it by $\Psi(t)$. Then since

$$t\sum_{n=0}^{\infty} f_{n+1}t^n = t\,(f_1 + f_2 t + \ldots) = \left(f_0 + f_1 t + f_2 t^2 + \ldots\right) - f_0 = \Psi(t) - f_0,$$

and

$$t^2\sum_{n=0}^{\infty} f_{n+2}t^n = t^2\,(f_2 + f_3 t + \ldots) = \Psi(t) - f_0 - f_1 t,$$

we see from (3.41) that

$$t^2\sum_{n=0}^{\infty} f_{n+2}t^n = t^2\sum_{n=0}^{\infty} f_{n+1}t^n + t^2\sum_{n=0}^{\infty} f_n t^n,$$

or

$$\Psi(t) - f_0 - f_1 t = t(\Psi(t) - f_0) + t^2\Psi(t). \tag{3.42}$$

From (3.42), we may solve for $\Psi(t)$, and obtain

$$\Psi(t) = \frac{-t}{t^2+t-1}.$$

Since the coefficients of the power series $\Psi(t)$ are the Fibonacci numbers, there are at least two ways to uncover them. First, recall from Calculus that

$$\Psi(0) = f_0, \Psi'(0) = f_1, \Psi''(0) = 2!f_2, ..., \Psi^{(n)}(0) = n!f_n, ...,$$

thus

$$f_n = \frac{1}{n!}\Psi^{(n)}(0),\ n \in N.$$

In our case, this method seems to be tedious. So we turn to another method. Note that by the method of partial fractions,

$$\frac{-t}{t^2+t-1} = -\frac{1}{2}\left\{\frac{1}{t+\gamma_+} + \frac{1}{t-\gamma_-}\right\},$$

where

$$\gamma_\pm = -\frac{1}{2} \pm \frac{1}{2}\sqrt{5}.$$

By expanding the rational functions $1/(t \pm \gamma_\pm)$, we see further that

$$\frac{-t}{t^2+t-1} = \sum_{n=0}^{\infty}(-1)^{n+1}\frac{\gamma_+^n - \gamma_-^n}{\sqrt{5}}t^n.$$

By comparing coefficients of the power series $\Psi(t)$ and $-t/(t^2+t-1)$, we finally end up with

$$f_n = (-1)^{n+1}\frac{\gamma_+^n - \gamma_-^n}{\sqrt{5}},\ n \in N. \tag{3.43}$$

We remark that the above method seems all right except for the convergence questions of the power series involved. We therefore return to our symbolic calculus for a rigorous procedure. We treat $f_0 = 0, f_1 = 1$ as the operator conditions $\overline{f_0} = \overline{0}, \overline{f_1} = \overline{1}$ respectively, and treat equation (3.40) as

$$\overline{f_{n+1}} = \overline{f}_{n+1} + \overline{f}_n,\ n = 0, 1, 2, ... \tag{3.44}$$

Convoluting both sides of (3.44) with $\hbar^n$, then summing from $n = 0$ to ∞, and then taking the convolution product of the resulting equation with $\hbar^2$, we obtain

$$\hbar^2 * \sum_{n=0}^{\infty}\overline{f_{n+2}} * \hbar^n = \hbar^2 * \sum_{n=0}^{\infty}\overline{f_{n+1}} * \hbar^n + \hbar^2 * \sum_{n=0}^{\infty}\overline{f_n} * \hbar^n,$$

or

$$\hbar^2 * E^2 f = \hbar^2 * Ef + \hbar^2 * f.$$

Recall the formula in Theorem 28,

$$(\overline{1} - \delta)^n(E^n f) = f - \overline{f_0} - (\overline{1} - \delta)f_1 - ... - (\overline{1} - \delta)^{n-1}f_{n-1},$$

or

$$\hbar^n(E^n f) = f - \overline{f_0} - \hbar\overline{f_1} - \ldots - \hbar^{n-1}\overline{f_{n-1}}, \tag{3.45}$$

where $n \in Z^+$, and $f = \{f_k\} \in l^N$. We see further that

$$f - \overline{f_0} - \overline{f_1} * \hbar = \hbar * (f - \overline{f_0}) + \hbar^2 * f,$$

which shows that

$$\frac{f}{\overline{1}} = \frac{-\hbar}{\hbar * \hbar + \hbar - \overline{1}}.$$

By the method of partial fractions again,

$$\frac{f}{\overline{1}} = \frac{\overline{-1}}{\overline{2}}\left\{\frac{\overline{1}}{\hbar + \overline{\gamma_+}} + \frac{\overline{1}}{\hbar - \overline{\gamma_-}}\right\}. \tag{3.46}$$

In view of the formula

$$\{\alpha^k\} = \frac{1}{1 - \alpha\hbar},$$

we may "expand" the right hand side of (3.46) to obtain

$$\frac{f}{\overline{1}} = \sum_{n=0}^{\infty} \frac{\overline{(-1)^{n+1}(\gamma_+^n - \gamma_-^n)/\sqrt{5} * \hbar^n}}{\overline{1}},$$

so that the same formula (3.43) holds.

This example motivates a general procedure for solving difference equations (and possibly others) of the form (3.39). We first write (3.39) in the form

$$\alpha_0 E^n f + \alpha_1 E^{n-1} f + \ldots + \alpha_n f = g \tag{3.47}$$

and then multiply both sides by $\hbar^n$ or by $(1 - \delta)^n$. Then an algebraic equation involving the unknown operator f is obtained. After solving f from this equation, we need only to find the sequence which represents this operator. If this can be done, we obtain the desired sequence solution, otherwise, a generalized operator solution is obtained.

Example 42 *Consider the difference equation*

$$f_{k+2} + 3f_{k+1} + 2f_k = 1,\ k \in N, \tag{3.48}$$

subject to the conditions $f_0 = 1$ and $f_1 = 0$. We write equation (3.48) in the form

$$E^2 f + 3Ef + 2f = \sigma,$$

and the initial conditions as $\overline{f_0} = \overline{1}$, $\overline{f_1} = \overline{0}$ respectively. Then multiplying both sides of the above equation by $\hbar^2$, and then applying (3.45), we see that

$$\overline{2}f + \frac{\overline{3}}{\hbar}(f - \overline{1}) + \frac{\overline{1}}{\hbar^2}(f - \overline{1}) = \frac{\overline{1}}{\overline{1} - \hbar}.$$

Thus

$$f = \frac{\bar{1} + \bar{2}\hbar - \bar{2}\hbar^2}{(\bar{1} - \hbar)(\bar{1} + \hbar)(\bar{1} + \bar{2}\hbar)}$$
$$= \frac{\bar{1}}{\bar{6}}\frac{\bar{1}}{\bar{1} - \hbar} + \frac{\bar{3}}{\bar{2}}\frac{\bar{1}}{\bar{1} + \hbar} - \frac{\bar{2}}{\bar{3}}\frac{\bar{1}}{\bar{1} + \bar{2}\hbar}$$
$$= \sum_{k=0}^{\infty} \left(\frac{1}{6} + \frac{3}{2}(-1)^k - \frac{2}{3}(-2)^k\right)\hbar^k,$$

which implies that

$$f_k = \frac{1}{6} + \frac{3}{2}(-1)^k - \frac{2}{3}(-2)^k, \ k \in N.$$

In general, from (3.47) and (3.45), we see that

$$\frac{\alpha_0}{\hbar^n}\left(f - f_0 ... - f_{n-1}\hbar^{n-1}\right) + ... + \frac{\alpha_{n-1}}{\hbar}(f - f_0) + \alpha_n f = g,$$

so that

$$f = \frac{\beta_{n-1}\hbar^{n-1} + ... + \beta_0}{\alpha_n \hbar^n + ... + \alpha_0} + \frac{\hbar^n g}{\alpha_n \hbar^n + ... + \alpha_0}, \tag{3.49}$$

where

$$\beta_i = \alpha_0 f_i + \alpha_1 f_{i-1} + ... + \alpha_i f_0.$$

If we can expand the right hand side of (3.49) into a power series of the translation operator $\hbar$, then we will obtain the sequence $\{f_k\}$. We remark that there is at least one case in which this can be done, namely, when

$$g = \frac{v_p \hbar^p + ... + v_1 \hbar + v_0}{\mu_q \hbar^q + ... + \mu_1 \hbar + \mu_0}$$

is a rational operator in $\hbar$. This is due to the fact that a rational operator can be decomposed into the sum of a polynomial in $\hbar$ and simple fractions in $\hbar$, which have already been discussed in the previous sections. We remark further that by means of the algebraic derivatives, we may also proceed, as in Example 39, to derive the k-th term of the solution. This procedure may, however, be quite tedious.

3.3 Semi-Infinite Bivariate Sequences

3.3.1 Ring of Double Sequences

Let $l^{N \times N}$ be the set of all complex bivariate sequences of the form $f = \{f_{ij}\}_{i,j \in N}$. Such a bivariate sequence f is a function defined on the set of all nonnegative lattice points $N \times N$ and a natural way to view a bivariate sequence is to regard it as an infinite matrix of the form

$$\begin{bmatrix} f_{00} & f_{01} & \cdots \\ f_{10} & f_{11} & \cdots \\ \cdots & \cdots & \cdots \end{bmatrix}.$$

For the sake of convenience, we will also write $\{f_{ij}\}$ instead of $\{f_{ij}\}_{i,j \in N}$ if no confusion is caused. The number f_{ij} will be called the (i, j)-th component of the bivariate sequence f, while the sequences $\{f_{i0}, f_{i1}, ...\}$ and $\{f_{0j}, f_{1j}, ...\}$ will be called its i-th row and j-th column. The set of all bivariate sequences with null rows except the 0-th ones will be denoted by $X_0(l^{N\times N})$, and the set of bivariate sequences with null columns except the 0-th columns by $Y_0(l^{N\times N})$. A typical element in $X_0(l^{N\times N})$ is of the form

$$\begin{bmatrix} * & * & * & \cdots \\ 0 & 0 & 0 & \cdots \\ 0 & 0 & 0 & \cdots \\ \cdots & \cdots & \cdots & \cdots \end{bmatrix}.$$

For any complex number α and $f = \{f_{ij}\}, g = \{g_{ij}\} \in l^{N\times N}$, we define $-f, \alpha f$ and $f + g$ by $\{-f_{ij}\}, \{\alpha f_{ij}\}, \{f_{ij} + g_{ij}\}$ as usual.

It will be necessary to list the components of a bivariate sequence in $l^{N\times N}$ in a linear order. This can be done once we introduce a linear ordering for $N \times N$. One such ordering $\preccurlyeq$ has been introduced in the first section of Chapter 2. Under such an ordering, we may enumerate the components of $\{f_{ij}\}$ in $l^{N\times N}$ as $f_{00}; f_{10}, f_{01}; f_{2,0}, f_{11}, f_{02}; \cdots$.

There are some common sequences in $l^{N\times N}$ which deserve special notations. First of all, let α be a complex number, the sequence whose $(0,0)$-th component is α and others are zero will be denoted by $\overline{\alpha}$ and is called a *scalar bivariate sequence*. In particular, the sequence with all zero components will be denoted by $\overline{0}$. Note that we have used a similar notation for scalar sequence of one integral variable. However, confusion cannot arise if the number of independent variables is indicated. The bivariate sequence whose $(1,0)$-th component is 1 and others are zero will be denoted by $\hbar_x$, while the sequence whose $(0,1)$-th component is 1 and others are zero will be denoted by $\hbar_y$:

$$\hbar_x = \begin{bmatrix} 0 & 0 & 0 & \cdots \\ 1 & 0 & 0 & \cdots \\ 0 & 0 & 0 & \cdots \\ \cdots & \cdots & \cdots & \cdots \end{bmatrix}, \quad \hbar_y = \begin{bmatrix} 0 & 1 & 0 & \cdots \\ 0 & 0 & 0 & \cdots \\ 0 & 0 & 0 & \cdots \\ \cdots & \cdots & \cdots & \cdots \end{bmatrix},$$

while the bivariate sequences $\sigma_x, \sigma_y, \delta_x$ and δ_y are defined by

$$\sigma_x = \begin{bmatrix} 1 & 0 & 0 & \cdots \\ 1 & 0 & 0 & \cdots \\ 1 & 0 & 0 & \cdots \\ \cdots & \cdots & \cdots & \cdots \end{bmatrix}, \quad \sigma_y = \begin{bmatrix} 1 & 1 & 1 & \cdots \\ 0 & 0 & 0 & \cdots \\ 0 & 0 & 0 & \cdots \\ \cdots & \cdots & \cdots & \cdots \end{bmatrix},$$

$$\delta_x = \begin{bmatrix} 1 & 0 & 0 & \cdots \\ -1 & 0 & 0 & \cdots \\ 0 & 0 & 0 & \cdots \\ \cdots & \cdots & \cdots & \cdots \end{bmatrix}, \quad \delta_y = \begin{bmatrix} 1 & -1 & 0 & \cdots \\ 0 & 0 & 0 & \cdots \\ 0 & 0 & 0 & \cdots \\ \cdots & \cdots & \cdots & \cdots \end{bmatrix}$$

respectively. It is important to note that $\delta_x + \hbar_x = \overline{1}$ and $\delta_y + \hbar_y = \overline{1}$.

The bivariate sequence obtained from the bivariate sequence $f = \{f_{ij}\}$ by annihilating all its rows except the i-th row is denoted by $X_i f$. $Y_j f$ is similarly defined.

For example,

$$X_1 \begin{bmatrix} f_{00} & f_{01} & f_{02} & \cdots \\ f_{10} & f_{11} & f_{12} & \cdots \\ f_{20} & f_{21} & f_{22} & \cdots \\ \cdots & \cdots & \cdots & \cdots \end{bmatrix} = \begin{bmatrix} 0 & 0 & 0 & \cdots \\ f_{10} & f_{11} & f_{12} & \cdots \\ 0 & 0 & 0 & \cdots \\ \cdots & \cdots & \cdots & \cdots \end{bmatrix},$$

and

$$Y_0 \begin{bmatrix} f_{00} & f_{01} & f_{02} & \cdots \\ f_{10} & f_{11} & f_{12} & \cdots \\ f_{20} & f_{21} & f_{22} & \cdots \\ \cdots & \cdots & \cdots & \cdots \end{bmatrix} = \begin{bmatrix} f_{00} & 0 & 0 & \cdots \\ f_{10} & 0 & 0 & \cdots \\ f_{20} & 0 & 0 & \cdots \\ \cdots & \cdots & \cdots & \cdots \end{bmatrix}.$$

More generally, if we annihilate all the components of f except those on m distinct rows and n columns, the resulting sequence is called an annihilated bivariate sequence of f. For example, the matrix

$$\begin{bmatrix} f_{00} & f_{01} & f_{02} & \cdots \\ f_{10} & 0 & 0 & \cdots \\ f_{20} & 0 & 0 & \cdots \\ \cdots & \cdots & \cdots & \cdots \end{bmatrix}$$

represents an annihilated bivariate sequence and can be written as $X_0 f + Y_0 f - X_0 Y_0 f$. In general, if $\{i_1, ..., i_m\}$ and $\{j_1, ..., j_n\}$ are two sets of mutually distinct nonnegative integers, then

$$\sum_{s=1}^{m} X_{i_s} f + \sum_{t=1}^{n} Y_{j_t} f - \sum_{u=1}^{m} \sum_{v=1}^{n} X_{i_u} Y_{j_v} f$$

is the bivariate sequence whose components are annihilated except those on the $i_1, ..., i_m$-th rows and $j_1, ..., j_n$-th columns.

The bivariate sequence $\{f_{i+m,j+n}\}_{i,j \in Z}$ will be denoted by $E_x^m E_y^n \{f_{ij}\}$, where $m, n \in N$. The sequence $E_x^m E_y^n f$ is called a *translated sequence* of f. For the sake of convenience, $E_x^0 E_y^n f$ and $E_x^m E_y^0 f$ are also denoted by $E_y^n f$ and $E_x^m f$ respectively.

The bivariate sequence $\{f_{i+1,j} - f_{ij}\}$ obtained by taking the partial differences of the bivariate sequence $f = \{f_{ij}\}$ will be denoted by $\Delta_x f$ and is called the first partial difference of f with respect to the first variable. The partial differences $\Delta_y f$, $\Delta_x \Delta_y f$, $\Delta_x^2 f$, $\Delta_x^m \Delta_y^n f$, etc. are similarly defined. In particular,

$$\begin{aligned} \Delta_x \Delta_y \{f_{ij}\} &= \{f_{i+1,j+1} - f_{i+1,j} - f_{i,j+1} + f_{ij}\} \\ &= E_x E_y f - E_x f - E_y f + f, \end{aligned}$$

and

$$\Delta_x^m \Delta_y^n \{f_{ij}\} = \sum_{i=0}^{m} \sum_{j=0}^{n} (-1)^{m+n-i-j} C_i^{(m)} C_j^{(n)} E_x^i E_y^j f, \quad m, n \in Z^+.$$

For any complex numbers λ and μ, the sequence $\{\lambda^i \mu^j f_{ij}\}$ obtained by scaling each term of the bivariate sequence $\{f_{ij}\}$ with a growth factor $\lambda^i \mu^j$ is called an attenuated bivariate sequence of f and is denoted by $T_x^\lambda T_y^\mu f$. For the sake of convenience, $T_x^1 T_y^\mu f$ and $T_x^\lambda T_y^1 f$ are also denoted by $T_y^\mu f$ and $T_x^\lambda f$ respectively. It is easily seen that $T_x^0 T_y^0 f = \overline{f_{00}}$, $T_x^1 T_y^1 f = f$ and $(T_x^\lambda T_y^\mu)(T_x^\rho T_y^\tau) f = T_x^{\lambda\rho} T_y^{\mu\tau} f$.

For any $f = \{f_{ij}\}, g = \{g_{ij}\} \in l^{N\times N}$, we define the convolution product $f * g$, by

$$(f * g)_{ij} = \sum_{u=0}^{i}\sum_{v=0}^{j} f_{uv}g_{i-u,j-v},\ i,j \geq 0.$$

Under the ordering $\preccurlyeq$, we may evaluate the components of $f*g$ in an orderly manner as follows:

$$f_{00}g_{00};\ f_{10}g_{00} + f_{00}g_{10},\ f_{01}g_{00} + f_{00}g_{01};$$

$$f_{20}g_{00} + f_{10}g_{10} + f_{00}g_{20},\ f_{11}g_{00} + f_{01}g_{10} + f_{10}g_{01} + f_{00}g_{11}, \cdots.$$

For the sake of convenience, we will also use the simpler notation fg for the product $f * g$. Note that $f * f$, $f * (f * f)$, ..., will also be written as $f^2, f^3, ...$, respectively.

For example, $\overline{0}*f = \overline{0}$, $\overline{1}*f = f$, $\overline{\alpha}*\overline{\beta} = \overline{\alpha\beta}$, and $\overline{\alpha}*f = (\alpha\overline{1})*f = \alpha(\overline{1}*f) = \alpha f$. More complicated examples can also be given. First of all, $\hbar_x^m$ (or $\hbar_y^m$) is a bivariate sequence whose $(m,0)$-th component (respectively $(0,m)$-th component) is 1 and others are zero, while $\hbar_x^m\hbar_y^n$ is a bivariate sequence whose (m,n)-th component is 1 and others are zero. It is also interesting to note that $\hbar_x^m\hbar_y^n\{f_{ij}\} = \{g_{ij}\}$ where

$$g_{ij} = \begin{cases} f_{i-m,j-n} & i \geq m, j \geq n \\ 0 & \text{otherwise} \end{cases}.$$

For instance, the matrix representation of the bivariate sequence $\hbar_x^2\hbar_y^1\{f_{ij}\}$ is

$$\begin{bmatrix} 0 & 0 & 0 & 0 & \cdots \\ 0 & 0 & 0 & 0 & \cdots \\ 0 & f_{00} & f_{01} & f_{02} & \cdots \\ 0 & f_{10} & f_{11} & f_{12} & \cdots \\ \cdot & \cdots & \cdots & \cdots & \cdots \end{bmatrix},$$

while $f_{21}\hbar_x^2\hbar_y^1$ is

$$\begin{bmatrix} 0 & 0 & 0 & \cdots \\ 0 & 0 & 0 & \cdots \\ 0 & f_{21} & 0 & \cdots \\ 0 & 0 & 0 & \cdots \\ \cdots & \cdots & \cdots & \cdots \end{bmatrix}.$$

Next, δ_x^m is the bivariate sequence whose $(i,0)$-th component is $(-1)^i m^{[i]}/i!$ for $i \geq 0$ and others are zero, δ_y^n is the bivariate sequence whose $(0,j)$-th component is $(-1)^j n^{[j]}/j!$ for $j \geq 0$ and others are zero, while

$$\delta_x^m\delta_y^n = \left\{(-1)^{i+j}\frac{m^{[i]}n^{[j]}}{i!j!}\right\}_{i,j=0}^{\infty},\ m,n \in Z^+.$$

In particular, the matrix representations of $\delta_x\delta_y$ and $\delta_x^2\delta_y^3$ are

$$\begin{bmatrix} 1 & -1 & 0 & \cdots \\ -1 & 1 & 0 & \cdots \\ 0 & 0 & 0 & \cdots \\ \cdot & \cdot & \cdot & \cdots \end{bmatrix},$$

and

$$\begin{bmatrix} 1 & -3 & 3 & -1 & 0 & \dots \\ -2 & 6 & -6 & 2 & 0 & \dots \\ 1 & -3 & 3 & -1 & 0 & \dots \\ 0 & 0 & 0 & 0 & 0 & \dots \\ \cdot & \cdot & \cdot & \cdot & \cdot & \dots \end{bmatrix}$$

respectively. Next, σ_x^m is the bivariate sequence whose $(i,0)$-th component is

$$\frac{(i+m-1)^{[m-1]}}{(m-1)!}$$

for $i \geq 0$ and others are zero, σ_y^n the bivariate sequence whose $(0,j)$-th component is

$$\frac{(j+n-1)^{[n-1]}}{(n-1)!}$$

for $j \geq 0$ and others are zero, while

$$\sigma_x^m \sigma_y^n = \left\{ \frac{(i+m-1)^{[m-1]}}{(m-1)!} \frac{(j+n-1)^{[n-1]}}{(n-1)!} \right\}_{i,j=0}^{\infty}, \quad m,n \in Z^+.$$

Although the formulas for $\delta_x^m \delta_y^n$ and $\sigma_x^m \sigma_y^m$ can be proved directly by induction, we will present their alternate proofs in a later section.

There are several elementary facts related to the convolution product of bivariate sequences. First of all, we may show that for any bivariate sequences $f = \{f_{ij}\}$, $g = \{g_{ij}\}$ and $h = \{h(i,j)\}$, we have $fg = gf$ and $f(gh) = (fg)h$. Indeed, these are due to the fact that the convolution product of sequences of a single integral variable are commutative and associative:

$$\sum_{k=0}^{i} x_k y_{i-k} = \sum_{k=0}^{i} x_{i-k} y_k,$$

and

$$\sum_{k=0}^{m} \left(\sum_{i=0}^{k} x_i y_{k-i} \right) z_{m-k} = \sum_{i=0}^{m} \sum_{k=i}^{m} x_i y_{k-i} z_{m-k} = \sum_{i=0}^{m} x_i \sum_{j=0}^{m-i} y_j z_{m-i-j}.$$

Next, we show that when $f \neq \bar{0}$ and $g \neq \bar{0}$, then $fg \neq \bar{0}$. Indeed, suppose the components of f and g are ordered by the ordering $\preccurlyeq$, then

$$f_{00} = f_{10} = f_{01} = \dots = f_{m+1,n-1} = 0, \ f_{mn} \neq 0,$$

and

$$g_{00} = g_{10} = g_{01} = \dots = g_{s+1,t-1} = 0, \ g_{st} \neq 0,$$

where we may assume without loss of generality that $(m,n) \preccurlyeq (s,t)$. Since when $s+t \geq m+n$,

$$(fg)_{m+s,n+t} = f_{00} g_{m+s,n+t} + \dots + f_{mn} g_{st} + \dots + f_{m+s,n+t} g_{00} = f_{mn} g_{st} \neq 0.$$

we see that $fg \neq 0$.

Theorem 37 *Let $f = \{f_{ij}\}$ and $g = \{g_{ij}\}$ be bivariate sequences in $l^{N\times N}$. If $g_{00} \neq 0$, then there is a unique bivariate sequence $x = \{x_{ij}\}$ such that $g * x = f$.*

The proof is elementary. We write the component equations of $g * x = f$ in the following orderly manner:

$$g_{00}x_{00} = f_{00},$$

$$g_{00}x_{10} + g_{10}x_{00} = f_{10},$$

$$g_{00}x_{01} + g_{01}x_{00} = f_{01},$$

$$g_{00}x_{20} + g_{10}x_{10} + g_{20}x_{00} = f_{20},$$

$$g_{00}x_{11} + g_{10}x_{01} + g_{01}x_{10} + g_{11}x_{00} = f_{11},$$

and so on, and then obtain $x_{00} = f_{00}/g_{00}$, $x_{10} = (f_{10} - g_{10}x_{00})/g_{00}$, ..., successively in a unique manner.

Theorem 38 *Let $f = \{f_{ij}\} \in l^{N\times N}$. If $f_{00} = 0$, then $f_{ij}^n = 0$ for all $(i,j) \in \{(i,j) \in N^2 |\, i+j \le n-1\}$, where $n \in Z^+$.*

Proof. Let $Q_k = \{(i,j) \in N^2 |\, i+j = k\}$ for $k \in N$. Assume by induction that $f_{ij}^k = 0$ for $(i,j) \in Q_{k-1}$ where k is a positive integer. Then for $(i,j) \in Q_0 + \cdots + Q_{k-1}$,

$$f_{ij}^{k+1} = \sum_{u=0}^{i}\sum_{v=0}^{j} f_{uv}^k f_{i-u,j-v} = \sum_{u=0}^{i}\sum_{v=0}^{j} 0 \cdot f_{i-u,j-v} = 0.$$

For $(i,j) \in Q_k$, let $S = \{0,1,...,i\} \times \{0,1,...,j\}$, then

$$f_{ij}^{k+1} = \sum_{(u,v)\in S\setminus\{(i,j)\}} f_{uv}^k f_{i-u,j-v} + f_{ij}^k f_{00} = 0.$$

The proof is complete. ■

As an example, when $f_{00} = 0$, the matrix representation of f^4 is of the form

$$\begin{bmatrix} 0 & 0 & 0 & * & \dots \\ 0 & 0 & * & . & \dots \\ 0 & * & . & . & \dots \\ * & . & . & . & \dots \\ . & . & . & . & \dots \end{bmatrix}$$

Theorem 39 *Let $f = \{f_{ij}\}, g = \{g_{ij}\}$ be bivariate sequences in $l^{N\times N}$. Then $T_x^\lambda T_y^\mu (f * g) = (T_x^\lambda T_y^\mu f) * (T_x^\lambda T_y^\mu g)$ for $\lambda, \mu \in C$.*

Indeed,

$$\begin{aligned} T_x^\lambda (fg) &= \left\{ \sum_{u=0}^{i}\sum_{v=0}^{j} \lambda^i f_{uv} g_{i-u,j-v} \right\} \\ &= \left\{ \sum_{u=0}^{i}\sum_{v=0}^{j} \lambda^u f_{uv} \lambda^{i-u} g_{i-u,j-v} \right\} = (T_x^\lambda f) * (T_x^\lambda g). \end{aligned}$$

The general case is similarly proved.

3.3.2 Operators

It is easily verified that under the above defined addition and convolution product, $l^{N\times N}$ is a commutative ring with no zero divisor, and the additive and multiplicative identities are $\bar{0}$ and $\bar{1}$ respectively. Consequently, we can construct from $l^{N\times N}$ a field $l^{N\times N}/l^{N\times N}$ of quotients in a routine manner. Briefly, on the set

$$\{(f,g)|\ f,g \in l^{N\times N},\ g \neq 0\},$$

we define a relation $\sim$ by

$$(f,g) \sim (p,q) \Longleftrightarrow fq = pg.$$

Then it is easily verified that $\sim$ is an equivalence relation. Each equivalence class can be represented by an ordered pair (f,g), which is written in the form of a quotient f/g in analogy with the rationals. The set of all such quotients is denoted by $l^{N\times N}/l^{N\times N}$. Hence, two quotients f/g and p/q are equal if, and only if, $fq = pg$. We define the product and addition of quotients by

$$\frac{f}{g}\frac{p}{q} = \frac{fp}{gq},$$

and

$$\frac{f}{g} + \frac{p}{q} = \frac{fq+pg}{gq}$$

respectively. Under these definitions, $l^{N\times N}/l^{N\times N}$ is a field with the unique additive identity $\bar{0}/\bar{1}$ and the unique multiplicative identity $\bar{1}/\bar{1}$. For convenience, we will call a quotient f/g an *operator*. As usual, the additive inverse of an operator f/g is $-f/g$, which is also denoted by $-(f/g)$. Operators can be divided by one another. More specifically, if $\phi = p/q$ and $\psi = f/g$ are operators where $f,g,p,q \in l^{N\times N}$ with $f \neq 0$, then the quotient ϕ/ψ of the operators ϕ and ψ is pg/qf. In particular, when ψ is a nonzero operator, $\bar{1}/\psi$, which is also denoted by ψ^{-1}, is called the multiplicative inverse of the operator ψ. Note that a quotient of the form f/g may have different meanings. For definiteness, by a quotient operator, we will always mean an operator of the form f/g where $f,g \in l^{N\times N}$.

Every bivariate sequence f can be regarded as an operator since it can be identified with $f/\bar{1}$, and in such a case, the corresponding operator is said to be *ordinary*. In particular, an ordinary operator of the form $\bar{\alpha}/\bar{1}$, where $\alpha \in C$, will be called a *scalar operator*. If no confusion arises, we will denote an ordinary operator $f/\bar{1}$ by f.

Not all operators are ordinary. For example, the operator $\bar{1}/(\sigma_x\sigma_y - \bar{1})$ cannot be ordinary. For otherwise there would exist a bivariate sequence $f = \{f_{ij}\}$ such that $\sigma_x\sigma_y f - f = \bar{1}$. But then $f_{00} - f_{00} = 1$, which is a contradiction. However, any quotient operator f/g is ordinary when the $(0,0)$-th component of the sequence g is nontrivial. This is true in view of Theorem 37.

As is customary, for any operators ϕ and ψ, we denote $\phi+\phi$ by 2ϕ, $\phi+\psi+\psi$ by $\phi+2\psi$, $\phi\phi$ by ϕ^2, etc. The notation 2ϕ, for example, is actually a definition. Fortunately, $\phi+\phi = \bar{2}\phi$ so that we may identify 2ϕ as $\bar{2}\phi$. In the sequel, for a complex number α, a positive integer n and an operator ϕ, the notation $\alpha\phi$ will be taken to mean $\bar{\alpha}\phi$, ϕ^0 to mean $\bar{1}$, and ϕ^n to mean $\phi\phi\cdots\phi$. In an expression involving operators, the notation α will be taken to mean the ordinary scalar operator $\overline{\alpha}$. For instance, the following notations

$$\frac{\bar{1}}{\delta_x - \overline{\alpha}}, \frac{\bar{1}}{\delta_x - \alpha}, \frac{1}{\delta_x - \overline{\alpha}}, \frac{1}{\delta_x - \alpha}$$

are all equivalent but the last one is much simpler. We will use any one of the above notations whenever needed.

At this point, we recall that, in the previous section, a symbolic calculus has already been established for the set l^N of all complex sequences under the usual addition and convolution. We have also defined $X_0(l^{N\times N})$ to be the set of all bivariate sequences with null rows except the 0-th row. If we identify a bivariate sequence f in $X_0(l^{N\times N})$ by means of its first row, we are then setting up an isomorphism Φ between $X_0(l^{N\times N})$ with l^N so that

$$\Phi(f+g) = \Phi f + \Phi g,$$

and

$$\Phi(f * g) = (\Phi f) * (\Phi g).$$

Clearly then, for each formula which is valid for the sequences and/or operators in the calculus for l^N, there is a corresponding one for bivariate sequences and/or operators in the calculus for $X_0(l^{N\times N})$. For example, it is known that

$$\{2^k\}_{k=0}^{\infty} = \frac{\bar{1}}{\bar{2}\delta - \bar{1}} = \frac{\bar{1}}{\bar{1} - \bar{2}\hbar},$$

and

$$\{k+1\}_{k=0}^{\infty} = \sigma^2,$$

where $\delta = \{1, -1, 0, ...\}$, $\sigma = \{1, 1, ...\}$, $\hbar = \{0, 1, 0, ...\}$ and $\overline{\alpha} = \{\alpha, 0, 0...\}$. By means of the isomorphism between $X_0(l^{N\times N})$ and l^N, we identify the operators $\delta, \sigma, \hbar$ and $\overline{\alpha}$ with $\delta_y, \sigma_y, \hbar_y$ and $\overline{\alpha}$ in $X_0(l^{N\times N})$, and to $\delta_x, \sigma_x, \hbar_x$ and $\overline{\alpha}$ in $Y_0(l^{N\times N})$ and obtain

$$X_0\{2^j\}_{i,j\in Z} = \frac{\bar{1}}{\bar{2}\delta_y - \bar{1}} = \frac{\bar{1}}{\bar{1} - \bar{2}\hbar_y},$$

$$X_0\{j+1\}_{i,j\in Z} = \sigma_y^2,$$

and

$$Y_0\{2^i\} = \frac{\bar{1}}{\bar{2}\delta_x - \bar{1}}.$$

3.3.3 Separable Double Sequences

A bivariate sequence of the form $\{f_i g_j\}_{i,j\in Z}$ is said to be separable. As examples, $\{ij\}$, $\{2^{i+j}\}$, $\{2^{i-j}\}$, $\{a^i b^j\}$ are separable sequences. A separable bivariate sequence has the following important property.

Theorem 40 *Let $f = \{g_i h_j\}_{i,j\in Z}$. Then*

$$f = (Y_0\{g_i\}_{i,j\in Z}) * (X_0\{h_j\}_{i,j\in Z}).$$

Indeed, this follows by directly verifying that

$$\begin{bmatrix} g_0h_0 & g_0h_1 & \cdots \\ g_1h_0 & g_1h_1 & \cdots \\ \cdots & \cdots & \cdots \\ & & \end{bmatrix}$$

$$= \begin{bmatrix} g_0 & 0 & \cdots \\ g_1 & 0 & \cdots \\ g_2 & 0 & \cdots \\ \cdots & \cdots & \cdots \end{bmatrix} * \begin{bmatrix} f_0 & f_1 & f_2 & \cdots \\ 0 & 0 & 0 & \cdots \\ \cdots & \cdots & \cdots & \cdots \end{bmatrix}.$$

As an example, $\{1\}_{i,j\in Z} = \sigma_x\sigma_y$ as we have already seen. As another example, we have,

$$\{2^{i+j}\} = Y_0\{2^i\} * X_0\{2^j\} = \frac{1}{(2\delta_x - 1)(2\delta_y - 1)}.$$

More complicated examples can be given, for example,

$$\{2^{i+j} + j + 1\} = \frac{1}{(2\delta_x - 1)(2\delta_y - 1)} + \sigma_y^2.$$

Once the principle behind the above examples is fully understood, we may then derive a number of operator representations of some common bivariate sequences. A list follows.

Recall that

$$\frac{\bar{1}}{(\delta - \bar{\alpha})^\lambda} = \left\{\frac{\Gamma(k+\lambda)}{\Gamma(\lambda)\Gamma(k)}(1-\alpha)^{-k-\lambda}\right\},\ \alpha \neq 1, \lambda > 0. \tag{3.50}$$

Thus

$$\frac{\bar{1}}{(\delta_x - \bar{a})^\lambda(\delta_y - \bar{b})^\mu}$$
$$= \left\{\frac{\Gamma(i+\lambda)\Gamma(j+\mu)}{\Gamma(\lambda)\Gamma(i)\Gamma(\mu)\Gamma(j)}(1-a)^{-i-\lambda}(1-b)^{-j-\mu}\right\},\ a, b \neq 1,\ \lambda, \mu > 0.$$

In particular,

$$\sigma_x^m\sigma_y^n = \left\{\frac{(i+m-1)^{[m-1]}}{(m-1)!}\frac{(j+n-1)^{[n-1]}}{(n-1)!}\right\},\ m, n \in Z^+,$$

$$\begin{aligned} \sigma_x^m\sigma_y^n\{f_{ij}\} &= \left\{\sum_{u=0}^{i}\sum_{v=0}^{j}\frac{(u+m-1)^{[m-1]}}{(m-1)!}\frac{(v+n-1)^{[n-1]}}{(n-1)!}f_{i-u,j-v}\right\} \\ &= \left\{\sum_{u=0}^{i}\sum_{v=0}^{j}f_{uv}\frac{(i-u+m-1)^{[m-1]}}{(m-1)!}\frac{(j-v+n-1)^{[n-1]}}{(n-1)!}\right\}, \end{aligned}$$

for $m, n \in Z^+$, and

$$\sigma_x\sigma_y\{f_{ij}\} = \left\{\sum_{u=0}^{i}\sum_{v=0}^{j}f_{ij}\right\}.$$

In view of the above formulas, it is natural to call the corresponding ordinary operator $(\delta_x - \overline{a})^{-\lambda}(\delta_y - \overline{b})^{-\mu}$ a fractional summation operator.

Substituting $a = \overline{1}/(\overline{1} - \overline{\alpha})$ and $b = \overline{1}/(\overline{1} - \overline{\beta})$ into the fractional summation operator, we obtain

$$\{\alpha^i\beta^j\} = \frac{\overline{1}}{(\overline{\alpha}\delta_x - \overline{\alpha} + \overline{1})(\overline{\beta}\delta_y - \overline{\beta} + \overline{1})}.$$

Next, recall that

$$\delta^n = \left\{(-1)^k \frac{n^{[k]}}{k!}\right\}, \; n \in Z^+.$$

Thus

$$\delta_x^m \delta_y^n = \left\{(-1)^{i+j} \frac{m^{[i]} n^{[j]}}{i!j!}\right\}_{i,j=0}^{\infty}, \; m, n \in Z^+,$$

which we have already mentioned. By multiplying δ_x^m with σ_y^n, we also obtain

$$\delta_x^m \sigma_y^n = \left\{(-1)^i \frac{m^{[i]}}{i!} \frac{(j+n-1)^{[n-1]}}{(n-1)!}\right\}, \; m, n \in Z^+.$$

There are several variations of the above formulas. These are based on the fact that $\delta + \hbar = \overline{1}$, and $\delta_x + \hbar_x = \delta_y + \hbar_y = \overline{1}$. For instance, from (3.50), we have

$$\frac{\overline{1}}{(\overline{\alpha} - \hbar)^\lambda} = \left\{\frac{\Gamma(k+\lambda)}{\Gamma(\lambda)\Gamma(k)} \alpha^{-k-\lambda}\right\}, \; \alpha \neq 0, \lambda > 0.$$

Thus

$$\frac{\overline{1}}{(\overline{a} - \hbar_x)^\lambda (\overline{b} - \hbar_y)^\mu} = \left\{\frac{\Gamma(i+\lambda)\Gamma(j+\mu)}{\Gamma(\lambda)\Gamma(\mu)\Gamma(i)\Gamma(j)} a^{-i-\lambda} b^{-j-\mu}\right\}, \quad a, b \neq 0; \lambda, \mu > 0.$$

Next, recall that

$$\{\sin k\theta\} = \frac{\hbar \overline{\sin\theta}}{\overline{1} - \overline{2}\hbar\overline{\cos\theta} + \hbar^2},$$

$$\{\cos k\theta\} = \frac{\overline{1} - \hbar\overline{\cos\theta}}{\overline{1} - \overline{2}\hbar\overline{\cos\theta} + \hbar^2},$$

$$\left\{\frac{\sin((1+k)\phi)}{\beta(\alpha^2+\beta^2)^{(1+k)/2}}\right\} = \frac{\overline{1}}{(\hbar - \overline{\alpha})^2 + \overline{\beta}^2}, \quad \alpha, \beta \in R, \beta \neq 0,$$

and

$$\left\{\frac{-\cos((1+k)\phi)}{(\alpha^2+\beta^2)^{(1+k)/2}}\right\} = \frac{\hbar - \overline{\alpha}}{(\hbar - \overline{\alpha})^2 + \overline{\beta}^2}, \quad \alpha, \beta \in R, \beta \neq 0,$$

where ϕ is the argument of the number $\alpha + i\beta$. Thus

$$\begin{aligned}\{\sin(i+j)\theta\} &= \{\sin i\theta \cos j\theta + \cos i\theta \sin j\theta\} \\ &= \frac{\overline{\sin\theta}(\hbar_x + \hbar_y - \overline{2}\hbar_x\hbar_y\overline{\cos\theta})}{(\overline{1} - \overline{2}\hbar_x\overline{\cos\theta} + \hbar_x^2)(\overline{1} - \overline{2}\hbar_y \cos\theta + \hbar_y^2)}\end{aligned}$$

and

$$\left\{\frac{-\sin(2(1+k)\phi)}{\beta(\alpha^2+\beta^2)^{1+k}}\right\}$$
$$= \frac{\hbar_y - \overline{\alpha}}{((\hbar_x-\overline{\alpha})^2+\overline{\beta}^2)((\hbar_y-\overline{\alpha})^2+\overline{\beta}^2)}, \quad \alpha,\beta \in R, \beta \neq 0,$$

where ϕ is the argument of the number $\alpha + i\beta$. Other similar formulae for trigonometric sequence can be found later.

3.3.4 Basic Relations Between Operators

We have defined the sequence $\delta_x, \delta_y, \hbar_x$ and $\hbar_y$. We have also defined partial differences and annihilated sequences of a bivariate sequence. Several elementary observations related to these bivariate sequences can be made here:

$$X_m(\Delta_y f) = \Delta_y(X_m f),\ Y_n(\Delta_x f) = \Delta_x(Y_n f),\ m,n \in N,$$
$$\Delta_x(\hbar_y f) = \hbar_y(\Delta_x f),\ \Delta_y(\hbar_x f) = \hbar_y(\Delta_y f),$$
$$X_m(\hbar_y f) = \hbar_y(X_0 f),\ Y_n(\hbar_x f) = \hbar_x(Y_n f),\ m,n \in N$$
$$\hbar_x^n(X_m f) = X_{m+n}(\hbar_x^n f),\ \hbar_y^n(Y_m f) = Y_{m+n}(\hbar_y^n f),\quad m \in N, n \in Z^+,$$
$$X_0(\hbar_x f) = \overline{0} = Y_0(\hbar_y f),$$
$$\Delta_x \hbar_x f = \hbar_x \Delta_x f + X_0 f,\ \Delta_y \hbar_y f = \hbar_y \Delta_y f + Y_0 f,$$
$$\hbar_x \Delta_x X_0 f = -\hbar_x X_0 f,\ \hbar_y \Delta_y Y_0 f = -\hbar_y Y_0 f.$$

Now for any bivariate sequence $f = \{f_{ij}\}$, we have

$$\delta_x f = \begin{bmatrix} f_{00} & f_{01} & \cdots \\ \Delta_1 f_{00} & \Delta_1 f_{01} & \cdots \\ \Delta_1 f_{10} & \Delta_1 f_{11} & \cdots \\ \cdots & \cdots & \cdots \end{bmatrix},$$

so that $\delta_x f$ cannot be identified as $\Delta_x f$, but instead, we have

$$\delta_x f = \hbar_x(\Delta_x f) + X_0 f.$$

In a similar manner, we have

$$\delta_y f = \hbar_y(\Delta_y f) + Y_0 f.$$

As an example, note that

$$\delta_x\{2^{i+j}\} = \hbar_x\{2^{i+j}\} + X_0\{2^j\} = (\overline{1} - \delta_x)\{2^{i+j}\} + \frac{\overline{1}}{2\delta_y - \overline{1}},$$

or,

$$\{2^{i+j}\} = \frac{\overline{1}}{(2\delta_x - \overline{1})(2\delta_y - \overline{1})},$$

which have already been derived.

In view of the above relationships between δ_x, δ_y and the differences of a bivariate sequence, it is natural to call δ_x and δ_y the partial difference operators. Convolution products of the partial difference operators can be very complicated. As an example,

$$\begin{aligned}
\delta_x\delta_y f &= \delta_x\left[\hbar_y(\Delta_y f) + Y_0 f\right] \\
&= \hbar_x\Delta_x\left[\hbar_y(\Delta_y f) + Y_0 f\right] + X_0\left[\hbar_y(\Delta_y f) + Y_0 f\right] \\
&= \hbar_x\Delta_x(\hbar_y(\Delta_y f)) + \hbar_x(\Delta_x(Y_0 f)) + X_0(\hbar_y(\Delta_y f)) + X_0 Y_0 f \\
&= \hbar_x\hbar_y(\Delta_x\Delta_y f) + \hbar_x(Y_0(\Delta_x f)) + \hbar_y(X_0(\Delta_y f)) + \overline{f_{00}},
\end{aligned}$$

which can also be seen from direct calculation:

$$\delta_x\delta_y f = \begin{bmatrix} f_{00} & \Delta_2 f_{00} & \Delta_2 f_{01} & \cdots \\ \Delta_1 f_{00} & \Delta_x\Delta_y f_{00} & \Delta_x\Delta_y f_{01} & \cdots \\ \Delta_1 f_{10} & \Delta_x\Delta_y f_{10} & \Delta_x\Delta_y f_{11} & \cdots \\ \cdots & \cdots & \cdots & \cdots \end{bmatrix}.$$

As another example,

$$\begin{aligned}
\delta_x^2 f &= \delta_x\left[\hbar_x\Delta_x f + X_0 f\right] \\
&= \hbar_x\left[\Delta_x\left(\hbar_x\Delta_x f + X_0 f\right)\right] + X_0\left[\hbar_x\Delta_x f + X_0 f\right] \\
&= \hbar_x[\Delta_x(\hbar_x(\Delta_x f))] + \hbar_x\Delta_x X_0 f + X_0\hbar_x\Delta_x f + X_0 X_0 f \\
&= \hbar_x\left[\hbar_x\Delta_x(\Delta_x f) + X_0\Delta_x f\right] + \hbar_x\Delta_x X_0 f + X_0 f \\
&= \hbar_x^2\Delta_x^2 f + \hbar_x X_0\Delta_x f - \hbar_x X_0 f + X_0 f \\
&= \hbar_x^2\Delta_x^2 f + X_1\hbar_x(\Delta_x f - f) + X_0 f.
\end{aligned}$$

Next, we note that there are several elementary relationship between translated sequences, the translated operators and the annihilated sequences:

$$\hbar_x^m(E_x^m f) = f - X_0 f - X_1 f - \ldots - X_{m-1} f,$$

and

$$\hbar_y^n(E_y^n f) = f - Y_0 f - Y_1 f - \ldots - Y_{n-1} f,$$

and for $m, n \geq 1$,

$$\hbar_x^m\hbar_y^n(E_x^m E_y^n f) = f - \sum_{s=0}^{m-1} X_s f - \sum_{t=0}^{n-1} Y_t f + \sum_{u=0}^{m-1}\sum_{v=0}^{n-1} f_{uv}\hbar_x^u\hbar_y^v. \tag{3.51}$$

3.3.5 Attenuation Operators

Recall that the attenuated bivariate sequence of $f = \{f_{ij}\}$ is $T_x^\lambda T_y^\mu f = \{\lambda^i\mu^j f_{ij}\}$. Given a quotient operator f/g, we will call the operators

$$T_x^\lambda\left(\frac{f}{g}\right) = \frac{T_x^\lambda f}{T_x^\lambda g}, \ \lambda \in C, \lambda \neq 0,$$

and

$$T_y^\mu\left(\frac{f}{g}\right) = \frac{T_y^\mu f}{T_y^\mu g}, \ \mu \in C, \mu \neq 0,$$

the attenuated operators of the quotient operator f/g. Note that

$$T_x^\lambda\left(T_y^\mu\left(\frac{f}{g}\right)\right) = T_x^\lambda\left(\frac{T_y^\mu f}{T_y^\mu g}\right) = \frac{T_x^\lambda T_y^\mu f}{T_x^\lambda T_y^\mu g}.$$

It is also easily seen that for any $\alpha \in C$, $T_x^\lambda \overline{\alpha} = T_y^\mu \overline{\alpha} = \overline{\alpha}$, and for any quotient operator ϕ,

$$T_x^\lambda(T_x^\rho \phi) = T_x^{\lambda\rho}\phi,\ T_y^\mu(T_y^\zeta \phi) = T_y^{\mu\zeta}\phi.$$

Furthermore, for quotient operators $\phi = f/g$ and $\psi = p/q$,

$$T_x^\lambda(\phi+\psi) = \frac{T_x^\lambda(fq+pg)}{T_x^\lambda(gq)} = \frac{(T_x^\lambda f)(T_x^\lambda q) + (T_x^\lambda p)(T_x^\lambda g)}{(T_x^\lambda g)(T_x^\lambda q)} = T_x^\lambda \phi + T_x^\mu \psi,$$

$$T_x^\lambda(\phi\psi) = T_x^\lambda\left(\frac{fp}{gq}\right) = \frac{(T_x^\lambda f)(T_x^\lambda p)}{(T_x^\lambda g)(T_x^\lambda q)} = (T_x^\lambda \phi)(T_x^\lambda \psi),$$

and similarly,

$$T_y^\mu(\phi+\psi) = T_y^\mu \phi + T_y^\mu \psi,$$
$$T_y^\mu(\phi\psi) = (T_y^\mu \phi)(T_y^\mu \psi).$$

Consequently, we have

$$T_x^\lambda(\overline{\alpha}\phi) = (T_x^\lambda \overline{\alpha})(T_x^\lambda \phi) = \overline{\alpha} T_x^\lambda \phi,\ T_y^\mu(\overline{\alpha}\psi) = \overline{\alpha} T_y^\mu \psi,$$

$$T_x^\lambda\left(\frac{f}{g}\right) = \frac{\overline{1}}{T_x^\lambda(g/f)},\ T_y^\mu\left(\frac{f}{g}\right) = \frac{\overline{1}}{T_y^\mu(g/f)},\ f, g \neq 0,$$

$$T_x^\lambda \delta_x^m = (\overline{1} - \overline{\lambda}\hbar_x)^m,\ T_y^\mu \delta_y^n = (\overline{1} - \overline{\mu}\hbar_y)^n,\ m, n \in Z^+,$$

and

$$T_x^\lambda \hbar_x^m = \overline{\lambda}^m \hbar_x^m,\ T_y^\mu \hbar_y^n = \overline{\mu}^m \hbar_y^m,\ m, n \in Z^+.$$

3.3.6 Sequences and Series of Operators

Given a sequence $\{f^{(k)}\} = \{\{f_{ij}^{(k)}\}\}$ of bivariate sequences, we say that $\{f^{(k)}\}$ converges pointwise to $f = \{f_{ij}\}$ in $l^{N\times N}$ if

$$\lim_{k\to\infty} f_{ij}^{(k)} = f_{ij},\ i, j \in N.$$

A sequence of operators $\{\phi^{(k)}\}$ is said to converge to an operator f/g if there are two sequences $\{f^{(k)}\}$ and $\{g^{(k)}\}$ of bivariate sequences in $l^{N\times N}$ such that $\phi^{(k)} = f^{(k)}/g^{(k)}$ for each k, that $\{f^{(k)}\}$ and $\{g^{(k)}\}$ converge pointwise to f and g respectively, and that $g \neq \overline{0}$. Clearly, these two definitions are compatible in the sense that if a sequence $\{f^{(k)}\}$ of bivariate sequences converges to f, then the corresponding operator sequence $\{f^{(k)}/\overline{1}\}$ converges to the operator $f/\overline{1}$.

Let $\{\psi^{(k)}\}$ be a sequence of operators. The series $\psi^{(0)} + \psi^{(1)} + \ldots$ is said to converge to the sum ψ if the sequence of partial sums $\{\psi^{(0)} + \psi^{(1)} + \ldots + \psi^{(k)}\}$ converges to ψ. We will use the notation

$$\sum_{k=0}^{\infty} \psi^{(k)} \in l^{Z\times Z}/l^{Z\times Z}$$

to denote convergence. Note that if each $\psi^{(k)}$ is an ordinary operator, i.e. $\psi^{(k)} = \{\psi_{ij}^{(k)}\}$, then we have

$$\sum_{k=0}^{\infty} \psi^{(k)} = \sum_{k=0}^{\infty} \{\psi_{ij}^{(k)}\} = \left\{ \sum_{k=0}^{\infty} \psi_{ij}^{(k)} \right\}, \tag{3.52}$$

that is, the (i, j)-th component of the series is obtained by adding all the (i, j)-th components of the individual sequences.

A number of elementary properties of the limits of operator sequences can be obtained. For instance, we have the following result.

Theorem 41 *The limit of a sequence $\{\phi^{(k)}\}$ of operators is unique. If two operator sequences $\{\phi_k\}$ and $\{\psi_k\}$ converge to ϕ and ψ respectively, then $\lim_{k\to\infty}(\phi^{(k)} + \psi^{(k)}) = \phi + \psi$ and $\lim_{k\to\infty} \phi^{(k)}\psi^{(k)} = \phi\psi$. If the sequence $\{f^{(k)}/g^{(k)}\}$ of quotient operators converges to a nonzero operator ϕ, and $f^{(k)} \neq 0$ for each k, then $\lim_{k\to\infty} g^{(k)}/f^{(k)} = \bar{1}/\phi$.*

Given a doubly indexed sequence $\{\psi^{(mn)}\}_{m,n\in N}$ of operators, we may reorder it into a sequence $\{\phi_k\}_{k\in N}$ of operators by means of a linear ordering $\ll$ of $N \times N$. The corresponding series

$$\sum_{k=0}^{\infty} \phi_k$$

will be denoted by

$$\sum_{(m,n)\in(N\times N,\ll)} \psi^{(mn)}.$$

It is also possible that for each $m \in N$,

$$\sum_{n=0}^{\infty} \psi^{(mn)} = \psi^{(m)} \in l^N,$$

then the series $\psi^{(0)} + \psi^{(1)} + ...$ will also be written as

$$\sum_{m=0}^{\infty} \left(\sum_{n=0}^{\infty} \psi^{(mn)} \right).$$

For instance, if $\{\alpha_{ij}\}$ is a bivariate sequence of complex numbers, then for each $m \in N$,

$$\sum_{n=0}^{\infty} \overline{\alpha_{mn}} h_x^m h_y^n = X_m\{\alpha_{ij}\} \in l^N,$$

and

$$\sum_{m=0}^{\infty} \sum_{n=0}^{\infty} \overline{\alpha_{mn}} h_x^m h_y^n = \{\alpha_{ij}\} \in l^{N\times N}. \tag{3.53}$$

This example motivates the following two results.

Theorem 42 *Let $\{\alpha_{ij}\}$ be a bivariate sequence of complex numbers and $\{\beta_k\}$, $\{\gamma_k\}$ be strictly increasing integral sequences that tend to infinity. Let $N \times N$ be linearly ordered by the ordering $\leq$. Then the series*

$$\sum_{(m,n)\in(N\times N,\leq)} \overline{\alpha_{mn}}\hbar_x^{\beta_m}\hbar_y^{\gamma_n} = \overline{\alpha_{00}}\hbar_x^{\beta_0}\hbar_y^{\gamma_0} + \overline{\alpha_{10}}\hbar_x^{\beta_1}\hbar_y^{\gamma_0} + \ldots$$

is convergent.

Proof. Note that when β_0 and γ_0 are nonnegative, the above series is the bivariate sequence whose (β_m, γ_n)-th component is α_{mn}. Thus our theorem is clear if β_0 and γ_0 are nonnegative. If $\beta_0 < 0$ but $\gamma_0 \geq 0$, we can divide the series by $h_x^{\beta_0}$ to obtain the convergent series

$$\sum_{(m,n)\in(N\times N,\leq)} \overline{\alpha_{mn}}\hbar_x^{\beta_m-\beta_0}\hbar_y^{\gamma_n}.$$

The other two cases are similarly proved. ■

Theorem 43 *Let $\{\alpha_{ij}\}$ be a bivariate sequence of complex numbers and $\{\beta_k\}$, $\{\gamma_k\}$ be strictly increasing integral sequences that tend to infinity. Then for each $m \in N$,*

$$\sum_{n=0}^{\infty} \overline{\alpha_{mn}}\hbar_x^{\beta_m}\hbar_y^{\gamma_n}$$

is convergent and

$$\sum_{m=0}^{\infty}\sum_{n=0}^{\infty} \overline{\alpha_{mn}}\hbar_x^{\beta_m}\hbar_y^{\gamma_n}$$

is convergent.

The proof of this result is similar to that of the previous theorem and hence is omitted.

We remark that when the ordering $\leq$ in Theorem 42 is replaced with another linear ordering, the resulting limit will not be changed. This is because we have adopted a pointwise convergence. For instance, when $\{\beta_m\} = \{m\}_{m\in N}$, $\{\gamma_n\} = \{n\}_{n\in N}$, we have

$$\{\alpha_{mn}\} = \sum_{(m,n)\in(N\times N,\leq)} \overline{\alpha_{mn}}\hbar_x^m\hbar_y^n = \sum_{m=0}^{\infty}\sum_{n=0}^{\infty} \overline{\alpha_{mn}}\hbar_x^m\hbar_y^n.$$

Series of translation operators enjoy several interesting properties. First of all, in addition to (3.53), we have

$$\{f_{ij}\} = \overline{f_{00}} + \overline{f_{01}}\hbar_x + \overline{f_{01}}\hbar_y + \overline{f_{20}}\hbar_x^2 + \overline{f_{11}}\hbar_x\hbar_y + \ldots,$$

where the right hand side is a series of translation operators that is always convergent, and converges to the left hand side. Next, we have the following uniqueness theorem.

Theorem 44 *Let $\{\alpha_{ij}\}$ be a bivariate sequence of complex numbers and $\{\beta_k\}$, $\{\gamma_k\}$ are strictly increasing integral sequence that tends to infinity. Let $N \times N$ be linearly ordered by the ordering $\leq$. If*

$$\sum_{(m,n)\in(N\times N,\leq)} \overline{\alpha_{mn}} \hbar_x^{\beta_m} \hbar_y^{\gamma_n} = \overline{0},$$

or

$$\sum_{m=0}^{\infty}\sum_{n=0}^{\infty} \overline{\alpha_{mn}} \hbar_x^{\beta_m} \hbar_y^{\gamma_n} = \overline{0},$$

then $\alpha_{mn} = 0$ for $m, n \in N$.

Proof. If β_0 and γ_0 are nonnegative, then the above series is the bivariate sequence whose (β_m, γ_n)-th component is α_{mn}. Thus $\alpha_{mn} = 0$ for $m, n \geq 0$. If $\beta_0 < 0$ and $\gamma_0 \geq 0$, we can divide the series by $\hbar_x^{\beta_0}$ and obtain

$$\sum_{(m,n)\in N\times N} \overline{\alpha_{mn}} \hbar_x^{\beta_m-\beta_0} \hbar_y^{\gamma_n} = \overline{0}$$

so that the same argument applies. The other cases are similarly proved. ∎

We now deal with more general series of operators of the form

$$\overline{\alpha_0} + \overline{\alpha_1} f + \overline{\alpha_2} f^2 + ...,$$

where $\{\alpha_k\}$ is a complex sequence and f is an ordinary operator. Recall that f^n stands for $f * f * f ... * f$.

Theorem 45 *Let $\{\alpha_k\}_{k=0}^{\infty}$ be a complex sequence and $f = \{f_{ij}\}$ an ordinary operator. If $f_{00} = 0$ or if the series $\alpha_0 + \alpha_1 z + \alpha_2 z^2 + ...$ in the complex variables z is entire, then the series $\overline{\alpha_0} + \overline{\alpha_1} f + \overline{\alpha_2} f^2 + ...$ is convergent.*

Proof. In view of Theorem 4.38, if $f_{00} = 0$, then $f_{ij}^n = 0$ for all $(i,j) \in Q_0 + ... + Q_{n-1}$. Thus for $(i,j) \in Q_0 + ... + Q_{n-1}$,

$$\alpha_0 + \alpha_1 f_{ij} + \alpha_2 f_{ij}^2 + ... = \alpha_0 + \alpha_1 f_{ij} + ... + \alpha_{n-1} f_{ij}^{n-1} < \infty.$$

The first case is proved. Next, let J be an arbitrary nonnegative integer. Let $M = \max_{0\leq u,v\leq J} |f_{uv}|$. For a fixed lattice point (i,j) satisfying $0 \leq i,j \leq K$, we have

$$|f_{ij}^2| = \left|\sum_{u=0}^{i}\sum_{v=0}^{j} f_{uv} f_{i-u,j-v}\right| \leq M^2(K+1)^2,$$

$$|f_{ij}^3| = \left|\sum_{u=0}^{i}\sum_{v=0}^{j} f_{uv}^2 f_{i-u,j-v}\right| \leq M^3(K+1)^4,$$

and in general,

$$|f_{ij}^k| \leq M^k(J+1)^{2k-2}, \ k \in Z^+.$$

Thus

$$|\alpha_0 + \alpha_1 f_{ij} + \alpha_2 f_{ij}^2 + ...| \leq |\alpha_0| + \sum_{k=1}^{\infty} |\alpha_k| M^k (K+1)^{2k-2} < \infty$$

in view of the fact that the power series $\alpha_0 + \alpha_1 z + \ldots$ is entire. The proof is complete. ■

We remark that if the power series $\alpha_0 + \alpha_1 z + \alpha_1 z^2 + \ldots$ is not entire, the above result may not hold. For instance, the geometric series $1 + z + z^2 + \ldots$ is analytic in the unit disk, nevertheless, the series $\overline{1} + \overline{1} + \ldots$ is not convergent.

By means of the above theorem, we see that for any bivariate sequence $f = \{f_{ij}\}$, the series

$$\sum_{k=0}^{\infty} \frac{f^n}{\overline{n!}}, \ f \in l^{N\times N}$$

is convergent. We will denote this series by e^f or $\exp f$ and call it the exponential of the ordinary operator f. By means of (3.52), we see that e^f is an ordinary operator and the (i,j)-th component of it is given by

$$f_{ij} + \frac{1}{2}\sum_{u=0}^{i}\sum_{v=0}^{j} f_{uv} f_{i-u,j-v} + \ldots, \ i,j \in Z^+$$

that $e^f \neq 0$ since the $(0,0)$-th component is equal to $e^{f_{00}}$, and that e^f is a scalar operator when f is a scalar bivariate sequence. We may also show that

$$e^f e^g = e^{f+g}, \ f,g \in l^{N\times N}.$$

Indeed, this follows from

$$\sum_{k=0}^{\infty}\frac{a^n}{k!}\sum_{k=0}^{\infty}\frac{b^k}{k!} = \sum_{k=0}^{\infty}\sum_{s=0}^{k}\frac{a^j}{j!}\frac{b^{k-j}}{(k-j)!} = \sum_{k=0}^{\infty}\frac{(a+b)^k}{k!}$$

which holds for any complex number a and b.

3.3.7 Algebraic Derivatives

Given a bivariate sequence $f = \{f_{ij}\}$, we denote the sequences $\{(i+1)f_{i+1,j}\}$ and $\{(j+1)f_{i,j+1}\}$ by $D_x f$ and $D_y f$ respectively and call them the algebraic derivatives of f. The higher algebraic and mixed derivatives are defined recursively. Thus we have

$$D_x\overline{\alpha} = D_y\overline{\alpha} = \overline{0}, \ \alpha \in C,$$

and

$$\begin{aligned} D_x^m D_y^n f &= \{[(i+1)\cdots(i+m)][(j+1)\cdots(j+n)]f_{i+m,j+n}\} \\ &= D_y^n D_x^m f, \ m,n \in Z^+. \end{aligned}$$

It is easily verified that for any $\alpha, \beta \in C$, and $f,g \in l^{N\times N}$,

$$D_x(\alpha f + \beta g) = \alpha D_x f + \beta D_x g, \ D_y(\alpha f + \beta g) = \alpha D_y f + \beta D_y g$$

and

$$D_x(f * g) = f * D_x g + g * D_x f, \ D_y(f * g) = f * D_y g + g * D_y f.$$

Given a quotient operator f/g, we denote the operators

$$\frac{gD_xf - fD_xg}{g^2}, \frac{gD_yf - fD_yg}{g^2}$$

by $D_x(f/g)$ and $D_y(f/g)$ respectively, and again we call them the algebraic derivatives of the quotient operator f/g. Clearly, if f is an ordinary operator, then $D_x(f/\bar{1}) = (D_xf)/\bar{1}$. It is also easy to verify that for any $\alpha, \beta \in C$ and quotient operators ϕ, ψ,

$$D_x(\overline{\alpha}\phi + \overline{\beta}\psi) = \overline{\alpha}D_x\phi + \overline{\beta}D_x\psi,$$

and

$$D_x(\phi\psi) = \phi D_x\psi + \psi D_x\phi.$$

Algebraic derivatives of some common operators can easily be found. More complicated derivatives can be obtained by employing the following list of useful formulas:

$$D_x\hbar_y = D_y\hbar_x = \bar{0},$$

$$D_x(X_0f) = \bar{0},$$

$$\hbar_xD_x\{f_{ij}\} = \{if_{ij}\},$$

$$D_x\hbar_x^m = m\hbar_x^{m-1}, \; m \in Z$$

$$D_x\delta_x^m = -m\delta_x^{m-1}, \; m \in Z,$$

$$\begin{aligned} D_x\phi^m &= D_x(\phi^{m-1}\phi) \\ &= \phi^{m-1}D_x\phi + \phi D_x\phi^{m-1} \\ &= \ldots \\ &= m\phi^{m-1}D_x\phi, \; \phi \in l^{N\times N}/l^{N\times N}, m \in Z^+, \end{aligned}$$

$$D_x\sum_{k=0}^{\infty}\phi_k = \sum_{k=0}^{\infty}D\phi_k,$$

where $\phi_0 + \phi_1 + \ldots$ is a convergent series of operators,

$$D_xe^f = e^fD_xf, \; f \in l^{N\times N},$$

$$D_x^n(\hbar_x^m\{f_{ij}\}) = \hbar_x^{m-n}\left\{(m+i)^{[n]}f_{ij}\right\}, \; m \geq n \geq 1,$$

and finally

$$\hbar_x^mD_x^n\{f_{ij}\} = \left\{(i+n-m)^{[n]}f(i+n-m,j)\right\}, \; n \geq m \geq 0.$$

Example 43 *Let us calculate*

$$D_x\frac{\bar{1}}{(\bar{1}-\bar{3}\hbar_x)(\bar{1}-\bar{3}\hbar_y)}.$$

Since

$$D_x\frac{\bar{1}}{\bar{1}-\bar{3}\hbar_y} = \frac{\bar{3}D_x\hbar_y}{(\bar{1}-\bar{3}\hbar_y)^2} = \bar{0},$$

and

$$D_x \frac{\bar{1}}{(\bar{1}-\bar{3}\hbar_x)} = \frac{\bar{3}D_x\hbar_x}{(\bar{1}-\bar{3}\hbar_x)^2} = \frac{\bar{3}}{(\bar{1}-\bar{3}\hbar_x)^2},$$

thus

$$D_x \frac{\bar{1}}{(\bar{1}-\bar{3}\hbar_x)(\bar{1}-\bar{3}\hbar_y)} = \frac{\bar{1}}{\bar{1}-\bar{3}\hbar_x} D_x \frac{\bar{1}}{\bar{1}-\bar{3}\hbar_y} + \frac{\bar{1}}{\bar{1}-\bar{3}\hbar_y} D_x \frac{\bar{1}}{\bar{1}-\bar{3}\hbar_x}$$
$$= \frac{\bar{3}}{(\bar{1}-\bar{3}\hbar_y)(\bar{1}-\bar{3}\hbar_x)^2}.$$

We conclude this section by remarking that iterated algebraic integrals can be introduced. They are just the primitives of partial algebraic derivatives and thus their properties follow from those of algebraic integrals defined in Section 3.2.8.

3.4 Notes and Remarks

The calculus presented here has origins in several works by Berg [6], Mikusinski [121], Moore [126], Niven [131], Fenyes and Kosik [66], and a series of papers by Jentsch [87], [88], [90], [91], [92], [93].

Berg [6] contains a chapter on discrete operational calculus, but his definition of convolution product is different from ours. Niven [131] discusses most of his calculus in terms of formal power series, but in the last section of his paper, the idea of an extension field is introduced to facilitate discussions of the cosecant and cotangent of a sequence. The presentation in the papers by Jentsch are close to ours, and contain some additional results.

One purpose of our calculus is to provide tools for finding explicit solutions to ordinary as well as partial difference equations. As mentioned in the Introduction, the Z-transform is also a well-established method for such a purpose. However, the Z-transform of a sequence requires subexponentiality. In the case of an ordinary linear difference equation with constant coefficients, it is well known that all its solutions are subexponential. Hence the Z-transform is applicable. However, for linear partial difference equations with constant coefficients, its solutions may not be subexponential as will be seen in Chapter 6. Thus the Z-transform will become a formal method and additional care has to be taken.

Chapter 4

Monotonicity and Convexity

4.1 Introduction

Functions which satisfy a certain functional relation in a domain Ω and, because of this, achieve their maxima on the boundary of Ω are said to possess a maximum principle. An elementary example is that any smooth function $f(x)$ which satisfies the inequalities $f'(x) \geq 0$ or $f''(x) \geq 0$ on a real closed interval achieves its maximum value at one of its end points. Furthermore, if $f''(x) = 0$ on this interval, additional properties of f can be said. In this chapter, we will be concerned with univariate sequences $\{u_k\}$ or bivariate sequences $\{u_{ij}\}$ which satisfy "monotone" or "convex" functional relations of the form

$$\Delta u_k \geq 0,\ \Delta^2 u_k \geq 0,\ \Xi u_{ij} \geq 0,\ \Xi u_{ij} + q_{ij} u_{ij} = 0,\ \ldots,$$

etc., where Ξ is the discrete Laplacian defined in Example 22. Maximum principles and the so called discrete Wirtinger's inequalities will be derived. These principles and inequalities will be useful in yielding various qualitative properties of several types of partial difference equations.

4.2 Univariate Maximum Principles

In this section, we first obtain several elementary maximum principles for solutions of second order functional difference inequalities. Let $u = \{u_k\}$ be a sequence defined on a set of consecutive integers of the form $\{a-1, a, \ldots, b, b+1\}$. Suppose for some $c \in \{a, \ldots, b\}$, $\Delta u_{c-1} \geq 0$ and $\Delta u_c \leq 0$, then for any nonnegative sequences $\{q_k\}_{k=a}^{b}$ and $\{r_k\}_{k=a}^{b}$,

$$-q_c \Delta u_{c-1} + r_c \Delta u_c \leq 0.$$

Consequently, if u is known to satisfy a functional difference inequality of the form

$$-q_k \Delta u_{k-1} + r_k \Delta u_k > 0,\ a \leq k \leq b, \tag{4.1}$$

then it is clear that relations $\Delta u_{c-1} \geq 0$ and $\Delta u_c \leq 0$ cannot be satisfied at any point $c \in \{a, \ldots, b\}$. In other words, whenever (4.1) holds, the maximum of u cannot be attained anywhere in $\{a-1, \ldots, b+1\}$ except at the endpoints $a-1$ or $b+1$.

An essential feature of the above argument is the requirement that the inequality (4.1) be strict. This we can remove, but at the expense of admitting other requirements.

Theorem 46 *Suppose $q_k > 0$ and $r_k > 0$ for $k = a, a+1, ..., b$ and $u = \{u_k\}_{k=a-1}^{b+1}$ satisfies the functional inequality*

$$-q_k \Delta u_{k-1} + r_k \Delta u_k \geq 0, \ k = a, a+1, ..., b.$$

If the maximum M of u is attained in $\{a, ..., b\}$, then $u_k \equiv M$ for $k = a-1, a, ..., b+1$. If the maximum M of u is attained at $a-1$ (or $b+1$), then $\Delta u_{a-1} < 0$ (respectively $\Delta u_b > 0$) or $u_k \equiv M$ for $k = a-1, ..., b+1$.

Suppose that $u_c = M$ for some integer c in $\{a, ..., b\}$, then we have

$$-q_c \Delta u_{c-1} + r_c \Delta u_c = 0.$$

Consequently,

$$0 \geq r_c \Delta u_c = q_c \Delta u_{c-1} \geq 0.$$

Since $r_c > 0$ and $q_c > 0$, we must have $\Delta u_c = \Delta u_{c-1} = 0$ so that $u_{c-1} = u_c = u_{c+1} = M$. If $c+1 \neq b$, we may repeat the same argument to deduce the fact that $u_{c+2} = M$. The proof may now be completed by finite induction.

Suppose $u_{a-1} = M$, then $\Delta u_{a-1} \leq 0$. If $\Delta u_{a-1} = 0$, then $u_a = M$, hence by what we have just proved, $u_k = M$ for $k = a-1, ..., b+1$. The proof is complete.

We remark that the assumptions that $q_a > 0$ and $r_b > 0$ can be relaxed to $q_a \geq 0$ and $r_a \geq 0$ respectively. This fact may be useful in some cases, but we will not emphasize it for the sake of simplicity. A variant of the above theorem is as follows.

Theorem 47 *Suppose in the above theorem, the condition that $r_k > 0$ for $k = a, ..., b$ is replaced by $r_k \geq 0$ for $k = a, ..., b$. Then assuming $u_c = M$ for some c in $\{a, ..., b\}$, we have $u_{a-1} = u_a = ... = u_{c-1} = u_c$. Furthermore, if we assume that $u_{b+1} = M$, then $\Delta u_b < 0$ unless $u_k = M$ for $k = a-1, ..., b+1$.*

Another variant of the above theorem is as follows.

Theorem 48 *Suppose $q_k > 0$, $r_k > 0$ and $s_k \geq 0$ for $k = a, a+1, ..., b$ and $u = \{u_k\}_{k=a-1}^{b+1}$ satisfies the functional inequality*

$$-q_k \Delta u_{k-1} + r_k \Delta u_k - s_k u_k \geq 0, \ k = a, a+1, ..., b.$$

If the maximum M of u is nonnegative and attained in $\{a, ..., b\}$, then $u_k \equiv M$ for $k = a-1, a, ..., b+1$. If the maximum M of u is nonnegative and attained at $a-1$ (or $b+1$), then $\Delta u_{a-1} < 0$ (respectively $\Delta u_b > 0$) or $u_k \equiv M$ for $k = a-1, ..., b+1$. If $u_{a-1} \leq 0$ and $u_{b+1} \leq 0$, then $u_k < 0$ for $k = a, a+1, ..., b$, or $u_k \equiv 0$ for $k = a-1, a, ..., b+1$.

The first two assertions are proved in manners similar to that in the proof of the previous theorem. As for the last assertion, note that it holds when the maximum M of u is negative. If M is positive, then it must be attained in $\{a, ..., b\}$, hence $u_k = M$ for $k = a-1, ..., b+1$, which is contrary to our assumptions that

$u_{a-1}, u_{b+1} \leq 0$. Suppose $M = 0$. If it is attained in $\{a, ..., b\}$, then $u_k = M = 0$ for $k = a-1, a, ..., b+1$. If it is not attained in $\{a, .., b\}$, that is, $u_k \neq M$ for $k = a, ..., b$, then $u_k < M = 0$ for $k = a, ..., b$. The proof is complete.

In the above theorem, the sequence $\{s_k\}$ is assumed to be nonnegative. We will derive a generalized maximum principle in which the above assumption is relaxed. For this purpose, let us employ the notation

$$(Lu)_k = -q_k \Delta u_{k-1} + r_k \Delta u_k - s_k u_k.$$

Assume that $v = \{v_k\}_{k=a-1}^{b+1}$ is a positive sequence which satisfies the functional inequality

$$(Lv)_k \leq 0, \quad k = a, ..., b.$$

Define $g_k = u_k/v_k$ for $k = a-1, ..., b+1$. A simple computation yields

$$\begin{aligned}
(Lu)_k &= (Lgv)_k = -q_k \Delta(g_{k-1} v_{k-1}) + r_k \Delta(g_k v_k) - s_k g_k v_k \\
&= -q_k \{v_{k-1} \Delta g_{k-1} + g_k \Delta v_{k-1}\} + r_k \{g_k \Delta v_k + v_{k+1} \Delta g_k\} - s_k g_k v_k \\
&= -q_k v_{k-1} \Delta g_{k-1} + r_k v_{k+1} \Delta g_k + (Lv)_k g_k,
\end{aligned}$$

for $k = a, ..., b$. Consequently, if we assume that $(Lu)_k \geq 0$ for $k = a, ..., b$, then the previous maximum principles can be applied.

Theorem 49 *Suppose $\{s_k\}_{k=a}^{b}$ is a real sequence, $q_k > 0$ and $r_k > 0$ for $k = a, ..., b$, and $u = \{u_k\}_{k=a-1}^{b+1}$ satisfies the functional inequality $(Lu)_k \geq 0$ for $k = a, ..., b$. Suppose further that $v = \{v_k\}_{k=a-1}^{b+1}$ is a positive sequence which satisfies the functional inequality $(Lv)_k \leq 0$ for $k = a, ..., b$. Let $g = \{u_k/v_k\}_{k=a-1}^{b+1}$. If the maximum M of g is nonnegative and attained in $\{a, ..., b\}$, then $g_k \equiv M$ for $k = a-1, a, ..., b+1$. If the maximum M of g is nonnegative and attained at $a-1$ (or $b+1$), then $\Delta g_{a-1} < 0$ (respectively $\Delta g_b > 0$) or $g_k \equiv M$ for $k = a-1, ..., b+1$. If $g_{a-1} \leq 0$ and $g_{b+1} \leq 0$, then $g_k < 0$ for $k = a, a+1, ..., b$, or $g_k \equiv 0$ for $k = a-1, a, ..., b+1$.*

We shall illustrate our results in this section by several examples. Consider the steady state equation

$$\Delta^2 u_{k-1} - s_k u_k = p_k, \; k = 1, ..., n,$$

subject to the boundary conditions

$$u_0 = \alpha, \; u_{n+1} = \beta.$$

If $u = \{u_k\}_{k=0}^{n+1}$ and $\{v_k\}_{k=0}^{n+1}$ are any two solutions, then $w = u - v$ satisfies

$$\Delta^2 w_{k-1} - s_k w_k = 0, \; k = 1, ..., n, \tag{4.2}$$

and

$$w_0 = 0 = w_{n+1}.$$

According to Theorem 48, if $s_k \geq 0$ for $k = 1, ..., n$, then $w_k = u_k - v_k \leq 0$ for $k = 0, 1, ..., n+1$. If we repeat the above argument with w substituted by $v - u$, we may conclude that $v_k - u_k \leq 0$ for $k = 0, 1, ..., n+1$. Therefore, $u_k = v_k$ for

$k = 0, ..., n+1$. In other words, we have shown the uniqueness theorem for our steady state equation.

As another example, consider the boundary value problem

$$\Delta^2 u_{k-1} = g_k, \ k = 1, 2, ..., n,$$

$$u_0 = 0 = u_{n+1}.$$

which can be considered as a system of linear equation of the form

$$\mathbf{J}\,\mathrm{col}(u_1, ..., u_n) = -\mathrm{col}(g_1, ..., g_n),$$

where the Jacobi matrix $\mathbf{J}$ is defined by (2.20). When $g = \mathrm{col}(g_1, ..., g_n) = 0$, our problem has the trivial solution. It is the only solution in view of our previous example. Thus our boundary value problem has a unique solution when $g \neq 0$. Furthermore, if $g \geq 0$ and $g \neq 0$, then the corresponding solution $\{u_0, u_1, ..., u_n, u_{n+1}\}$ will satisfy $u_1, ..., u_n < 0$ in view of Theorem 48. As an interesting consequence, the inverse of the Jacobi matrix $\mathbf{J}$ exists and each of its components is positive.

Next, consider the nonlinear steady state equation

$$\Delta^2 u_{k-1} - u_k^3 = 0, \ k = 1, ..., n,$$

and the functional difference inequality

$$\Delta^2 v_{k-1} - v_k^3 \geq 0, \ k = 1, 2, ..., n.$$

If $u = \{u_k\}_{k=0}^{n+1}$ and $v = \{v_k\}_{k=0}^{n=1}$ are their respective solutions, then the function $w = v - u$ will satisfy

$$\Delta^2 w - 3(w_k^*)^2 w_k \geq 0, \ k = 1, 2, ..., n,$$

where $w_k^* = u_k + \alpha_k(v_k - u_k)$ and $0 < \alpha_k < 1$ for $k = 1, 2, ..., n$. Thus if w has a nonnegative M maximum attained in $\{1, ..., n\}$, then $v_k = M + u_k$ for $k = 0, 1, ..., n+1$.

4.3 Bivariate Maximum Principles

The maximum principles obtained in the previous section can easily be modified to suit functional difference inequalities involving bivariate sequences. Let Ω be a finite domain in the lattice plane and $\partial\Omega$ its exterior boundary. We introduce the following abbreviations:

$$(L_1 u)_{ij} = -p_{ij}\Delta_1 u_{i-1,j} + q_{ij}\Delta_1 u_{ij},$$

$$(L_2 u)_{ij} = -r_{ij}\Delta_2 u_{i,j-1} + s_{ij}\Delta_2 u_{ij},$$

and

$$(Lu)_{ij} = (L_1 u)_{ij} + (L_2 u)_{ij}.$$

Let $u = \{u_{ij}\}$ be a real sequence defined on $\Omega + \partial\Omega$. Suppose for some point (m, n) in Ω, the value of u at (m, n) is not less than its values at the four neighboring points of (m, n), i.e.,

$$\Delta_1 u_{m-1,n} \geq 0, \ \Delta_1 u_{mn} \leq 0, \ \Delta_2 u_{m,m-1} \geq 0, \ \Delta_2 u_{mn} \leq 0,$$

then for any functions p, q, r, s which are nonnegative on Ω,

$$(L_1u)_{mn} \le 0, \ (L_2u)_{mn} \le 0,$$

and

$$(Lu)_{mn} \le 0. \tag{4.3}$$

Consequently, if u is known to satisfy a **partial difference inequality** of the form

$$(Lu)_{ij} > 0, \ (i,j) \in \Omega, \tag{4.4}$$

then it is clear that relation (4.3) cannot be satisfied at any point (m,n) in Ω. In other words, whenever (4.4) holds for all $(i,j) \in \Omega$, the maximum of u cannot be attained anywhere in $\Omega + \partial\Omega$ except at the exterior boundary points.

Theorem 50 *Let Ω be finite domain in Z^2. Suppose the functions p, q, r and s are positive on $\Omega + \partial\Omega$, and suppose $u = \{u_{ij}\}_{(i,j)\in\Omega+\partial\Omega}$ satisfies the functional inequality*

$$(Lu)_{ij} \ge 0, \ (i,j) \in \Omega. \tag{4.5}$$

If the maximum $M = \max_{(i,j)\in\Omega+\partial\Omega} u_{ij}$ is attained in Ω, then $u_{ij} \equiv M$ for all $(i,j) \in \Omega + \partial\Omega$.

The proof is as follows. Suppose that $u_{mn} = M$ for some point (m,n) in Ω. We assert that for any point (a,b) in $\Omega + \partial\Omega$, $u_{ab} = M$. Indeed, let $z_1 = (m,n), z_2, ..., z_n = (a,b)$ be a chain of points contained in $\Omega + \partial\Omega$. Since $u_{mn} = M$, the values of u at the neighboring points of (m,n) are not greater than M. Consequently, (4.3) must hold. But in view of (4.5), we must have

$$(L_1u)_{mn} + (L_2u)_{mn} = 0$$

and

$$0 \ge (L_1u)_{mn} = -(L_2u)_{mn} \ge 0,$$

so that

$$0 \le p_{mn}\Delta_1 u_{m-1,m} = q_{mn}\Delta_1 u_{mn} \le 0$$

and

$$0 \le r_{mn}\Delta_2 u_{m,n-1} = s_{mn}\Delta_2 u_{mn} \le 0.$$

In view of the positivity of the functions p, q, r and s at (m,n), we conclude that the values of u at the four neighbors of (m,n) are equal to M, so that $u_{z_2} = M$. If z_2 is not equal to (a,b), we may repeat the above argument to deduce the fact that $u_{z_3} = M$. The proof can now be completed by finite induction.

We remark that under the same assumptions of the above theorem, if the maximum M of u is attained at an exterior boundary point (m,n), then for any interior boundary point (x,y) in Ω which is adjacent to (m,n), we must have $u_{mn} - u_{xy} > 0$ unless u is identically equal to M.

We remark further that a dual statement can be made and easily proved: Suppose the functions p, q, r and s are positive on $\Omega + \partial\Omega$, and suppose u satisfies the inequality

$$(Lu)_{ij} \le 0, \ (i,j) \in \Omega.$$

Let $K = \min_{(i,j)\in\Omega+\partial\Omega} u_{ij}$. Then $u_{ij} > K$ for all $(i,j) \in \Omega$ unless $u \equiv K$.

By modifying the above arguments slightly, it is easy to see that the following result will hold.

Theorem 51 *Let Ω be a finite domain in Z^2. Suppose p, q, r, s are positive and h is nonnegative on Ω. Suppose $u = \{u_{ij}\}_{(i,j)\in\Omega+\partial\Omega}$ satisfies*

$$(Lu)_{ij} - h_{ij}u_{ij} \geq 0,\ (i,j) \in \Omega.$$

Suppose the maximum M of u over $\Omega + \partial\Omega$ is nonnegative and is attained in Ω, then $u_{ij} \equiv M$ for all $(i,j) \in \Omega + \partial\Omega$.

As an immediate consequence, we see that under the above assumptions, if $u_{ij} \leq 0$ for all $(i,j) \in \partial\Omega$, we have $u_{ij} < 0$ for all (i,j) in Ω, unless $u \equiv 0$. Indeed, we may suppose $M = 0$. If M is attained in Ω, then $u \equiv M = 0$. If M is not attained in Ω, then by definition, $u_{ij} < M = 0$ for $(i,j) \in \Omega$.

In the above theorem, the function h is assumed to be nonnegative. We can derive a generalized maximum principle in which this assumption is removed. For this purpose, we assume that $u = \{u_{ij}\}$ is a real sequence which satisfies the inequalities

$$u_{ij} > 0,\ (i,j) \in \Omega + \partial\Omega,$$

and

$$(Lu)_{ij} \leq 0,\ (i,j) \in \Omega.$$

We define the new dependent variables

$$v_{ij} = \frac{w_{ij}}{u_{ij}}.$$

A simple computation yields

$$\begin{aligned}(Lw)_{ij} &= (Luv)_{ij} \\ &= -p_{ij}u_{i-1,j}\Delta_1 v_{i-1,j} + q_{ij}u_{i+1,j}\Delta_1 v_{ij} \\ &\quad -r_{ij}u_{i,j-1}\Delta_2 v_{i,j-1} + s_{ij}u_{i,j+1}\Delta_2 v_{ij} + (Lu)_{ij}v_{ij}.\end{aligned}$$

Consequently, if we assume that $(Lw)_{ij} \geq 0$ for $(i,j) \in \Omega$, then we may apply the above theorem to conclude that v satisfies the maximum principle as given there. We summarize these in the following result.

Theorem 52 *Let Ω be a finite domain in Z^2. Suppose h is a real function on Ω, p, q, r and s are positive functions in Ω, and suppose w satisfies $(Lw)_{ij} \geq 0$ for $(i,j) \in \Omega$. Suppose $u = \{u_{ij}\}_{(i,j)\in\Omega+\partial\Omega}$ is positive and $(Lu)_{ij} \leq 0$ for $(i,j) \in \Omega$. Then (i) when the maximum M of $v = w/u$ over $\Omega + \partial\Omega$ is nonnegative, v cannot attain the value M in Ω unless $v \equiv M$; (ii) when v attains the value M at a boundary point (m,n), then the values of v at the neighboring points (m,n) in Ω are strictly less than M unless $v \equiv M$; and (iii) when $v_{ij} \leq 0$ for all (i,j) in $\partial\Omega$, then $v_{ij} < 0$ for all $(i,j) \in \Omega$ unless $v \equiv 0$.*

To see how maximum principles can be applied, let Ω be a finite domain in Z^2, and let $h_{mn} \geq 0$ for $(m,n) \in \Omega$. Suppose the (real) sequence $v = \{v_{mn}\}_{(m,n)\in\Omega+\partial\Omega}$ satisfies the functional inequality

$$\Delta_1^2 v_{m-1,n} + \Delta_2^2 v_{m,n-1} - h_{mn}v_{mn} \geq 0,\ (m,n) \in \Omega. \tag{4.6}$$

Suppose further that the maximum M of v over $\Omega + \partial\Omega$ is nonnegative. Then $v_{mn} < M$ for $x \in \Omega$ unless v is a constant bivariate sequence.

As an immediate application, let us consider the following homogeneous linear system

$$\Delta_1^2 v_{m-1,n} + \Delta_2^2 v_{m,n-1} - \Gamma v_{mn} = 0,\ (m,n) \in \Omega,$$
$$v_{mn} = 0,\ (m,n) \in \partial\Omega,$$

where Ω is a finite domain, and Γ is a nonnegative constant. We assert that this system can have the trivial solution only. Indeed, if u is a nontrivial solution, we may assume that its maximum over $\Omega + \partial\Omega$ is positive. But since the maximum must occur at some point of the exterior boundary, u must be identically zero. This contradiction completes the proof of our assertion. The following is now clear.

Theorem 53 *Let Ω be a finite domain and $\Gamma \geq 0$. Then the nonhomogeneous linear system*

$$\Delta_1^2 v_{m-1,n} + \Delta_2^2 v_{m,n-1} - \Gamma v_{mn} = w_{mn},\ (m,n) \in \Omega,$$
$$v_{mn} = g_{mn},\ x \in \partial\Omega,$$

has a unique solution.

Next, let us consider the equation

$$(Lu)_{ij} - h_{ij}u_{ij} = f_{ij},\ (i,j) \in \Omega$$

under the boundary condition

$$u_{ij} = \psi_{ij},\ (i,j) \in \partial\Omega.$$

For simplicity, we will assume that the sequences p, q, r, s are positive and h is nonnegative. Suppose we can find a function $v = \{v_{ij}\}$ with the properties

$$(Lv)_{ij} - h_{ij}v_{ij} \leq f_{ij},\ (i,j) \in \Omega,$$

and

$$v_{ij} \geq \psi_{ij},\ (i,j) \in \partial\Omega.$$

Then the function

$$x_{ij} = u_{ij} - v_{ij},\ (i,j) \in \Omega + \partial\Omega$$

satisfies

$$(Lx)_{ij} - h_{ij}x_{ij} \geq 0,\ (i,j) \in \Omega,$$

and

$$x_{ij} \leq 0,\ (i,j) \in \partial\Omega.$$

The maximum principle stated above may be applied to x_{ij}, and we conclude that $x_{ij} \leq 0$ on $\Omega + \partial\Omega$. That is, $u_{ij} \leq v_{ij}$ for $(i,j) \in \Omega + \partial\Omega$. Similarly, a lower bound for u may be obtained by finding a bivariate sequence $w = \{w_{ij}\}$ with the properties

$$(Lw)_{ij} - h_{ij}w_{ij} \geq f_{ij},\ (i,j) \in \Omega,$$

and

$$w_{ij} \leq \psi_{ij},\ (i,j) \in \partial\Omega.$$

The previous development can be modified when one or more of the coefficient sequences p, q, r or s vanish. For instance, we have the following results.

Theorem 54 *Let Ω be a finite domain in Z^2. Suppose p, q, r are positive and s is nonnegative on Ω. Suppose further that $u = \{u_{ij}\}_{(i,j)\in\Omega+\partial\Omega}$ satisfies the following functional relation*

$$(Lu)_{ij} \leq 0, \ (i,j) \in \Omega.$$

Let

$$M = \max_{(i,j)\in\Omega+\partial_L\Omega+\partial_D\Omega+\partial_R\Omega} u_{ij}.$$

If M is attained at some point (m,n) in Ω, then $u_{ij} \equiv M$ for all (i,j) in $(\Omega+\partial\Omega)\cap \{(i,j) \in Z^2 | \ j \leq n\}$.

Theorem 55 *Let Ω be a finite domain in Z^2. Suppose p, q, r are positive, s is nonnegative and h is nonnegative on Ω. Suppose $u = \{u_{ij}\}_{(i,j)\in\Omega+\partial\Omega}$ satisfies*

$$(Lu)_{ij} - h_{ij}u_{ij} \geq 0, \ (i,j) \in \Omega.$$

Suppose the maximum M of u over $\Omega + \partial_L\Omega + \partial_D\Omega + \partial_R\Omega$ is nonnegative and is attained at some point (m,n) in Ω, then $u_{ij} \equiv M$ for all $(i,j) \in (\Omega+\partial\Omega)\cap\{(i,j) \in Z^2 | \ j \leq n\}$.

Theorem 56 *Let Ω be a finite domain in Z^2. Suppose h is a real function on Ω, p, q and r are positive functions, and s is nonnegative in Ω, and suppose w satisfies $(Lw)_{ij} - h_{ij}w_{ij} \geq 0$ for $(i,j) \in \Omega$. Suppose $u = \{u_{ij}\}_{(i,j)\in\Omega+\partial\Omega}$ is positive and $(Lu)_{ij} - h_{ij}u_{ij} \leq 0$ for $(i,j) \in \Omega$. If the maximum M of $v = w/u$ over $\Omega + \partial_L\Omega + \partial_D\Omega + \partial_R\Omega$ is nonnegative and is attained at some point (m,n) in Ω, then $v_{ij} \equiv M$ for all $(i,j) \in (\Omega+\partial\Omega)\cap\{(i,j) \in Z^2 | \ j \leq n\}$.*

4.4 Univariate Wirtinger's Inequalities

Recall that when $p_k > 0$ and $s_k \geq 0$, then a solution of the functional relation

$$-p_{k-1}\Delta u_{k-1} + p_k\Delta u_k - s_k u_k = 0, \ k = 1, 2, ..., n,$$

will satisfy certain maximum principles. In case $\{s_k\}$ takes on variable signs, we can still obtain some useful information regarding the corresponding solutions. More precisely, let us consider the so called self-adjoint ordinary difference equations

$$\Delta(p_{k-1}\Delta u_{k-1}) + q_k u_k = 0, \ k = 1, 2, ..., n, \tag{4.7}$$

where $p = \{p_0, p_1, ..., p_n\}$ is a real sequence with nonzero components, and $q = \{q_1, ..., q_n\}$ an arbitrary real sequence. Let a and b be any two fixed real numbers. A real sequence $y = \{y_0, y_1, ..., y_{n+1}\}$ is said to be admissible (with respect to the pair (a,b)) if it is nontrivial and satisfies

$$y_0 + ay_1 = 0 = y_{n+1} + by_n. \tag{4.8}$$

For any sequence $y = \{y_0, ..., y_{n+1}\}$, we define the functional

$$J[y] = (1+a)p_0y_1^2 + \sum_{k=1}^{n-1} p_k(\Delta y_k)^2 - \sum_{k=1}^{n} q_k y_k^2 + (1+b)p_n y_n^2.$$

It is not difficult to see that if u is an admissible solution of (4.7), then $J[u] = 0$. Indeed, we multiply (4.7) by u_k and rearrange to obtain

$$-\Delta(p_{k-1}u_{k-1}\Delta u_{k-1}) + p_{k-1}(\Delta u_{k-1})^2 - q_k u_k^2 = 0, \ k = 1, ..., n.$$

Then summing from $k = 1$ to $k = n$, we will have

$$-p_n u_n \Delta u_n + p_0 u_0 \Delta u_0 + \sum_{k=0}^{n-1} p_k (\Delta u_k)^2 - \sum_{k=1}^{n} q_k u_k^2 = 0.$$

If we now substitute $\Delta u_n = -(1+b)u_n$ and $\Delta u_0 = u_1(1+a)$ into the above equality, we obtain $J[u] = 0$.

Next we let $u = \{u_0, ..., u_{n+1}\}$ be a solution of (4.7) such that $u_k \neq 0$ for $1 \leq k \leq n$. Let y be any admissible sequence. If we define

$$v_k = \frac{y_k^2}{u_k}, \ 1 \leq k \leq n,$$

then in view of (2.5),

$$\begin{aligned} & p_k \Delta u_k \Delta\left(\frac{y_k^2}{u_k}\right) + \left(\frac{y_k^2}{u_k}\right) \Delta(p_{k-1}\Delta u_{k-1}) \\ = \ & \Delta\left(\frac{p_{k-1}u_k y_k^2}{u_k}\right) - \Delta\left(\frac{p_{k-1}u_{k-1}y_k^2}{u_k}\right). \end{aligned}$$

Hence by (2.6) and (4.7)

$$\begin{aligned} & p_k (\Delta y_k)^2 - p_k u_k u_{k+1} \left(\Delta\left(\frac{y_k}{u_k}\right)\right)^2 - q_k y_k^2 \\ = \ & \Delta\left(p_{k-1} y_k^2\right) - \Delta\left(\frac{p_{k-1}u_{k-1}y_k^2}{u_k}\right) \end{aligned}$$

for $1 \leq k \leq n-1$. Summing from $k = 1$ to $k = n-1$, we obtain

$$\begin{aligned} & \sum_{k=1}^{n-1} \left\{ p_k (\Delta y_k)^2 - p_k u_k u_{k+1} \left(\Delta\left(\frac{y_k}{u_k}\right)\right)^2 - q_k y_k^2 \right\} \\ = \ & p_{n-1} y_n^2 - p_0 y_1^2 - \frac{p_{n-1}u_{n-1}y_n^2}{u_n} + \frac{p_0 u_0 y_1^2}{u_1} \\ = \ & \frac{y_n^2}{u_n} \{[p_{n-1}u_n - p_{n-1}u_{n-1} - p_n u_{n+1} + p_n u_n] + p_n u_{n+1} - p_n u_n\} \\ & -p_0 y_1^2 + \frac{p_0 u_0 y_0^2}{u_1} \\ = \ & q_n y_n^2 + p_n y_n^2 \left(\frac{u_{n+1}}{u_n} - 1\right) + p_0 y_1^2 \left(\frac{u_0}{u_1} - 1\right). \end{aligned} \tag{4.9}$$

Theorem 57 *Suppose $p_k > 0$ for $k = 1, ..., n-1$. Suppose further that $u = (u_0, u_1, ..., u_{n+1})$ is an admissible solution of (4.7) and $u_k > 0$ for $1 \leq k \leq n$. Then for any admissible vector $y = (y_0, ..., y_{n+1})$, we have $J[y] \geq 0$. Moreover, $J[y] = 0$ if, and only if, y is a constant multiple of u.*

Proof. By substituting $u_0 = -au_1$ and $u_{n+1} = -bu_n$ into (4.9) and rearranging, we have

$$J[y] = \sum_{k=1}^{n-1} p_k u_k u_{k+1} \left(\Delta\left(\frac{y_k}{u_k}\right)\right)^2 \geq 0.$$

Moreover, $J[y] = 0$ only if

$$\Delta\left(\frac{y_k}{u_k}\right) = 0,\ 1 \leq k \leq n-1.$$

As a consequence, $y_k = Du_k$ for $1 \leq k \leq n$, where D is some constant. Furthermore, since y and u are both admissible, y is a constant multiple of u. If $y = Du$, then $J[y] = D^2 J[u] = 0$. The proof is complete. ■

Note that the positivity assumptions on p and u are not essential [13]. All we need is to assume that $p_k u_k u_{k+1} > 0$ for $k = 1, ..., n-1$. Similar remarks hold for the following theorems in this section.

Theorem 58 *Suppose $p_k > 0$ for $k = 0, ..., n$. Suppose $u = (u_0, ..., u_{n+1})$ is a solution of (4.7) such that $u_k > 0$ for $1 \leq k \leq n$. Suppose further that $u_0 + au_1 \geq 0$ as well as $u_{n+1} + bu_n \geq 0$. Then for any admissible vector y, we have $J[y] \geq 0$. If in addition, one of the inequalities $u_0 + au_1 \geq 0$ or $u_{n+1} + bu_n \geq 0$ is strict, then $J[y] \geq 0$ can be strengthened to $J[y] > 0$.*

For the proof, we rearrange (4.9) to obtain

$$J[y] = \sum_{k=1}^{n-1} p_k u_k u_{k+1} \left(\Delta\left(\frac{y_k}{u_k}\right)\right)^2 + p_0 y_1^2 \left(\frac{u_0}{u_1} + a\right) + p_n \left(\frac{u_{n+1}}{u_n} + b\right) y_n^2,$$

which, in view of our assumptions, is nonnegative. Moreover, $J[y]$ cannot vanish, say, when $u_{n+1} + bu_n > 0$, for otherwise $y_n = 0$ and $\Delta(y_k/u_k) = 0$ for $1 \leq k \leq n-1$, would imply that y is trivial.

We remark that in order that Theorem 58 holds, we need $p_1, p_n > 0$, which are not needed in Theorem 57.

As applications, we consider comparison theorems for oscillations of ordinary difference equations. Comparison theorems for partial difference equations will be discussed later.

We first define the concept of a node. Let $u = \{u_k\}_{k=\alpha}^{\beta}$ be a real sequence. If the points $(k, u_k), \alpha \leq k \leq \beta$, are joined by straight line segments to form a broken line, then this broken line gives rise to a representation of a continuous function, henceforth denoted by $u^*(t)$, such that $u^*(k) = u_k$ for $k = \alpha, ..., \beta$. The zeros of $u^*(t)$ are called the nodes of u. Now let $u = \{u_k\}_{k=0}^{n+1}$ be an admissible vector. Note that when $a \geq 0$, the condition $u_0 + au_1 = 0$ holds if, and only if, $1 - 1/(1+a)$ is a node of u. Similarly, if $b \geq 0$, then the condition $u_{n+1} + bu_n = 0$ holds if, and only if, $n + 1/(1+b)$ is a node of u.

Suppose $p_k > 0$ for $k = 0, 1, ..., n$, $a, b \geq 0$, and $u = (u_0, ..., u_{n+1})$ has consecutive nodes $1 - 1/(1+a)$ and $n + 1/(1+b)$. If $J[u] \leq 0$, then every solution v of (4.7) has a node in $(1 - 1/(1+a), n + 1/(1+b))$, unless v is a constant multiple of u. Indeed, if v does not have a node in $(1 - 1/(1+a), n + 1/(1+b))$, then we may assume without loss of generality that $v_0 + av_1 \geq 0$, $v_{n+1} + bv_n \geq 0$, and $v_k > 0$ for

$k = 1, ..., n$. Thus $J[u] \geq 0$ in view of Theorem 58. The case $J[u] > 0$ is ruled out by our assumption. If $J[u] = 0$, then v must be a constant multiple of u in view of Theorem 57.

Let $(P_0, ..., P_n)$ be a real vector with nonzero components, and let $(Q_1, ..., Q_n)$ be an arbitrary real vector. In addition to (4.7), consider the equation

$$\Delta(P_{k-1}\Delta u_{k-1}) + Q_k u_k = 0, \; k = 1, 2, ..., n, \tag{4.10}$$

and its associated function $\tilde{J}$ which is the analog of the functional J defined above.

Theorem 59 *Suppose $p_k > 0$ for $k = 0, ..., n$. Suppose $a, b \geq 0$ and suppose further that $u = (u_0, ..., u_{n+1})$ is an admissible solution of (4.10). If $\tilde{J}[u] \geq J[u]$, then every solution v of (4.7) has a node in $(1 - 1/(1 + a), n + 1/(1 + b))$ unless v is a constant multiple of u.*

The proof follows from noting that $\tilde{J}[u] = 0$.

Note that for any admissible u,

$$\begin{aligned} \tilde{J}[u] - J[u] &= (P_0 - p_0)(1 + a)u_1^2 + (P_n - p_n)(1 - b)u_n^2 \\ &+ \sum_{k=1}^{n-1}(P_k - p_k)(\Delta u_k)^2 + \sum_{k=1}^{n}(q_k - Q_k)u_k^2. \end{aligned}$$

As a consequence, we have the following discrete analogue of the classical Sturm-Picone comparison theorem.

Theorem 60 *Suppose $a, b \geq 0$ and suppose $P_k \geq p_k > 0$ for $k = 0, 1, ..., n$ and $q_k \geq Q_k$ for $k = 1, ..., n$. If (4.10) has a nontrivial solution u which has nodes $1 - 1/(1+a)$ and $n+1/(1+b)$, then every solution of (4.7) has a node in $(1-1/(1+a), n+1/(1+b))$ unless it is a constant multiple of u.*

In the above result, various additional conditions can be imposed so that every solution of (4.7) has a node in $(1 - 1/(1 + a), n + 1/(1 + b))$. The obvious ones are (i) $p_0 < P_0$ or $p_n < P_n$, and (ii) $q_i > Q_i$ and $q_{i+1} > Q_{i+1}$ for some integer i in $\{1, ..., n - 1\}$. A less obvious one is (iii) $q_i > Q_i$ for some integer i in $\{1, ..., n\}$ and $P_k > p_k$ for $k = 1, ..., n - 1$. To see (iii), note that if $\tilde{J}[u] = J[u]$, then (iii) implies $y_1 = ... = y_n$. Since y_1 is not zero, we may assume that $y_1 > 0$ so that $\Delta y_0 < 0, \Delta y_n > 0$ and $y_k = y_1 > 0$ for $k = 1, ..., n$. If (4.7) has a solution which is a constant multiple of u, then

$$(P_i - p_i)\Delta u_i = (P_{i-1} - p_{i-1})\Delta u_{i-1} + (q_i - Q_i)u_i > 0.$$

But then $u_{i+1} > u_i$, which is a contradiction.

In the special case that equations (4.7) and (4.10) coincide, we obtain the well known separation theorem: The nodes of linearly independent solutions of (4.7) separate each other.

4.5 Bivariate Wirtinger's Inequalities

Wirtinger's inequalities for univariate sequences have been discussed in the previous section. We will also need similar inequalities for bivariate sequences. Again, we approach these inequalities via "self-adjoint" partial difference equations of the form

$$\Delta_1(\bar{p}_{i-1,j}\Delta_1 y_{i-1,j}) + \Delta_2(\tilde{p}_{i,j-1}\Delta_2 y_{i,j-1}) + q_{ij}y_{ij} = 0, \ (i,j) \in \Omega, \tag{4.11}$$

where Ω is a nonempty finite domain, $\bar{p}_{ij} \neq 0$ for $(i,j) \in \Omega + \partial_L\Omega$ and $\tilde{p}_{ij} \neq 0$ for $(i,j) \in \Omega + \partial_D\Omega$, and $\{q_{ij}\}_{(i,j)\in\Omega}$ is real. Let $\Upsilon(\Omega)$ denote the set of points (w,z) in $\partial\Omega \times \partial'\Omega$ such that w, z are neighbors. Also, let $\Psi = \Psi_w^{(z)}$ be a real function defined on $\Upsilon(\Omega)$. A real bivariate sequence $\{y_{ij}\}_{(i,j)\in\Omega+\partial\Omega}$ is said to be admissible (with respect to the function Ψ) if it is nontrivial and satisfies

$$y_{ij} + \Psi_{ij}^{(u,v)} y_{uv} = 0, \ ((i,j),(u,v)) \in \Upsilon(\Omega). \tag{4.12}$$

We can break up the function $\Psi_w^{(z)}$ by defining L, R, T and D as follows:

$$\begin{aligned}
L_w^{(z)} &= \Psi_w^z, \ w \in \partial_L\Omega, z \in \partial'_L\Omega, \\
R_w^{(z)} &= \Psi_w^z, \ w \in \partial_R\Omega, z \in \partial'_R\Omega, \\
T_w^{(z)} &= \Psi_w^z, \ w \in \partial_T\Omega, z \in \partial'_T\Omega, \\
D_w^{(z)} &= \Psi_w^z, \ w \in \partial_D\Omega, z \in \partial'_D\Omega,
\end{aligned}$$

Given a real function Ψ_w^z defined on $\Upsilon(\Omega)$, we define the functional $J[y] = J[y \mid \bar{p}, \tilde{p}, q]$ for any bivariate sequence $\{y_{ij}\}_{(i,j)\in\Omega+\partial\Omega}$ as follows:

$$\begin{aligned}
&J[y] \\
= &\sum_{(i,j)\in\Omega\setminus\partial'_R\Omega} \bar{p}_{ij}\left(\Delta_1 y_{ij}\right)^2 + \sum_{(i,j)\in\Omega\setminus\partial'_T\Omega} \tilde{p}_{ij}\left(\Delta_2 y_{ij}\right)^2 - \sum_{(i,j)\in\Omega} q_{ij}y_{ij}^2 \\
&+ \sum_{(i,j)\in\partial_R\Omega} \left(1 + R_{ij}^{(i-1,j)}\right)\bar{p}_{i-1,j}y_{i-1,j}^2 + \sum_{(i,j)\in\partial_L\Omega} \left(1 + L_{ij}^{(i+1,j)}\right)\bar{p}_{ij}y_{i+1,j}^2 \\
&+ \sum_{(i,j)\in\partial_T\Omega} \left(1 + T_{ij}^{(i,j-1)}\right)\tilde{p}_{i,j-1}y_{i,j-1}^2 + \sum_{(i,j)\in\partial_D\Omega} \left(1 + D_{ij}^{(i,j+1)}\right)\tilde{p}_{ij}y_{i,j+1}^2.
\end{aligned}$$

We assert that if $\{y_{ij}\}$ is a Ψ-admissible solution of (4.11), then $J[y] = 0$. Indeed, note that

$$\begin{aligned}
&\sum_{(i,j)\in\Omega} \bar{p}_{ij}(\Delta_1 y_{ij})^2 \\
= &\sum_{(i,j)\in\Omega\setminus\partial'_R\Omega} \bar{p}_{ij}(\Delta_1 y_{ij})^2 + \sum_{(i,j)\in\partial'_R\Omega} \bar{p}_{ij}(\Delta_1 y_{ij})^2 \\
= &\sum_{(i,j)\in\Omega\setminus\partial'_R\Omega} \bar{p}_{ij}(\Delta_1 y_{ij})^2 + \sum_{(i,j)\in\partial_R\Omega} \bar{p}_{i-1,j}(\Delta_1 y_{i-1,j})^2 \\
= &\sum_{(i,j)\in\Omega\setminus\partial'_R\Omega} \bar{p}_{ij}(\Delta_1 y_{ij})^2 + \sum_{(i,j)\in\partial_R\Omega} \left(1 + R_{ij}^{(i-1,j)}\right)^2 \bar{p}_{i-1,j}y_{i-1,j}^2,
\end{aligned}$$

where the last equality is obtained from

$$\Delta_1 y_{i-1,j} = -R_{ij}^{(i-1,j)} y_{i-1,j} - y_{i-1,j}.$$

Similarly,

$$\sum_{(i,j)\in\Omega} \tilde{p}_{ij}(\Delta_2 y_{ij})^2 = \sum_{(i,j)\in\Omega\backslash\partial_T'\Omega} \tilde{p}_{ij}(\Delta_2 y_{ij})^2 + \sum_{(i,j)\in\partial_T\Omega} \left(1+T_{ij}^{(i,j-1)}\right)^2 \tilde{p}_{i,j-1} y_{i,j-1}^2.$$

Next, if we multiply (4.11) by y_{ij} and then sum over Ω, we obtain

$$\sum_{(i,j)\in\Omega} y_{ij}\left\{\Delta_1(\bar{p}_{i-1,j}\Delta_1 y_{i-1,j}) + \Delta_2(\tilde{p}_{i,j-1}\Delta_2 y_{i,j-1})\right\} = -\sum_{(i,j)\in\Omega} q_{ij} y_{ij}^2.$$

Adding the three equalities obtained above, we have

$$\begin{aligned}
&\sum_{(i,j)\in\Omega} \tilde{p}_{ij}(\Delta_1 y_{ij})^2 + \sum_{(i,j)\in\Omega} \bar{p}_{ij}(\Delta_1 y_{ij})^2 \\
&+ \sum_{(i,j)\in\Omega} y_{ij}\left\{\Delta_1(\bar{p}_{i-1,j}\Delta_1 y_{i-1,j}) + \Delta_2(\tilde{p}_{i,j-1}\Delta_2 y_{i,j-1})\right\} \\
= &\sum_{(i,j)\in\Omega\backslash\partial_R'\Omega} \bar{p}_{ij}(\Delta_1 y_{ij})^2 + \sum_{(i,j)\in\partial_R\Omega} \left(1+R_{ij}^{(i-1,j)}\right)^2 \bar{p}_{i-1,j} y_{i-1,j}^2 \\
&+ \sum_{(i,j)\in\Omega\backslash\partial_T'\Omega} \tilde{p}_{ij}(\Delta_2 y_{ij})^2 + \sum_{(i,j)\in\partial_T\Omega} \left(1+T_{ij}^{(i,j-1)}\right)^2 \tilde{p}_{i,j-1} y_{i,j-1}^2 \\
&- \sum_{(i,j)\in\Omega} q_{ij} y_{ij}^2.
\end{aligned}$$

By Theorem 4, the left hand side is equal to

$$\begin{aligned}
&\sum_{(i,j)\in\partial_R\Omega} \bar{p}_{i-1,j} y_{ij}\Delta_1 y_{i-1,j} - \sum_{(i,j)\in\partial_L\Omega} \bar{p}_{ij} y_{i+1,j}\Delta_1 y_{ij} \\
&+ \sum_{(i,j)\in\partial_T\Omega} \tilde{p}_{i,j-1} y_{ij}\Delta_2 y_{i,j-1} - \sum_{(i,j)\in\partial_D\Omega} \tilde{p}_{ij} y_{i,j+1}\Delta_2 y_{ij},
\end{aligned}$$

which, in view of (4.12), is just

$$\begin{aligned}
&\sum_{(i,j)\in\partial_R\Omega} \bar{p}_{i-1,j} R_{ij}^{(i-1,j)}\left(1+R_{ij}^{(i-1,j)}\right) y_{i-1,j}^2 - \sum_{(i,j)\in\partial_L\Omega} \left(1+L_{ij}^{(i+1,j)}\right)\bar{p}_{ij} y_{i+1,j}^2 \\
&+ \sum_{(i,j)\in\partial_T\Omega} \tilde{p}_{i,j-1} T_{ij}^{(i,j-1)}\left(1+T_{ij}^{(i,j-1)}\right) y_{i,j-1}^2 - \sum_{(i,j)\in\partial_D\Omega} \left(1+D_{ij}^{(i,j+1)}\right)\tilde{p}_{ij} y_{i,j+1}^2
\end{aligned}$$

Our assertion now follows easily.

Theorem 61 *Suppose $\{u_{ij}\}$ is a solution of (4.11) such that $u_{ij} \neq 0$ for $(i,j) \in \Omega$. Then for any Ψ-admissible function $\{y_{ij}\}_{(i,j)\in\Omega+\partial\Omega}$, we have*

$$\begin{aligned}
J[y] = &\sum_{(i,j)\in\Omega\backslash\partial_R'\Omega} \bar{p}_{ij} u_{ij} u_{i+1,j}\left\{\Delta_1\left(\frac{y_{ij}}{u_{ij}}\right)\right\}^2 \\
&+ \sum_{(i,j)\in\Omega\backslash\partial_T'\Omega} \tilde{p}_{ij} u_{ij} u_{i,j+1}\left\{\Delta_2\left(\frac{y_{ij}}{u_{ij}}\right)\right\}^2
\end{aligned}$$

$$+\sum_{(i,j)\in\partial_R\Omega}\left\{\frac{u_{ij}}{u_{i-1,j}}+R_{ij}^{(i-1,j)}\right\}\bar{p}_{i-1,j}y_{i-1,j}^2$$
$$+\sum_{(i,j)\in\partial_L\Omega}\left\{\frac{u_{ij}}{u_{i+1,j}}+L_{ij}^{(i+1,j)}\right\}\bar{p}_{ij}y_{i+1,j}^2$$
$$+\sum_{(i,j)\in\partial_T\Omega}\left\{\frac{u_{ij}}{u_{i,j-1}}+T_{ij}^{(i,j-1)}\right\}\tilde{p}_{i,j-1}y_{i,j-1}^2$$
$$+\sum_{(i,j)\in\partial_D\Omega}\left\{\frac{u_{ij}}{u_{i,j+1}}+D_{ij}^{(i,j+1)}\right\}\tilde{p}_{ij}y_{i,j+1}^2. \tag{4.13}$$

Proof. Note first that

$$\sum_{(i,j)\in\partial_R'\Omega}\bar{p}_{i-1,j}\left(\frac{y_{ij}^2}{u_{ij}}\right)\Delta_1 u_{i-1,j}$$
$$=\sum_{(i,j)\in\partial_R'\Omega}\left\{\left(\frac{y_{ij}^2}{u_{ij}}\right)[\bar{p}_{i-1,j}\Delta_1 u_{i-1,j}-\bar{p}_{ij}\Delta_1 u_{ij}]+\frac{y_{ij}^2\bar{p}_{ij}u_{i+1,j}}{u_{ij}}-y_{ij}^2\bar{p}_{ij}\right\}$$
$$=\sum_{(i,j)\in\partial_R'\Omega}\left\{-\left(\frac{y_{ij}^2}{u_{ij}}\right)\Delta_1(\bar{p}_{i-1,j}\Delta_1 u_{i-1,j})+\left(\frac{u_{i+1,j}}{u_{ij}}-1\right)\bar{p}_{ij}y_{ij}^2\right\}$$
$$=-\sum_{(i,j)\in\partial_R'\Omega}\left(\frac{y_{ij}^2}{u_{ij}}\right)\Delta_1(\bar{p}_{i-1,j}\Delta_1 u_{i-1,j})+\sum_{(i,j)\in\partial_R\Omega}\left(\frac{u_{ij}}{u_{i-1,j}}-1\right)\bar{p}_{i-1,j}y_{i-1,j}^2.$$

Note further that, in view of Theorem 2.4,

$$\sum_{(i,j)\in\Omega\backslash\partial_R'\Omega}\left\{\bar{p}_{ij}\Delta_1 u_{ij}\Delta_1\left(\frac{y_{ij}^2}{u_{ij}}\right)+\left(\frac{y_{ij}^2}{u_{ij}}\right)\Delta_1(\bar{p}_{i-1,j}\Delta_1 u_{i-1,j})\right\}$$
$$=\sum_{(i,j)\in\partial_R'\Omega}\bar{p}_{i-1,j}\left(\frac{y_{ij}^2}{u_{ij}}\right)\Delta_1 u_{i-1,j}-\sum_{(i,j)\in\partial_L\Omega}\bar{p}_{ij}\left(\frac{y_{i+1,j}^2}{u_{i+1,j}}\right)\Delta_1 u_{ij}$$
$$=-\sum_{(i,j)\in\partial_R'\Omega}\left(\frac{y_{ij}^2}{u_{ij}}\right)\Delta_1(\bar{p}_{i-1,j}\Delta_1 u_{i-1,j})+\sum_{(i,j)\in\partial_R\Omega}\left(\frac{u_{ij}}{u_{i-1,j}}-1\right)\bar{p}_{i-1,j}y_{i-1,j}^2$$
$$-\sum_{(i,j)\in\partial_L\Omega}\bar{p}_{ij}y_{i+1,j}^2\left(1-\frac{u_{ij}}{u_{i+1,j}}\right).$$

Thus,

$$\sum_{(i,j)\in\partial_R\Omega}\left(\frac{u_{ij}}{u_{i-1,j}}-1\right)\bar{p}_{i-1,j}y_{i-1,j}^2+\sum_{(i,j)\in\partial_L\Omega}\left(\frac{u_{ij}}{u_{i+1,j}}-1\right)\bar{p}_{ij}y_{i+1,j}^2$$
$$=\sum_{(i,j)\in\Omega\backslash\partial_R'\Omega}\bar{p}_{ij}\Delta_1 u_{ij}\Delta_1\left(\frac{y_{ij}^2}{u_{ij}}\right)+\sum_{(i,j)\in\Omega}\left(\frac{y_{ij}^2}{u_{ij}}\right)\Delta_1(\bar{p}_{i-1,j}\Delta_1 u_{i-1,j})$$

$$= \sum_{(i,j)\in\Omega\backslash\partial_R'\Omega} \left\{ \bar{p}_{ij}(\Delta_1 y_{ij})^2 - \bar{p}_{ij}u_{ij}u_{i+1,j}\left[\Delta_1\left(\frac{y_{ij}}{u_{ij}}\right)\right]^2 \right\}$$

$$+ \sum_{(i,j)\in\Omega} \left(\frac{y_{ij}^2}{u_{ij}}\right)\Delta_1(\bar{p}_{i-1,j}\Delta_1 u_{i-1,j}).$$

Substituting the boundary condition (4.12) and rearranging the subsequent equality, we obtain

$$\sum_{(i,j)\in\Omega\backslash\partial_R'\Omega} \bar{p}_{ij}(\Delta_1 y_{ij})^2 + \sum_{(i,j)\in\Omega}\left(\frac{y_{ij}^2}{u_{ij}}\right)\Delta_1(\bar{p}_{i-1,j}\Delta_1 u_{i-1,j})$$

$$+ \sum_{(i,j)\in\partial_R\Omega}\left(1+R_{ij}^{(i-1,j)}\right)\bar{p}_{i-1,j}y_{i-1,j}^2 + \sum_{(i,j)\in\partial_L\Omega}\left(1+L_{ij}^{(i+1,j)}\right)\bar{p}_{ij}y_{i+1,j}^2$$

$$= \sum_{(i,j)\in\Omega\backslash\partial_R'\Omega} \bar{p}_{ij}u_{ij}u_{i+1,j}\left[\Delta_1\left(\frac{y_{ij}}{u_{ij}}\right)\right]^2$$

$$+ \sum_{(i,j)\in\partial_R\Omega}\left(\frac{u_{ij}}{u_{i-1,j}} + R_{ij}^{(i-1,j)}\right)\bar{p}_{i-1,j}y_{i-1,j}^2$$

$$+ \sum_{(i,j)\in\partial_L\Omega}\left(\frac{u_{ij}}{u_{i+1,j}} + L_{ij}^{(i+1,j)}\right)\bar{p}_{ij}y_{i+1,j}^2.$$

By symmetric arguments, we may also obtain

$$\sum_{(i,j)\in\Omega\backslash\partial_T'\Omega} \tilde{p}_{ij}(\Delta_2 y_{ij})^2 + \sum_{(i,j)\in\Omega}\left(\frac{y_{ij}^2}{u_{ij}}\right)\Delta_2(\tilde{p}_{i,j-1}\Delta_1 u_{i,j-1})$$

$$+ \sum_{(i,j)\in\partial_T\Omega}\left(1+T_{ij}^{(i,j-1)}\right)\tilde{p}_{i,j-1}y_{i,j-1}^2 + \sum_{(i,j)\in\partial_D\Omega}\left(1+D_{ij}^{(i,j+1)}\right)\tilde{p}_{ij}y_{i,j+1}^2$$

$$= \sum_{(i,j)\in\Omega\backslash\partial_T'\Omega} \tilde{p}_{ij}u_{ij}u_{i,j+1}\left[\Delta_2\left(\frac{y_{ij}}{u_{ij}}\right)\right]^2$$

$$+ \sum_{(i,j)\in\partial_T\Omega}\left(\frac{u_{ij}}{u_{i,j-1}} + T_{ij}^{(i,j-1)}\right)\tilde{p}_{i,j-1}y_{i,j-1}^2$$

$$+ \sum_{(i,j)\in\partial_D\Omega}\left(\frac{u_{ij}}{u_{i,j+1}} + D_{ij}^{(i,j+1)}\right)\tilde{p}_{ij}y_{i,j+1}^2.$$

Our assertion now follows easily by adding the last two equalities. ■

We are now ready for the following two dimensional analog of Theorem 57.

Theorem 62 *Suppose $\bar{p}_{ij} > 0$ for $(i,j) \in \Omega\backslash\partial_R'\Omega$ and $\tilde{p}_{ij} > 0$ for $(i,j) \in \Omega\backslash\partial_T'\Omega$. Suppose $u = \{u_{ij}\}_{(i,j)\in\Omega+\partial\Omega}$ is an admissible solution of (4.11) such that $u_{ij} > 0$ for $(i,j) \in \Omega$. Then for any admissible function $y = \{y_{ij}\}_{(i,j)\in\Omega+\partial\Omega}$, we have $J[y] \geq 0$. Moreover, $J[y] = 0$ if, and only if, y is a constant multiple of u.*

Proof. In view of (4.12) and Theorem 3.61,

$$\begin{aligned} J[y] &= \sum_{(i,j)\in\Omega\backslash\partial'_R\Omega} \bar{p}_{ij}u_{ij}u_{i+1,j}\left(\Delta_1\left(\frac{y_{ij}}{u_{ij}}\right)\right)^2 \\ &\quad + \sum_{(i,j)\in\Omega\backslash\partial'_T\Omega} \tilde{p}_{ij}u_{ij}u_{i,j+1}\left(\Delta_2\left(\frac{y_{ij}}{u_{ij}}\right)\right)^2 \\ &\geq 0. \end{aligned}$$

Moreover, $J[y] = 0$ only if

$$\Delta_1\left(\frac{y_{ij}}{u_{ij}}\right) = 0,\ (i,j)\in\Omega\backslash\partial'_R\Omega, \tag{4.14}$$

$$\Delta_2\left(\frac{y_{ij}}{u_{ij}}\right) = 0,\ (i,j)\in\Omega\backslash\partial'_T\Omega. \tag{4.15}$$

Since Ω is connected, $y_{ij} = Du_{ij}$ for $(i,j)\in\Omega$, where D is a constant. Furthermore, since u and y are both admissible, $y_{ij} = Du_{ij}$ for $(i,j)\in\partial\Omega$. ∎

The above theorem can be rephrased as follows: If there is an admissible function y such that $J[y] < 0$, then no admissible solution $u = \{u_{ij}\}$ satisfies $u_{ij} > 0$ for $(i,j)\in\Omega$; if there is an admissible function z such that $J[z] = 0$, then any admissible solution is a constant multiple of z.

We remark that the conditions $\bar{p}_{ij}, \tilde{p}_{ij}, u_{ij} > 0$ in the above theorem are not essential [18]. All we need is to assume that $\bar{p}_{ij}u_{ij}u_{i+1,j} > 0$ for $(i,j)\in\Omega\backslash\partial'_R\Omega$ and $\tilde{p}_{ij}u_{ij}u_{i,j+1} > 0$ for $(i,j)\in\Omega\backslash\partial'_T\Omega$. For instance, if we assume that $\bar{p}_{ij} > 0$ for $(i,j)\in\Omega\backslash\partial'_R\Omega$ and $\tilde{p}_{ij} < 0$ for $(i,j)\in\Omega\backslash\partial'_T\Omega$, as well as an admissible solution u such that $u_{ij}u_{i+1,j} > 0$ and $u_{ij}u_{i,j+1} < 0$ for $(i,j)\in\Omega$, the conclusion of the above theorem remains true. Similar remarks hold for the rest of the results in this section.

Theorem 63 *Suppose $\bar{p}_{ij} > 0$ for $(i,j)\in\Omega+\partial_L\Omega$, and $\tilde{p}_{ij} > 0$ for $(i,j)\in\Omega+\partial_D\Omega$. Suppose $u = \{u_{ij}\}$ is a solution of (4.11) such that $u_{ij} > 0$ for $(i,j)\in\Omega$. If*

$$u_{ij} + \Psi_{ij}^{(u,v)}u_{uv} \geq 0,\ ((i,j),(u,v))\in\Upsilon(\Omega), \tag{4.16}$$

and if there is some $((\alpha,\beta),(\gamma,\delta))\in\Upsilon(\Omega)$ such that

$$u_{\alpha\beta} + \Psi_{\alpha\beta}^{(\gamma,\delta)}u_{\gamma\delta} > 0,$$

then for any admissible function $y = \{y_{ij}\}$, we have $J[y] > 0$.

Indeed, in view of (4.13), we see that $J[y] \geq 0$. If $(\alpha,\beta)\in\partial_R\Omega$, then $J[y] = 0$ only if (4.14),(4.15) hold as well as $y_{\alpha-1,\beta} = 0$. Since Ω is connected, $y_{ij} = 0$ for $(i,j)\in\Omega$. This contradicts our assumption that y is nontrivial.

The above theorem can be rephrased as follows: If there is an admissible function y such that $J[y] \leq 0$, then equation (4.11) cannot have a solution $u = \{u_{ij}\}$ which satisfies $u_{ij} > 0$ for $(i,j)\in\Omega$, and satisfies (4.16) with a strict inequality.

Some straightforward two dimensional discrete inequalities of Wirtinger's type will now be derived. Note first that the function

$$u_{ij} = \sin(2c_1 - d_1)\sin(2c_2 - d_2)$$

is a formal solution of

$$\Delta_1^2 u_{i-1,j} + \Delta_2^2 u_{i,j-1} + 4(\sin^2 c_1 + \sin^2 c_2)u_{ij} = 0,$$

for arbitrary real numbers c_1, c_2, d_1 and d_2.

Example 44 *Let* $\Omega = \{1, ..., n\} \times \{1, ..., m\}$. *For any* $y = \{y_{ij}\}_{(i,j)\in\Omega+\partial\Omega}$ *such that* $y_{ij} = 0$ *for* $(i, j) \in \partial\Omega$, *we have*

$$\sum_{(i,j)\in\Omega+\partial_L\Omega} (\Delta_1 y_{ij})^2 + \sum_{(i,j)\in\Omega+\partial_D\Omega} (\Delta_2 y_{ij})^2$$
$$\geq 4\left\{\sin^2 \frac{\pi}{2n+2} + \sin^2 \frac{\pi}{2m+2}\right\} \sum_{(i,j)\in\Omega} y_{ij}^2,$$

where equality holds only if,

$$y_{ij} = C \sin \frac{\pi i}{n+1} \sin \frac{\pi j}{m+1}, \ (i, j) \in \Omega + \partial\Omega.$$

Example 45 *Let* $\Omega = \{1, ..., n\} \times \{1, ..., m\}$. *For any* $y = \{y_{ij}\}_{(i,j)\in\Omega+\partial\Omega}$ *such that* $y_{ij} = 0$ *for* $(i, j) \in \partial_L\Omega + \partial_D\Omega$ *and*

$$y_{nj} = y_{n+1,j}, \ 1 \leq j \leq m,$$
$$y_{im} = y_{i,m+1}, \ 1 \leq i \leq n,$$

we have

$$\sum_{(i,j)\in\Omega+\partial_L\Omega\backslash\partial'_R\Omega} (\Delta_1 y_{ij})^2 + \sum_{(i,j)\in\Omega+\partial_D\Omega\backslash\partial'_T\Omega} (\Delta_2 y_{ij})^2$$
$$\geq 4\left\{\sin^2 \frac{\pi}{4n+2} + \sin^2 \frac{\pi}{4m+2}\right\} \sum_{\Omega} y_{ij}^2,$$

where equality holds only if,

$$y_{ij} = C \sin \frac{\pi i}{2n+1} \sin \frac{\pi j}{2m+1}, \ (i, j) \in \Omega + \partial\Omega.$$

Next note that the function

$$u_{ij} = \sin(c(i+j) - b_1)\sin(c(i-j) - b_2)$$

is a formal solution of

$$\Delta_1^2 u_{i-1,j} + \Delta_2^2 u_{i,j-1} + 4(\sin^2 c)u_{ij} = 0,$$

where c, b_1 and b_2 are arbitrary real numbers.

Example 46 *Let* $\Omega = \{(i,j) \in Z^2 | \, 0 < i-j < n, 0 < i+j < n\}$. *For any* $y = \{y_{ij}\}_{(i,j)\in\Omega+\partial\Omega}$ *such that* $y_{ij} = 0$ *for* $(i,j) \in \Omega$, *we have*

$$\sum_{(i,j)\in\Omega+\partial_L\Omega} (\Delta_1 y_{ij})^2 + \sum_{(i,j)\in\Omega+\partial_D\Omega} (\Delta_2 y_{ij})^2 \geq 4\sin^2\frac{\pi}{n}\sum_{(i,j)\in\Omega} y_{ij}^2,$$

where equality holds only if,

$$y_{ij} = C\sin\frac{\pi(i+j)}{n}\sin\frac{\pi(i-j)}{n},\ (i,j)\in\Omega+\partial\Omega.$$

Example 47 *Note that*

$$u_{ij} = \sin\frac{im\pi}{m+1}\sin\frac{j\pi}{n+1},\ 0\leq i\leq m+1, 0\leq j\leq n+1,$$

defines a solution $u = \{u_{ij}\}$ *of the boundary value problem*

$$u_{0j} = 0 = u_{m+1,j},\ 1\leq j\leq n,$$
$$u_{i0} = 0 = u_{i,n+1},\ 1\leq i\leq m,$$

and

$$\Delta_2^2 u_{i,j-1} - \Delta_1^2 u_{i-1,j} + 4\left\{\sin^2\frac{\pi}{2n+2} - \sin^2\frac{m\pi}{2m+2}\right\} u_{ij} = 0,$$

for $1\leq i\leq m$ *and* $1\leq j\leq n$. *Since* $u_{ij}u_{i+1,j} < 0$ *for* $1\leq i\leq m-1$ *and* $1\leq j\leq n$ *and* $u_{ij}u_{i,j+1} > 0$ *for* $1\leq i\leq m$ *and* $1\leq j\leq n-1$. *Thus in view of the remark following Theorem 62, for any bivariate sequence* $y = \{y_{ij}\}$ *which satisfies the same boundary conditions that* u *satisfies, we have*

$$\sum_{i=1}^{m}\sum_{j=0}^{n}(\Delta_2 y_{ij})^2 - \sum_{i=0}^{n}\sum_{j=1}^{n}(\Delta_1 y_{ij})^2$$
$$\geq 4\left\{\sin^2\frac{\pi}{2n+2} - \sin^2\frac{m\pi}{2m+2}\right\}\sum_{i=1}^{m}\sum_{j=1}^{n} y_{ij}^2,$$

and equality holds only if y *is a constant multiple of* u.

Let $\bar{P}_{ij} \neq 0$ for $(i,j)\in\Omega+\partial_L\Omega$, $\tilde{P}_{ij}\neq 0$ for $(i,j)\in\Omega+\partial_D\Omega$ and Q_{ij} is real for $(i,j)\in\Omega$. In addition to equation (4.11), consider the partial difference equation

$$\Delta_1(\bar{P}_{i-1,j}\Delta_1 u_{i-1,j}) + \Delta_2(\tilde{P}_{i,j-1}\Delta_2 u_{i,j-1}) + Q_{ij}u_{ij} = 0,\ (i,j)\in\Omega, \tag{4.17}$$

and its associated functional $\hat{J}$ which is the analog of the functional J defined for equation (4.11). Note that for any $y = \{y_{ij}\}_{(i,j)\in\Omega+\partial\Omega}$,

$$\begin{aligned} &\hat{J}[y] - J[y] \\ = &\sum_{(i,j)\in\Omega\backslash\partial_R'\Omega} (\bar{P}_{ij} - \bar{p}_{ij})(\Delta_1 y_{ij})^2 + \sum_{(i,j)\in\Omega\backslash\partial_T'\Omega} (\tilde{P}_{ij} - \tilde{p}_{ij})(\Delta_2 y_{ij})^2 \\ &- \sum_{(i,j)\in\Omega}(Q_{ij} - q_{ij})y_{ij}^2 + \sum_{(i,j)\in\partial_R\Omega}\left(1 + R_{ij}^{(i-1,j)}\right)(\bar{P}_{i-1,j} - \bar{p}_{i-1,j})y_{i-1,j}^2 \end{aligned}$$

$$+ \sum_{(i,j)\in\partial_L\Omega} \left(1 + L_{ij}^{(i+1,j)}\right)(\bar{P}_{ij} - \bar{p}_{ij})y_{i+1,j}^2$$
$$+ \sum_{(i,j)\in\partial_T\Omega} \left(1 + T_{ij}^{(i,j-1)}\right)(\tilde{P}_{i,j-1} - \tilde{p}_{i,j-1})y_{i,j-1}^2$$
$$+ \sum_{(i,j)\in\partial_D\Omega} \left(1 + D_{ij}^{(i,j+1)}\right)(\tilde{P}_{ij} - \tilde{p}_{ij})y_{i,j+1}^2.$$

Thus we have the following two dimensional discrete analog of the classical Sturm-Picone comparison theorem.

Theorem 64 *Suppose $\bar{P}_{ij} \geq p_{ij} > 0$ for $(i,j) \in \Omega + \partial_L\Omega$, $\tilde{P}_{ij} \geq \tilde{p}_{ij} > 0$ for $(i,j) \in \Omega + \partial_D\Omega$, and $q_{ij} \geq Q_{ij}$ for $(i,j) \in \Omega$. If Ψ is nonnegative and y is a Ψ-admissible solution of (4.17), then (4.11) cannot have a Ψ-admissible solution $u = \{u_{ij}\}_{(i,j)\in\Omega+\partial\Omega}$ such that $u_{ij} > 0$ for $(i,j) \in \Omega$, unless u is a constant multiple of y.*

4.6 Notes and Remarks

Discrete maximum principles are known to many authors. See for example Cheng [11], Kuo and Trudinger [101], Heilbronn [80], and others. Our presentations are motivated by those in Protter and Weinberger [141]. Discrete harmonic functions are functions that satisfy the discrete Laplace equations and hence our maximum principles can be applied. Discrete harmonic functions have been studied by a number of authors, see for example Heilbronn [80], Allen and Murdoch [3], Streater [150], and Duffin [60].

Discrete Wirtinger's inequalities involving sequences of one variable are also known to many authors. However, those that involve double sequences are probably first announced by Cheng and Lu [22]. Our discussions of these inequalities are based on Cheng [12], [13], [18], and Cheng and Lu [22].

While Wirtinger's inequalities are associated with linear selfadjoint difference equations, Hardy's inequalities are associated with nonlinear equations. For such inequalities, the reader may consult Cheng and Lu [23].

Chapter 5

Explicit Solutions

5.1 Introduction

When a partial difference equation and its domain of definition is given, sometimes it is relatively easy to construct "solutions". Furthermore, under "well posed" auxiliary conditions, we may construct solutions that are unique. As an example, consider the discrete heat equation (1.3)

$$u_m^{(n+1)} = au_{m-1}^{(n)} + bu_m^{(n)} + cu_{m+1}^{(n)}, \ m \in Z, n \in N \tag{5.1}$$

where a, b, c are real numbers. It is easy to see that under the initial condition

$$u_m^{(0)} = f_m, \ m \in Z, \tag{5.2}$$

we can calculate

$$u_{01}; u_{-1,1}, u_{11}, u_{02}; u_{-2,1}, u_{-1,2}, u_{12}, u_{21}, u_{03}; \ldots$$

successively in a unique manner. Such a sequence is called a solution of the initial value problem (5.1),(5.2). An existence and uniqueness theorem for a solution of (5.1) subject to (5.2) is thus easily formulated and proved. Note that there are two interesting consequences of the computational scheme. First of all, the value $u_0^{(n)}$ is completely determined by the values $f_{-n}, f_{-n+1}, \ldots, f_n$; and similarly, $u_m^{(n)}$ by f_k, $m - n \le k \le m + n$. For this reason, we say that the integers $m - n, \ldots, m + n$ are the *domain of dependence* for the value $u_m^{(n)}$. Second, if $f_m = 0$ for $m \ge k$, then the values $u_m^{(n)}$ will vanish for $m \ge n + k$. As a special case, suppose $b = 0$ in (5.1). Let $f_m = 0$ when m is odd and $f_m = (-1)^{m/2}$ when m is even. Then by means of the computational scheme stated above, we obtain the sequence

$$0; -(a-c), (a-c), -(a-c)^2; 0, 0, 0, 0, 0; (a-c), (a-c)^2, (a-c)^3, \ldots,$$

which shows that $u_m^{(n)}$ is either equal to 0 or to $\pm(a-c)^n$.

As another example, consider the partial difference equation

$$u_{i+1,j+1} + au_{i+1,j} + bu_{ij} + cu_{ij} = f_{ij}, \ i, j \in N. \tag{5.3}$$

Since this equation can be written as

$$u_{i+1,j+1} = -au_{i+1,j} - bu_{i,j+1} - cu_{ij} + f_{ij}$$

it is clear that if the conditions

$$u_{0j} = \psi_j,\ j \in N, \tag{5.4}$$

$$u_{i0} = \phi_i,\ i \in N, \tag{5.5}$$

and the *compatibility condition*

$$\psi_0 = \phi_0, \tag{5.6}$$

are given, we can calculate

$$u_{11}; u_{12}, u_{21}; u_{11}, u_{22}, u_{31}; \ldots$$

successively in a unique manner. An existence and uniqueness theorem for solutions is thus easily formulated and proved.

In the previous examples, solutions can be constructed in a recursive manner. To see an example in which solutions cannot be obtained in such manners, consider the discrete time independent Poisson equation

$$u_{i-1,j} + u_{i+1,j} + u_{i,j-1} + u_{i,j+1} - 4u_{ij} = f_{ij},\ (i,j) \in \Omega, \tag{5.7}$$

where Ω is a finite domain. If we impose boundary conditions of the form

$$u_{ij} = 0,\ (i,j) \in \partial\Omega, \tag{5.8}$$

then it is not clear how a solution can be constructed. Fortunately, we may regard this boundary problem as a linear system of equations and then the existence problem can be dealt with by standard means in linear algebra. For example, let $\Omega = \{(1,2),(2,2),(1,1),(2,1)\}$. Then

$$\partial\Omega = \{(1,3),(2,3),(3,2),(3,1),(2,0),(1,0),(0,1),(0,2)\}.$$

The corresponding discrete Poisson equation together with the boundary condition can now be written as

$$\begin{aligned} 0 + 0 + u_{22} + u_{11} - 4u_{12} &= f_{12}, \\ u_{12} + 0 + 0 + u_{21} - 4u_{22} &= f_{22}, \\ u_{11} + u_{22} + 0 + 0 - 4u_{21} &= f_{21}, \\ 0 + u_{12} + u_{21} + 0 - 4u_{11} &= f_{11}. \end{aligned}$$

This is a linear system in the unknowns u_{12}, u_{22}, u_{11} and u_{21}, and by showing its corresponding determinant is nonzero, it is easy to see that it has a unique solution.

In general, the question of existence and/or uniqueness of solutions is much more involved. The previous examples, however, show that the solutions depend on the coefficient sequences, the initial conditions, the boundary data, the forcing terms and the control terms. Sometimes, there are additional parameters involved (e.g. the grid spacings or step lengths inherited from discretization of partial differential equations). Then the solutions also depend on these parameters. For the sake of convenience, we will call these factors (coefficient sequences, forcing terms, etc.) the accompanying *nonstructural variables* of our partial difference equation.

Example 48 *Consider the equation*

$$u_{i+1,j} + bu_{i,j+1} = 0,\ b \neq 0,\ i,j \in N$$

subject to the conditions

$$u_{i0} = i!,\ i \in N.$$

A solution can be found and is given by

$$u_{ij} = \left(-\frac{1}{b}\right)^j (i+j)!,\ i,j \in N.$$

Example 49 *The equation*

$$u_{i+1,j+1} + u_{i+1,j} + u_{i,j+1} + u_{ij} = 0,\ i,j \in N.$$

subject to the conditions

$$u_{i0} = (i+1)^i,\ i \in N,$$

and

$$u_{0j} = 2^{-j},\ j \in N,$$

has a solution

$$u_{ij} = (-1)^j (i+1)^i + (-1)^i 2^{-j} - (-1)^{i+j},\ i,j \in N.$$

Example 50 *The equation*

$$u_{i,j+1} - u_{i+1,j} - u_{ij} - u_{i-1,j} = 0,\ i,j \in N,$$

under the conditions

$$u_{i0} = (i+1)!,\ i = -1, -2, ...,$$

$$u_{-1,j} = 0,\ j \in N,$$

has a solution $u = \{u_{ij}\}$ *given by*

$$u_{ij} = \sum_{k=0}^{j} \sum_{s=0}^{j-k} C_k^{(j)} C_s^{(j-k)} (i - j + 2k + s + 1)!$$

for $i \geq j \geq 1$. *Note that*

$$u_{ij} \geq (i-j+1)! \sum_{k=0}^{j} \sum_{s=0}^{j-k} C_k^{(j)} C_s^{(j-k)} \geq 3^j (i-j+1)!,\ i \geq j.$$

In general, the exact dependence of a solution on the accompanying nonstructural variables is a difficult issue. Indeed, finding the exact dependence is the same as finding a *general solution*. In some rare cases, general solutions can indeed be found.

Example 51 *Consider the discrete heat equation*

$$u_i^{(j+1)} = au_{i-1}^{(j)} + bu_i^{(j)} + cu_{i+1}^{(j)} + g_i^{(j)},\ i = 1, ..., n;\ j \in N, \tag{5.9}$$

where $a, b, c \in R$, $g = \{g_i^{(j)}\}$ *is real function defined for* $i = 1, 2, ..., n$ *and* $j \in N$. *We will also assume that side conditions*

$$u_0^{(j)} = \phi_j,\ j \in N, \tag{5.10}$$

$$u_{n+1}^{(j)} = \psi_j,\ j \in N, \tag{5.11}$$

$$u_i^{(0)} = w_i,\ i = 1, 2, ..., n \tag{5.12}$$

are imposed, where $w = \text{col}(w_1, ..., w_n)$ *is a real vector, and* $\phi = \{\phi_j\}_{j=0}^{\infty}$ *and* $\psi = \{\psi_j\}_{j=0}^{\infty}$ *are real sequences. Let*

$$\Omega = \{(i,j) |\ i = 0, 1, ..., n+1;\ j \in N\}.$$

A solution of (5.9)–(5.12) is a discrete function $u = \{u_i^{(j)}\}_{(i,j)\in\Omega}$ *which satisfies the functional relation (5.9) and also the side conditions (5.10)–(5.12). It is easily seen that an existence and uniqueness theorem holds for solutions of the problem (5.9)–(5.12). It is also relatively easy to find the corresponding solutions. As an example, when* $w = 0$, $\phi = 0$, $\psi = 0$, *and* $g = 0$, *then* $\{0\}_{(i,j)\in\Omega}$ *is a solution of (5.9)–(5.12). As another example, when* $n = 1$, *we may show by induction that the unique solution* $\{u_i^{(j)}\}_{(i,j)\in\Omega}$ *of (5.9)–(5.12) satisfies*

$$u_1^{(j+1)} = \sum_{s=0}^{j} ab^{j-s}\phi_s + b^{j+1}w_1 + \sum_{s=0}^{j} b^{j-s}\psi_s + \sum_{s=0}^{j} b^{j-s}g_1^{(s)}$$

for $j \in Z^+$. *In general, by designating* $\text{col}(u_1^{(j)}, u_2^{(j)}, ..., u_n^{(j)})$ *as the real vector* $u^{(j)}$, *we see that the sequence* $\{u^{(j)}\}_{j=0}^{\infty}$ *will satisfy the vector equation*

$$u^{(j+1)} = Au^{(j)} + f^{(j)},\ j \in N, \tag{5.13}$$

and the initial condition

$$u^{(0)} = w \tag{5.14}$$

where

$$A = \begin{bmatrix} b & c & 0 & \dots & 0 \\ a & b & c & \dots & 0 \\ & . & . & . & \\ 0 & \dots & a & b & c \\ 0 & \dots & 0 & a & b \end{bmatrix}_{n\times n}, \tag{5.15}$$

$$f^{(j)} = \text{col}(g_1^{(j)}, ..., g_n^{(j)}) + \text{col}(a\phi_j, 0, ..., 0, c\psi_j). \tag{5.16}$$

Conversely, if $\{u^{(j)}\}_{j=0}^{\infty}$ *is a solution of (5.13)–(5.14), then by augmenting each* $u^{(j)} = \text{col}(u_1^{(j)}, ..., u_n^{(j)})$ *with the terms* ϕ_j *and* ψ_j *to form* $\left\{\phi_j, u_1^{(j)}, ..., u_n^{(j)}, \psi_j\right\}$, *we see that the resulting family forms a solution of (5.9)–(5.12). Now, by inductive arguments, it is easily seen that the unique solution* $\{u^{(j)}\}_{j=0}^{\infty}$ *of (5.13) subject to* $u^{(0)} = w$ *now satisfies*

$$u^{(j+1)} = A^{j+1}w + \sum_{k=0}^{j} A^{j-k} f^{(k)}, \ j \in N. \tag{5.17}$$

We have found a general solution in the above example. Other examples will be discussed in later sections. The methods for obtaining these solutions will also be explained.

5.2 Formal Methods

The method of trial and error is still one of the basic methods for obtaining explicit solutions. Next, there are formal methods which may also lead to plausible solutions. We will consider several simple examples in this section and find some or all their solutions. More systematic and rigorous approaches will be taken up later.

Example 52 *Consider the partial difference equation*

$$u_m^{(n+1)} = 4u_{m+1}^{(n)}. \tag{5.18}$$

If we guess that a solution is of the form $u_m^{(n)} = \mu^m \lambda^n$, *then substituting* $u_m^{(n)}$ *into (5.18), we are led to*

$$\lambda^{n+1}\mu^m - 4\mu^{m+1}\lambda^n = 0,$$

or

$$\lambda = 4\mu.$$

From $\lambda = 4\mu$, *it follows that a solution is given by* $u_m^{(n)} = 4^n \mu^{m+n}$. *Since sums of these solutions over any values of* μ *are solutions, we are then led to the plausible solutions*

$$\sum_{\mu} 4^m \mu^{m+n} = 4^m \sum_{\mu} \mu^{m+n} = 4^m v_{m+n},$$

where $\{v_k\}$ *is an arbitrary sequence.*

Example 53 *Consider the partial difference equation*

$$P_{ij} = pP_{i-1,j} + qP_{i,j-1}, \ i,j \in Z^+,$$

where $p,q > 0$ *and* $p+q=1$, *subject to the boundary conditions*

$$P_{i0} = 0, \ i \in N,$$
$$P_{0j} = 1, \ j \in Z^+.$$

For each $i \in N$, *let*

$$G_i(t) = P_{i0} + P_{i1}t + P_{i2}t^2 + \dots = \sum_{j=0}^{\infty} P_{ij}t^j$$

be the "generating function" of the sequence $\{P_{ij}\}_{j=0}^{\infty}$. Then multiplying the difference equation (1.7) by t^j and summing from $j = 1$ to ∞, we have

$$G_i(t) - P_{i0} = p\{G_{i-1}(t) - P_{i-1,0}\} + qtG_i(t).$$

In view of the boundary conditions, we further have

$$G_i(t) = pG_{i-1}(t) + qtG_i(t),$$

or

$$G_i(t) = \frac{p}{1-qt} G_{i-1}(t),$$

which can readily be solved to yield

$$G_i(t) = \left(\frac{p}{1-qt}\right)^i G_0(t).$$

Since

$$G_0(t) = P_{00} + P_{01}t + P_{02}t^2 + ... = t + t^2 + t^3 + ... = \frac{t}{1-t},$$

we see that

$$G_i(t) = \left(\frac{p}{1-qt}\right)^i \frac{t}{1-t}.$$

By expanding $G_i(t)$ into a series of powers of t, we see that

$$P_{ij} = p^i \sum_{k=0}^{j-1} C_k^{(i+k-1)} q^k.$$

Before embarking on the next example, let us introduce another formal approach based on the same idea of generating functions. Let $u = \{u_{ij}\}_{i,j\in Z}$ be a bivariate sequence. The generating function for the sequence u is formally defined by the bivariate infinite series

$$Y(s,t) = \sum_{i=0}^{\infty}\sum_{j=0}^{\infty} u_{ij} \frac{s^i}{i!}\frac{t^j}{j!}.$$

Assuming convergence of the series $Y(s,t)$ for $|s| < \alpha$ and $|t| < \beta$, we see that

$$Y(s,0) = \sum_{i=0}^{\infty} u_{i0} \frac{s^i}{i!},$$

$$Y(0,t) = \sum_{j=0}^{\infty} u_{0j} \frac{t^j}{j!},$$

and we may also calculate the partial derivatives within the region of convergence and obtain

$$\frac{\partial Y(s,t)}{\partial s} = \sum_{i=0}^{\infty}\sum_{j=0}^{\infty} u_{i+1,n} \frac{s^i}{i!}\frac{t^j}{j!},$$

$$\frac{\partial Y(s,t)}{\partial t} = \sum_{i=0}^{\infty}\sum_{j=0}^{\infty} u_{i,j+1}\frac{s^i}{i!}\frac{t^j}{j!},$$

and

$$\frac{\partial^{m+n} Y(s,t)}{\partial s^m \partial t^n} = \sum_{i=0}^{\infty}\sum_{j=0}^{\infty} u_{i+m,j+n}\frac{s^i}{i!}\frac{t^j}{j!}.$$

Furthermore, we have

$$u_{ij} = \left.\frac{\partial^{i+j} Y(s,t)}{\partial s^i \partial t^j}\right|_{s=0,t=0}.$$

Example 54 *Consider the partial difference equation*

$$u_{i+1,j+1} - u_{i,j+1} - u_{ij} = 0, \; i,j \in N, \tag{5.19}$$

under the auxiliary conditions

$$\begin{aligned} u_{00} &= 0, \\ u_{01} = u_{10} &= 1, \\ u_{i+1,0} = u_{i0} + u_{i-1,0}, &\; i \in Z^+, \\ u_{0,j+1} = u_{0,j} + u_{0,j-1}, &\; j \in Z^+. \end{aligned}$$

Multiplying equation (5.19) by $(s^i t^j)/(i!j!)$ and summing on m and n from 0 to ∞, we get the transformed equation

$$\frac{\partial^2 Y(s,t)}{\partial s \partial t} - \frac{\partial Y(s,t)}{\partial t} - Y(s,t) = 0. \tag{5.20}$$

Similarly, in view of the auxiliary conditions, we have

$$Y(0,0) = 0,$$

$$\frac{\partial Y(0,0)}{\partial s} = 1 = \frac{\partial Y(0,0)}{\partial t},$$

$$\frac{\partial^2 Y(s,0)}{\partial s^2} - \frac{\partial Y(s,0)}{\partial s} - Y(s,0) = 0,$$

$$\frac{\partial^2 Y(0,t)}{\partial t^2} - \frac{\partial Y(0,t)}{\partial t} - Y(0,t) = 0.$$

The functions $Y(s,0)$ and $Y(0,t)$ can readily be solved according to the theory of linear ordinary differential equations with constant coefficients and are given by

$$Y(s,0) = \frac{1}{\sqrt{5}}\left\{e^{s\gamma_+} - e^{s\gamma_-}\right\}, \tag{5.21}$$

$$Y(0,t) = \frac{1}{\sqrt{5}}\left\{e^{t\gamma_+} - e^{t\gamma_-}\right\}, \tag{5.22}$$

where $\gamma_\pm$ the distinct roots of the characteristic equation

$$x^2 - x - 1 = 0,$$

i.e.,

$$\gamma_{\pm} = \frac{1 \pm \sqrt{5}}{2}.$$

The solution of (5.20),(5.21) and (5.22) can be found and is given by

$$Y(s,t) = \frac{1}{\sqrt{5}} \left\{ e^{\gamma_+(t+s)} - e^{\gamma_-(t+s)} \right\},$$

thus

$$u_{ij} = \left. \frac{\partial^{i+j} Y(s,t)}{\partial s^i \partial t^j} \right|_{s=0,t=0} = \frac{1}{\sqrt{5}} \left\{ \gamma_+^{i+j} - \gamma_-^{i+j} \right\}.$$

5.3 The Method of Translation

Although we were able to derive some explicit solutions to several partial difference equations in the previous section, not all the methods employed are justified. Therefore, we need to do at least one of the following two things. Either we develope rigorous methods for finding solutions, or else we verify the correctness of our solutions by systematic means. One such rigorous and easy method is as follows. Let l^N be the set of all complex sequences of the form $\{f_k\}_{k\in N}$. Under the usual addition and scalar multiplication, l^N is a linear space. Let us define a family of translation mappings E^m, $m \in N$, as follows:

$$E^0 f = f,\ f \in l^N,$$

$$E^m \{f_k\}_{k\in N} = \{f_{m+k}\}_{k\in N},\ \{f_k\}_{k\in N} \in l^N.$$

These mappings have the following two simple properties:

$$E^m(\alpha E^n f) = \alpha E^{m+n} f,\ \alpha \in C;\ m,n \in N;\ f \in l^N,$$

and

$$\left(E^i + E^j\right)^n f = \sum_{m=0}^{n} C_m^{(n)} E^{im+jn-jm} f,\ f \in l^N;\ i,j,n \in N.$$

Example 55 *Consider the equation*

$$u_m^{(n+1)} = 4u_{m+1}^{(n)},\ m,n \in N$$

subject to the auxiliary condition

$$u_m^{(0)} = v_m,\ m \in N.$$

For each $n \in N$, let us view $\{u_m^{(n)}\}_{m\in Z}$ as a sequence $u^{(n)}$ in l^N. Then we can rewrite our problem as

$$u^{(n+1)} = 4Eu^{(n)},\ n \in N,$$
$$u^{(0)} = \{v_m\}_{m\in N}.$$

Thus

$$u^{(n+1)} = 4Eu^{(n)} = 4E\left(4Eu^{(n-1)}\right) = 4^2E^2u^{(n-1)} = \ldots = 4^{n+1}E^{n+1}u^{(0)},$$

so that

$$u_m^{(n)} = 4^n v_{m+n},\ m,n \in N.$$

Example 56 *Consider the partial difference equation*

$$u_m^{(n+1)} - bu_{m+1}^{(n)} - cu_m^{(n)} = 0, \ m, n \in N, \tag{5.23}$$

subject to the initial condition

$$u_m^{(0)} = f_m, \ m \in N. \tag{5.24}$$

As in the previous example, we can write equation (5.23) as

$$u^{(n+1)} = (bE + cE^0)u^{(n)},$$

then

$$u^{(t)} = (bE + cE^0)^t u^{(0)} = (bE + cE^0)^t \{f_j\} \\ = \sum_{k=0}^{t} C_k^{(t)} (bE)^{t-k} c^k \{f_j\}.$$

Thus the unique solution of the initial value problem (5.23)–(5.24) is given by

$$u_j^{(t)} = \sum_{k=0}^{t} C_k^{(t)} b^{t-k} c^k \psi_{j+t-k}, \ j, t \in N. \tag{5.25}$$

5.4 The Method of Operators

The method of operators as explained in Chapter 3 also provides rigorous means for solving certain linear partial difference equations of the form

$$\sum_{m=0}^{M} \sum_{n=0}^{K} c_{mn} f_{i+m,j+n} = g_{ij}, \ i, j \in N, \tag{5.26}$$

where $c_{00}, ..., c_{MK}$ are complex numbers such that $c_{MK} \neq 0$.

Example 57 *Consider the partial difference equation*

$$f_{i+1,j+1} - f_{i+1,j} - f_{i,j+1} - 3f_{ij} = 0, \ i, j \in N, \tag{5.27}$$

subject to the conditions

$$f_{i0} = 3^i, \ i \in N,$$

and

$$f_{0j} = 3^j, \ j \in N.$$

We may write the partial difference equation in the form

$$E_x E_y f - E_x f - E_y f - 3f = 0, \ f = \{f_{ij}\}, \tag{5.28}$$

and write the two side conditions as

$$Y_0 f = \frac{1}{1 - 3\hbar_x},$$

and

$$X_0 f = \frac{1}{1-3\hbar_y}$$

respectively. In view of (3.51), by multiplying (5.27) with $\hbar_x \hbar_y$, we obtain

$$f - X_0 f - Y_0 f + f_{00} - \hbar_y [f - X_0 f] - \hbar_x [f - Y_0 f] - 3\hbar_x \hbar_y f = 0,$$

so that

$$[1 - \hbar_x - \hbar_y - 3\hbar_x \hbar_y] f + 1 + \frac{\hbar_x - 1}{1-3\hbar_x} + \frac{\hbar_y - 1}{1 - 3\hbar_y} = 0.$$

Solving the above algebraic equation, we obtain

$$f = \frac{1}{(1-3\hbar_x)(1-3\hbar_y)} = \{3^{i+j}\}.$$

This example motivates a general procedure for solving difference equations of the form (5.26). We first write (5.26) in the form

$$\sum_{m=0}^{M} \sum_{n=0}^{K} E_x^m E_y^n f = g,$$

and then multiply both sides by $\hbar_x^M \hbar_y^K$. Then an algebraic equation involving the unknown operator f is obtained. After solving f from this equation, we need only to find the bivariate sequence which represents this operator. If this can be done, we obtain the desired solution, otherwise, a generalized operator solution is obtained.

Example 58 *Consider the partial difference equation*

$$f_{i+1,j+1} - \alpha f_{i,j+1} - \beta f_{i+1,j} + \alpha\beta f_{ij} = g_{ij},\ i,j \in N, \tag{5.29}$$

subject to the conditions

$$f_{i0} = \phi_i,\ i \in N,$$
$$f_{0j} = \psi_j,\ j \in N,$$

and the compatibility condition

$$\phi_0 = \psi_0.$$

Equation (5.29) can be written as

$$E_x E_y f - \alpha E_y f - \beta E_x f + \alpha\beta f = g,\ f = \{f_{ij}\},\ g = \{g_{ij}\},$$

which yields, after multiplying by $\hbar_x \hbar_y$, the algebraic equation

$$(1-\alpha\hbar_x)(1-\beta\hbar_y) f = (1-\beta\hbar_y) X_0 f + (1-\alpha\hbar_x) Y_0 f + \hbar_x \hbar_y g - f_{00}.$$

Since

$$\frac{1}{(1-\alpha\hbar_x)(1-\beta\hbar_y)} = \{\alpha^i \beta^j\},$$

$$\frac{Y_0 f}{1-\beta\hbar_y} = \{\phi_i \beta^j\},$$

and

$$\frac{X_0 f}{1-\alpha\hbar_x} = \{\alpha^i \psi_j\},$$

thus

$$f = \{\alpha^i \beta^j\} \hbar_x \hbar_y g + \{\phi_i \beta^j\} + \{\alpha^i \psi_j\} - \phi_0 \{\alpha^i \beta^j\}.$$

The above two equations (5.27) and (5.29) are particular cases of the following general partial difference equation

$$u_{i+1,j+1} = au_{i,j+1} + bu_{i+1,j} + cu_{ij},\ i,j \in N. \tag{5.30}$$

A solution of (5.30) is a real bivariate sequence $u = \{u_{ij}\}_{i,j\in Z}$ which satisfies (5.30). It is clear that if the conditions (5.4), (5.5) and (5.6) are imposed, we can calculate u_{11}; u_{12}, u_{21}; u_{13}, u_{22}, u_{31}; ... successively in a unique manner.

In view of the computing scheme just described, it is easily seen that the value u_{mn} is completely determined by the values $\psi_0, \psi_1, ..., \psi_n$; $\phi_1, ..., \phi_m$. Therefore, if we let the solutions of (5.30) corresponding to the following three sets of initial conditions

$$u_{00} = 1,\ u_{i0} = 0,\ u_{0j} = 0,\ i,j \in Z^+; \tag{5.31}$$

$$u_{00} = 0,\ u_{i0} = \delta_{i\alpha},\ u_{0j} = 0,\ i,j,\alpha \in Z^+,$$

and

$$u_{00} = 0,\ u_{i0} = 0,\ u_{0j} = \delta_{j\beta},\ i,j,\beta \in Z^+, \tag{5.32}$$

be denoted by $f = \{f_{ij}\}$, $g^{(\alpha)} = \{g_{ij}^{(\alpha)}\}$ and $h^{(\beta)} = \{h_{ij}^{(\beta)}\}$ respectively, then the general solution of (5.30) is given by

$$f_{ij} = \psi_0 f_{ij} + \sum_{\alpha=1}^{i} \psi_\alpha g_{mn}^{(\alpha)} + \sum_{\beta=1}^{j} \phi_\beta h_{mn}^{(\beta)}, \tag{5.33}$$

where $\delta_{mn} = 1$ if $m = n$ and $\delta_{mn} = 0$ otherwise. For this reason, it is natural to call the triplet $\{f, g, h\}$ the Green solution triplet of the initial value problem (5.30)–(5.32).

The methods described above can be employed to derive the solutions $f, g^{(\alpha)}$ and $h^{(\beta)}$, in spite of the fact that their explicit forms are awkward to express. Let us first follow the procedure described in Example 57, and derive the operator solution

$$f = \frac{1 - a\hbar_x - b\hbar_y}{1 - a\hbar_x - b\hbar_y - c\hbar_x\hbar_y}.$$

In order to find the components f_{ij}, first note that

$$\begin{aligned} & \frac{1}{1 - a\hbar_x - b\hbar_y - c\hbar_x\hbar_y} \\ = & \sum_{k=0}^{\infty} (a\hbar_x + b\hbar_y + c\hbar_x\hbar_y)^k \\ = & \sum_{k=0}^{\infty} \sum_{\alpha+\beta+\gamma=k;\alpha,\beta,\gamma\in N} \binom{k}{\alpha,\beta,\gamma} a^{\alpha} b^{\beta} c^{\gamma} \hbar_x^{\alpha+\gamma} \hbar_y^{\beta+\gamma}, \end{aligned}$$

where

$$\binom{n}{\alpha,\beta,\gamma} = \frac{n!}{\alpha!\beta!\gamma!},\ \alpha,\beta,\gamma \in N, \alpha+\beta+\gamma = n,$$

is the trinomial coefficients. Similarly,

$$\begin{aligned}&\frac{-a\hbar_x}{1-a\hbar_x-b\hbar_y-c\hbar_x\hbar_y}\\ =\ &-a\hbar_x\sum_{k=0}^{\infty}\sum_{\alpha+\beta+\gamma=k;\alpha,\beta,\gamma\in N}\binom{n}{\alpha,\beta,\gamma}a^{\alpha_1}b^{\beta}c^{\gamma}\hbar_x^{\alpha+\gamma}\hbar_y^{\beta+\gamma},\end{aligned}$$

and

$$\begin{aligned}&\frac{-b\hbar_y}{1-a\hbar_x-b\hbar_y-c\hbar_x\hbar_y}\\ =\ &-b\hbar_y\sum_{k=0}^{\infty}\sum_{\alpha+\beta+\gamma=k;\alpha,\beta,\gamma\in N}\binom{n}{\alpha,\beta,\gamma}a^{\alpha_1}b^{\beta}c^{\gamma}\hbar_x^{\alpha+\gamma}\hbar_y^{\beta+\gamma}.\end{aligned}$$

Adding the above series, we see that f_{ij} is just the coefficient of $\hbar_x^i\hbar_y^j$ in the sum series. The unique solutions $g^{(\alpha)}$ and $h^{(\beta)}$ can be obtained in similar manners. We remark that the explicit form of $f, g^{(\alpha)}$ and $h^{(\beta)}$ can be given as follows:

$$f_{ij}=\sum_{s=1}^{i}C_{s-1}^{(i-1)}C_{i-1}^{(i+j-1-s)}a^{i-s}b^{j-s}c^{s},\ 1\le i\le j,$$

$$f_{ij}=\sum_{s=1}^{i}(-1)^{s}C_{s-1}^{(j-1)}C_{j-1}^{(i+j-1-s)}a^{i-s}b^{j-s}c^{s},\ 1\le j\le i,$$

$$g_{ij}^{(\alpha)}=0,\ 0\le i\le\alpha-1,$$

$$g_{ij}^{(\alpha)}=\sum_{s=0}^{i-\alpha}C_{s}^{(j)}C_{j-1}^{(i+j-1-\alpha-s)}a^{i-\alpha-s}b^{j-s}c^{s},\ \alpha\le i\le j+\alpha,$$

$$g_{ij}^{(\alpha)}=\sum_{s=0}^{j}C_{s}^{(j)}C_{j-1}^{(i+j-1-\alpha-s)}a^{i-\alpha-s}b^{j-s}c^{s},\ i\ge j+\alpha,$$

$$h_{ij}^{(\beta)}=0,\ 0\le j\le\beta-1,$$

$$h_{ij}^{(\beta)}=\sum_{s=0}^{j-\beta}C_{s}^{(i)}C_{i-1}^{(i+j-1-\beta-s)}a^{i-s}b^{j-\beta-s}c^{s},\ \beta\le j\le i+\beta,$$

$$h_{ij}^{(\beta)}=\sum_{s=0}^{i}C_{s}^{(i)}C_{i-1}^{(i+j-1-\beta-s)}a^{i-s}b^{j-\beta-s}c^{s},\ j\ge i+\beta.$$

The verification of the above formulas are tedious and a direct proof of them can be found in Cheng and Lin [34]. The proofs, however, do not offer any new insights and are skipped.

5.5 The Method of Separable Solutions

In Example 52, we have seen that separable solutions of the form $u_i^{(j)} = \lambda^i \mu^j$ may sometimes be found. Finding separable solutions may lead to general solutions. We give two examples. Consider the following partial difference equation

$$\Delta_2 u_{jk} = a\Delta_1^2 u_{j-1,k+1} + b\Delta_1^2 u_{j-1,k},\ j = 1, ..., J;\ k \in N, \tag{5.34}$$

subject to the conditions

$$u_{0k} = 0 = u_{J+1,k} = 0,\ k \in N, \tag{5.35}$$

and

$$u_{j0} = \psi_j,\ j = 0, 1, ..., J+1. \tag{5.36}$$

Assuming that a solution of (5.34) is of the form

$$u_{jk} = v_j w_k,$$

we see from (5.34) that

$$\frac{\Delta^2 v_{j-1}}{v_j} = \frac{\Delta w_k}{a w_{k+1} + b w_k}.$$

Thus we "must" have

$$\Delta^2 v_{j-1} = \rho v_j,\ j = 1, 2, ..., J, \tag{5.37}$$

$$v_0 = 0 = v_{J+1}, \tag{5.38}$$

and

$$w_{k+1} = \frac{1+\rho b}{1-\rho a} w_k,\ k \in N. \tag{5.39}$$

By treating (5.37)–(5.38) as a linear eigenvalue problem, we may invoke Theorem 16 and conclude that the eigenvalues are given by

$$\rho_{(m)} = -4\sin^2 \frac{m\pi}{2(J+1)},\ m = 1, 2, ..., J,$$

and the corresponding eigenvectors are given by

$$v_j^{(m)} = \sin \frac{m\pi j}{J+1},\ j = 1, 2, ..., J-1;\ m = 1, ..., J.$$

Furthermore, for each $\rho = \rho_{(m)}$, the solution of (5.39) is given by

$$w_n^{(m)} = w_0^{(m)} \sigma_{(m)}^n,\ n \in Z^+,$$

where

$$\sigma_{(m)} = \frac{1+\rho_{(m)} b}{1-\rho_{(m)} a},\ m = 1, 2, ..., J.$$

Since the equation (5.34) is linear, the formal approach given above suggests, and as may be verified directly, that the general solution of (5.34)–(5.36) is given by

$$u_{jk} = \sum_{m=1}^{J} c_m \sigma_{(m)}^k \sin \frac{m\pi j}{J+1},\ j = 0, 1, ..., J+1;\ k \in N,$$

where the constants c_m are given by

$$c_m = \frac{2}{J+1}\sum_{j=1}^{J} \psi_j \sin\frac{m\pi j}{J+1}, \ m = 1, ..., J,$$

obtained by solving the system of equations (5.36), that is,

$$\psi_j = \sum_{k=1}^{J} c_m \sin\frac{m\pi j}{J+1}, \ j = 1, ..., J.$$

Next, let us try to find solutions to equation

$$u_m^{(t+1)} = au_{m-1}^{(t)} + bu_m^{(t)} + cu_{m+1}^{(t)} + du_m^{(t-1)}, \ m \in Z, t \in N, \tag{5.40}$$

in the following form

$$u_m^{(t)} = \alpha_t\beta_m, \ m \in Z, t \in N.$$

Substituting $u_m^{(t)}$ into (5.40), we obtain

$$\alpha_{t+1}\beta_m = a\alpha_t\beta_{m-1} + b\alpha_t\beta_m + c\alpha_t\beta_{m+1} + d\alpha_{t-1}\beta_m.$$

Proceeding formally, we divide this equation through by $\alpha_t\beta_m$, and obtain

$$\frac{\alpha_{t+1}}{\alpha_t} - d\frac{\alpha_{t-1}}{\alpha_t} = a\frac{\beta_{m-1}}{\beta_m} + b + c\frac{\beta_{m+1}}{\beta_m}.$$

In case

$$a\frac{\beta_{m-1}}{\beta_m} + b + c\frac{\beta_{m+1}}{\beta_m} = \lambda,$$

that is,

$$a\beta_{m-1} + b\beta_m + c\beta_{m+1} = \lambda\beta_m, \tag{5.41}$$

holds for some constant λ and some corresponding sequence $\{\beta_m\}$, then the solution $\{\alpha_t\}$ of the equation

$$\alpha_{t+1} = \lambda\alpha_t + d\alpha_{t-1} \tag{5.42}$$

coupled with the sequence $\{\beta_m\}$ will yield a desired solution $\{\alpha_t\beta_m\}$ of (5.40) which satisfies the initial conditions

$$u_m^{(0)} = \alpha_0\beta_m, \ m \in Z,$$

and

$$u_m^{(1)} = \alpha_1\beta_m, \ m \in Z.$$

At this point, we are not sure about the general solutions of the "infinite eigenvalue problem" (5.41). However, in case $\beta_m = \beta^m$ for some nontrivial number β, then from (5.41), we see that

$$\lambda = \frac{a}{\beta} + b + c\beta,$$

and (5.42) takes the form

$$\alpha_{t+1} = \left(\frac{a}{\beta} + b + c\beta\right)\alpha_t + d\alpha_{t-1}. \tag{5.43}$$

We summarize these in the following theorem, which can now be proved by direct verification.

Theorem 65 *The solution of (5.40) subject to the initial conditions*

$$u_m^{(0)} = \alpha_0\beta^m,\ m \in Z, \tag{5.44}$$

and

$$u_m^{(1)} = \alpha_1\beta^m,\ m \in Z, \tag{5.45}$$

where β is a nonzero number, is of the form $\{u_m^{(t)}\} = \{\alpha_t\beta^m\}$, where the first two terms α_0, α_1 of the sequence $\{\alpha_t\}_{t=0}^{\infty}$ are the numbers in (5.44) and (5.45) respectively, and α_t is defined by (5.43) for $t = 2, 3, \ldots$.

In other words, when given initial conditions of the form (5.44) and (5.45), we need to solve the three-term recurrence relation (5.43) to obtain $\{\alpha_t\}$ and then form the solution $\{\alpha_t\beta^m\}$. Solving linear three-term difference equations with constant coefficients is a routine matter and can be found in most text books or we can employ the theory of operators in Chapter 4 to solve them. We will therefore turn to two specific examples.

As our first example, consider the case where the coefficients in (5.40) satisfy $a = 1, b = 2, c = -2$ and $d = 1$. Furthermore, if we take $\beta = \alpha_0 = \alpha_1 = 1$ so that the initial conditions take on the form

$$u_m^{(0)} = 1 = u_m^{(1)},\ m \in Z,$$

then equation (5.43) reduces to

$$\alpha_{t+1} = \alpha_t + \alpha_{t-1},\ t \in Z^+.$$

and is subject to the initial conditions $\alpha_0 = \alpha_1 = 1$. The corresponding solution sequence is the well known sequence of Fibonacci numbers!

As another example, consider the Richardson's partial difference scheme

$$u_m^{(t+1)} = ru_{m-1}^{(t)} - 2ru_m^{(t)} + ru_{m+1}^{(t)} + u_m^{(t-1)},\ r > 0,\ t \in N, m \in Z, \tag{5.46}$$

subject to the initial conditions

$$u_m^{(0)} = (-1)^m,\ m \in Z, \tag{5.47}$$

and

$$u_m^{(1)} = (-1)^{m+1},\ m \in Z. \tag{5.48}$$

Equation (5.43) now takes the form

$$\alpha_{t+1} = -4r\alpha_t + \alpha_{t-1},\ t \in Z^+,$$

and is subject to the initial conditions $\alpha_0 = 1, \alpha_1 = -1$. The corresponding solution is given by

$$\begin{aligned}\alpha_t &= \left(\frac{1}{2} + \frac{(2r-1)\sqrt{4r^2+1}}{2(4r^2+1)}\right)\left(-2r + \sqrt{4r^2+1}\right)^t \\ &\quad + \left(\frac{1}{2} - \frac{(2r-1)\sqrt{4r^2+1}}{2(4r^2+1)}\right)\left(-2r - \sqrt{4r^2+1}\right)^t\end{aligned}$$

for $t \in N$. Thus the unique solution of the Richardson's partial difference scheme (5.46) subject to the initial conditions (5.47) and (5.48) is given by

$$u_m^{(t)} = (-1)^m \left(\frac{1}{2} + \frac{(2r-1)\sqrt{4r^2+1}}{2(4r^2+1)}\right)\left(-2r + \sqrt{4r^2+1}\right)^t$$

$$+(-1)^m \left(\frac{1}{2} - \frac{(2r-1)\sqrt{4r^2+1}}{2(4r^2+1)}\right)\left(-2r - \sqrt{4r^2+1}\right)^t$$

for $m \in Z$ and $t \in N$.

5.6 The Method of Convolution

When the domain of definition of a partial difference equation is the half plane, it is sometimes useful to think of the equation as a functional relation between sequences in l^Z. In this section, we will show by means of two examples how multi-level discrete heat equations can be viewed as convolution equations for sequences in l^Z, and how their general solutions can be obtained. We will also illustrate how qualitative as well as quantitative properties of our equations can be derived by analyzing their general solutions.

5.6.1 Two-Level Equations over the Upper Half Lattice Plane

Let us first consider the two-level discrete heat equation

$$u_m^{(n+1)} = au_{m-1}^{(n)} + bu_m^{(n)} + cu_{m+1}^{(n)},\ m \in Z, n \in N \tag{5.49}$$

subject to the initial condition

$$u_m^{(0)} = f_m,\ m \in Z, \tag{5.50}$$

and look for its general solution. Recall first that an existence and uniqueness theorem has been established by recursion, and that $u_m^{(n)}$ is completely determined by f_k, $m-n \le k \le m+n$. For each fixed $n \in N$, we will view $\left\{u_m^{(n)}\right\}_{m\in Z}$ as a vector in l^Z. This vector, denoted by $u^{(n)}$, will be called the n-th horizontal vector of the solution u. Next, note that we may also write (5.49) as

$$\{u_m\}^{(n+1)} = a\{u_{m-1}\}^{(n)} + b\{u_m\}^{(n)} + c\{u_{m+1}\}^{(n)},\ n \in N. \tag{5.51}$$

Recall that $\hat{h}^{[i]} = \{\hat{h}_m^{[i]}\}_{m\in Z}$ has been defined by $\hat{h}_m^{[i]} = 1$ if $i = m$ and $\hat{h}_m^{[i]} = 0$ if $i \neq m$. Recall further that for any two sequences $x = \{x_m\}$ and $y = \{y_m\}$ in l^Z, the convolution $x * y$ is given by

$$(x * y)_m = \sum_{i=-\infty}^{\infty} x_i y_{m-i},\ m \in Z,$$

whenever the infinite sums converge. Then since

$$\{u_{m+1}\} = \hat{h}^{[-1]} * \{u_m\}$$

and

$$\{u_{m-1}\} = \hat{h}^{[1]} * \{u_m\},$$

we can rewrite (5.51) as

$$u^{(n+1)} = (a\hat{h}^{[1]} + b\hat{h}^{[0]} + c\hat{h}^{[-1]}) * u^{(n)},\ n \in N, \tag{5.52}$$

where $u = \{u_m\}_{m\in Z}$ and (5.50) as

$$u^{(0)} = f, \tag{5.53}$$

where $f = \{f_m\}_{m\in Z}$. The solution of (5.52),(5.53) is easily obtained by induction and is given by

$$u^{(n)} = (a\hat{h}^{[1]} + b\hat{h}^{[0]} + c\hat{h}^{[-1]})^n * f,\ n \in Z^+, \tag{5.54}$$

which shows that $u_m^{(n)}$ is given by

$$u_m^{(n)} = \sum_{k=-\infty}^{\infty} (a\hat{h}^{[1]} + b\hat{h}^{[0]} + c\hat{h}^{[-1]})_k^n f_{m-k},\ n \in Z^+; m \in Z. \tag{5.55}$$

We remark that for each $n \geq 1$, since the convolution product $(a\hat{h}^{[1]} + b\hat{h}^{[0]} + c\hat{h}^{[-1]})^n$ has only a finite number of nonzero terms, $u^{(n)}$ is well defined. Thus, we have found the general solution of our equation. A special solution, however, that stands out is the Green's function. A solution of (5.49) will be called the Green's function of (5.49) if it satisfies the initial condition

$$u_m^{(0)} = \hat{h}_m^{[0]},\ m \in Z. \tag{5.56}$$

Such a solution (exists and is unique and) will be denoted by $G = \{G_m^{(n)}\}$. If we substitute f with $\hat{h}^{[0]}$ in (5.54), then we see that

$$G^{(n)} = (a\hat{h}^{[1]} + b\hat{h}^{[0]} + c\hat{h}^{[-1]})^n,\ n \in N. \tag{5.57}$$

In other words, we have the following.

Theorem 66 *The Green's function $G^{(n)}$ of (5.49) is equal to the convolution product $(a\hat{h}^{[1]} + b\hat{h}^{[0]} + c\hat{h}^{[-1]})^n$. The (m,n)-th component $G_m^{(n)}$ of the Green's function $G = \{G_m^{(n)}\}$ is the coefficient of the term $\hat{h}^{[m]}$ in the expansion of the convolution product $(a\hat{h}^{[1]} + b\hat{h}^{[0]} + c\hat{h}^{[-1]})^n$.*

In terms of the Green's function, we can rewrite (5.55) as

$$u_m^{(n)} = \sum_{k=-\infty}^{\infty} G_k^{(n)} f_{m-k} = \sum_{k=-n}^{n} G_k^{(n)} f_{m-k},\ n = 1, 2, ...; m \in Z.$$

In other words, the n-th horizontal vector of the solution u is the convolution of the n-th horizontal vector of the Green's function with the initial vector f.

Next, note that the algebraic operations involved in the expansion of the expression $(ax+b+cx^{-1})^n$ is "isomorphic" to the operations involved in the expansion of $(a\hat{h}^{[1]} + b\hat{h}^{[0]} + c\hat{h}^{[-1]})^n$. There is a certain advantage, however, in considering the algebraic expression $\left(ax + b + cx^{-1}\right)^n$, since we can formally "substitute" x with an arbitrary number other than zero and obtain several useful formulas.

Example 59 *If $b = 0$, then by means of*

$$\left(ax + cx^{-1}\right)^n = \sum_{k=0}^{n} C_k^{(n)} a^k c^{n-k} x^{-n+2k},$$

we see that $G_m^{(n)} = 0$ when $m + n$ is odd and

$$G_m^{(n)} = C_{(n-|m|)/2}^{(n)} a^{(n+m)/2} c^{(n-m)/2}, \ |m| \le n$$

when $m + n$ is even.

Example 60 *If $a = c$ and $b = 2a$, then by means of the equalities*

$$\left(ax + 2a + ax^{-1}\right)^n = a^n \left(x + 2 + x^{-1}\right)^n = a^n \left(\sqrt{x} + \frac{1}{\sqrt{x}}\right)^{2n},$$

it is easily seen that $G_m^{(n)} = C_{n-m}^{(2n)} a^n$, $|m| \le n$.

Example 61 *If $a = 0$, then*

$$G_m^{(n)} = C_{-m}^{(n)} b^{m+n} c^{-m}, \ m \le 0, \ |m| \le n,$$

and $G_m^{(n)} = 0$ otherwise.

Example 62 *If $c = 0$, $a = b = 1$, then*

$$G_m^{(n)} = C_m^{(n)}, \ |m| \le n.$$

This shows that the binomial coefficient is a Green's function of the initial value problem

$$u_m^{(n+1)} = u_{m-1}^{(n)} + u_m^{(n)}, \ (m, n) \in \Omega,$$
$$u^{(0)} = \hat{h}^{[0]}.$$

Example 63 *When $a = c \neq 0$ and $b = 1 - 2a$, then*

$$\begin{aligned}
\left(ax - 2a + ax^{-1} + 1\right)^n &= \left(a\left(x - 2 + x^{-1}\right) + 1\right)^n \\
&= \left(a\left(x^{1/2} - x^{-1/2}\right)^2 + 1\right)^n \\
&= \sum_{k=0}^{n} C_k^{(n)} a^k \left(x^{1/2} - x^{-1/2}\right)^{2k} \\
&= \sum_{k=0}^{n} \sum_{i=0}^{2k} (-1)^i C_k^{(n)} C_i^{(2k)} a^k x^{k-i},
\end{aligned}$$

which may be regarded as a sum of values each of which corresponds to a point in the subset $\{(i, k) \mid 0 \le i \le 2k, \ 0 \le k \le n\}$ of lattice points. To find $G_0^{(n)}$, we need *to look for the coefficient of the term x^0 in the last sum. But this is just the sum of the values corresponding to the subset $\{(i,k)|\ i=k, 0 \le k \le n\}$:*

$$G_0^{(n)} = \sum_{k=0}^{n} (-1)^k C_k^{(n)} C_k^{(2k)} a^k.$$

Similarly, we may calculate $G_m^{(n)}$ when $m \neq 0$ and obtain the Green's function

$$G_m^{(n)} = \sum_{k=|m|}^{n} (-1)^{k-m} C_k^{(n)} C_{k-m}^{(2k)} a^k, \ |m| \le n.$$

Theorem 67 *The Green's function $G = \{G_m^{(n)}\}$ of (5.49) is given by*

$$G_m^{(n)} = \sum_{i=|m|}^{n} \delta^{(i+m)} C_i^{(n)} C_{(i-|m|)/2}^{(i)} a^{(i+m)/2} b^{n-i} c^{(i-m)/2}, \qquad (5.58)$$

where

$$\delta^{(k)} = \frac{1+(-1)^k}{2}, \ k \in Z. \qquad (5.59)$$

The proof makes use of the expansions

$$\begin{aligned} \left(ax+b+cx^{-1}\right)^n &= \sum_{i=0}^{n} C_i^{(n)} b^{n-i} \left(ax+cx^{-1}\right)^i \\ &= \sum_{i=0}^{n} \sum_{k=0}^{i} C_i^{(n)} C_k^{(i)} a^{i-k} b^{n-i} c^k x^{i-2k}, \end{aligned}$$

and

$$\left(ax+b+cx^{-1}\right)^n = \sum_{m=-n}^{n} G_m^{(n)} x^m,$$

and then utilizes the same reasoning used in Example 63.

There are several immediate consequences of Theorem 66. First, we see that each horizontal vector of the Green's function is nontrivial when one of the coefficients a, b and c is nonzero. Second, if $a = c$, then since the coefficients of the terms x^m and x^{-m} are identical, we see that

$$G_m^{(n)} = G_{-m}^{(n)}, \ (m,n) \in \Omega.$$

Third, it is clear by substituting $x = 1$ and $x = -1$ into the rational function $(ax+b+cx^{-1})^n$ that

$$\sum_{k=-n}^{n} G_k^{(n)} = (a+b+c)^n, \qquad (5.60)$$

and

$$\sum_{k=-n}^{n} G_k^{(n)} (-1)^k = \sum_{k=-n}^{n} G_k^{(n)} (-1)^{-k} = (-1)^n (a-b+c)^n. \qquad (5.61)$$

In particular, if $a = c$ and $b = 1 - 2a$, then

$$\sum_{i=-n}^{n} G_i^{(n)} = 1.$$

This result may be regarded as a discrete analog of Theorem 2.D in Widder [161] which states that the integral of the source solution of the heat equation (1.10) over the space domain is one.

There are several immediate consequences of (5.55) also. For instance, if $f = \{1\}_{m\in Z}$, then the solution $\left\{u_m^{(n)}\right\}$ of (5.49) subject to (5.50) is given by

$$u_m^{(n)} = \sum_{k=-n}^{n} G_k^{(n)} = (a+b+c)^n.$$

If $f = \{(-1)^m\}_{m\in Z}$, then the solution $\left\{u_m^{(n)}\right\}$ of (5.49) subject to (5.50) is given by

$$u_m^{(n)} = \sum_{k=-n}^{n} G_k^{(n)}(-1)^{m-k} = (-1)^{m-n}(a-b+c)^n.$$

The next result states that the convolution of two different horizontal vectors of the Green's function is again a horizontal vector of the same Green's function. Its proof is a direct consequence of (5.57).

Theorem 68 *The Green's function $G = \{G_m^{(n)}\}$ of (5.49) satisfies the following "additivity" property:*

$$\{G_m^{(s)}\}_{m\in Z} * \{G_m^{(t)}\}_{m\in Z} = \{G_m^{(s+t)}\}_{m\in Z}.$$

As we will see below, there are several monotonicity properties of the Green's function.

Theorem 69 *Suppose $a = c \neq 0$. Then the Green's function $G = \{G_m^{(n)}\}$ satisfies*

$$G_{m+1}^{(n+1)} \geq G_m^{(n)}, \quad m, n \in N, \tag{5.62}$$

if, and only if, $a \geq 1$ and $b \geq 0$.

Proof. Assuming that $a = c \geq 1$ and $b \geq 0$, then for $m, n \geq 0$, we see that $G_{m+1}^{(n+1)} - G_m^{(n)}$ is equal to

$$\begin{aligned}
&\sum_{k=m+1}^{n+1} \delta^{(k+m+1)} C_k^{(n+1)} C_{(k-m-1)/2}^{(k)} a^k b^{n+1-k} \\
&- \sum_{k=m}^{n} \delta^{(k+m)} C_k^{(n)} C_{(k-m)/2}^{(k)} a^k b^{n-k} \\
= \ &\sum_{k=m}^{n} \delta^{(k+m)} C_{k+1}^{(n+1)} C_{(k-m)/2}^{(k+1)} a^{k+1} b^{n-k} - \sum_{k=m}^{n} \delta^{(k+m)} C_k^{(n)} C_{(k-m)/2}^{(k)} a^k b^{n-k}
\end{aligned}$$

$$\begin{aligned}
&= \sum_{k=m}^{n} \delta^{(k+m)} \left\{ C_k^{(n+1)} C_{(k-m)/2}^{(k+1)} a - C_k^{(n)} C_{(k-m)/2}^{(k)} \right\} a^k b^{n-k} \\
&\geq \sum_{k=m}^{n} \delta^{(k+m)} \left\{ C_k^{(n+1)} C_{(k-m)/2}^{(k+1)} - C_k^{(n)} C_{(k-m)/2}^{(k)} \right\} a^k b^{n-k} \\
&\geq 0.
\end{aligned}$$

Conversely, there are two cases to consider. Suppose first that $a < 1$. Then $G_m^{(m)} = a^m$ so that the statement $G_{m+1}^{(m+1)} \geq G_m^{(m)}$ cannot be satisfied for all $m \geq 0$. Next suppose $a \geq 1$ and $b < 0$. Then $G_m^{(m+1)} = (m+1)ba^m$ so that the statement that $G_{m+1}^{(m+2)} \geq G_m^{(m+1)}$ cannot be satisfied for all $m \geq 0$. The proof is complete. ■

We remark that when $a = c$, the Green's function $G_m^{(n)}$ satisfies $G_m^{(n)} = G_{-m}^{(n)}$. Thus the condition (5.62) in the above theorem can be replaced by the condition

$$G_{-m-1}^{(n+1)} \geq G_{-m}^{(n)}, \ m, n \in N.$$

As an immediate consequence, when $a = c \geq 1$ and $b \geq 0$, then the maximum of the Green's function over a lattice rectangle $\{\alpha, \alpha+1, ..., \beta\} \times \{\sigma, \sigma+1, ..., \tau\}$ occurs at the "lateral edges" of the rectangle. This result can be regarded as a maximum principle for equation (5.49).

It is clear from the representation (5.58) that G is nonnegative when the coefficients a, b and c are nonnegative. In case some of the coefficients a, b and c are negative, the Green's function will have negative values and a variety of oscillatory behaviors can also be shown.

Theorem 70 *The Green's function of (5.49) is nonnegative if, and only if,* $a, b, c \geq 0$.

Proof. If $b < 0$ and $a = c = 0$, then $G_0^{(n)} = b^n$ for $n \geq 0$ so that $G_0^{(1)} = b < 0$. If $b < 0$, $a > 0$ and $c \geq 0$, then $G_m^{(m+1)} = (m+1)ba^m < 0$ for $m \geq 0$. If $a < 0$, $b > 0$ and $c \geq 0$, then $G_m^{(m)} = a^m$ for $m \geq 0$ so that $G_1^{(1)} = a < 0$. The other cases are dealt with in a similar manner. ■

Next, we assert that when $a + b + c < 0$, adjacent horizontal vectors of the Green's function must have components with different signs.

Theorem 71 *If* $a + b + c < 0$, *then there exist integers* m_1 *and* m_2 *such that*

$$G_{m_1}^{(n)} G_{m_2}^{(n+1)} < 0, \ n \in N.$$

Proof. Since

$$(a+b+c)^n (a+b+c)^{n+1} < 0, \ n \in N,$$

thus by Theorem 5.66,

$$\left\{ \sum_{k=-n}^{n} G_k^{(n)} \right\} \left\{ \sum_{k=-(n+1)}^{n+1} G_k^{(n+1)} \right\} < 0,$$

which implies our assertion. ■

We may be more specific when $a, c < 0$ and $b \leq 0$.

Theorem 72 *Suppose $a, c < 0$ and $b \le 0$. If $G_m^{(n)} \neq 0$ for some lattice point (m, n), then $G_m^{(n)} > 0$ when n is even and $G_m^{(n)} < 0$ when n is odd.*

Proof. First suppose $a, b, c < 0$. Then by the representation (5.58),

$$G_m^{(n)} = (-1)^n \sum_{i=|m|}^{n} \delta^{(i+m)} C_i^{(n)} C_{(i-|m|)/2}^{(i)} |a|^{(i+m)/2} |b|^{n-i} |c|^{(i-m)/2}.$$

Thus if $G_m^{(n)}$ is not zero, then $G_m^{(n)}$ is positive when n is even and negative when n is odd. Next suppose $a, c < 0$ and $b = 0$. Since (see Example 5.59)

$$G_m^{(n)} = C_{(n-|m|)/2}^{(n)} a^{(n+m)/2} c^{(n-m)/2} = (-1)^n C_{(n-|m|)/2}^{(n)} |a|^{(n+m)/2} |c|^{(n-m)/2}$$

when $m + n$ is even and $|m| \le n$, and zero otherwise, the same conclusions may again be drawn. ■

We remark that for a slightly more general two-level equation of the form

$$u_m^{(t)} = a_{-\sigma} u_{m-\sigma}^{(t-\mu)} + a_{-\sigma+1} u_{m-\sigma+1}^{(t-\mu)} + \ldots + a_\tau u_{m+\tau}^{(t-\mu)} + g_m^{(t)}, \tag{5.63}$$

where μ, σ, τ are positive integers, $\tilde{a} = (a_{-\sigma}, a_{-\sigma+1}, \ldots, a_\tau)$ is a real vector, $g = \left\{g_m^{(t)}\right\}$ is a real bivariate sequence, $t \in N$ and $m \in Z$, the same technique of convolution will work. Indeed, when subsidiary initial conditions of the form

$$u_m^{(t)} = \phi_m^{(t)}, \quad m \in Z;\ t = -\mu, -\mu+1, \ldots, -1, \tag{5.64}$$

are given, we only need to note that equation (5.63) can be written as

$$u^{(t)} = (a_{-\sigma} \hat{h}^{[\sigma]} + \ldots + a_\tau \hat{h}^{[-\tau]}) * u^{(t-\mu)} + g^{(t)}, \quad t \in N,$$

and (5.64) as

$$u^{(t)} = \phi^{(t)}, \quad t = -\mu, -\mu+1, \ldots, -1,$$

where $u = \{u_m\}$, $g^{(t)} = \left\{g_m^{(t)}\right\}$ and $\phi^{(t)} = \left\{\phi_m^{(t)}\right\}$. In case $\mu = 1$, the corresponding solution is easily obtained by induction and is given by

$$u^{(t)} = U^{(t+1)} * \phi^{(-1)} + \sum_{i=0}^{t} U^{(i)} * g^{(t-i)}, \ t \in N,$$

where

$$U \equiv U^{(1)} = a_{-\sigma} \hat{h}^{[-\sigma]} + \ldots + a_\tau \hat{h}^{[-\tau]}, \ U^{(0)} = \hat{h}^{[0]}.$$

5.6.2 Three-Level Equations over the Upper Half Lattice Plane

When a "delayed" control mechanism of the form $du_m^{(t-1)}$ is introduced into our two-level discrete heat equation discussed in the last section, and assuming further that additional heat sources or sinks are provided at each integral point m, then the corresponding equation may look like

$$u_m^{(t+1)} = au_{m-1}^{(t)} + bu_m^{(t)} + cu_{m+1}^{(t)} + du_m^{(t-1)} + p_m^{(t)}, \tag{5.65}$$

where $t \in Z^+$ and $m \in Z$. To avoid trivial cases, we will assume throughout this section that $|a| + |b| + |c| \neq 0$. We will also make the assumption that $|a| + |c| \neq 0$. This is due to the fact that our equation will essentially be the same as the ordinary difference equation

$$u^{(t+1)} = bu^{(t)} + du^{(t-1)} + p^{(t)},\ t \in Z^+.$$

For the sake of convenience, in this section, we will denote the upper lattice plane by

$$\Omega = \{(i,j)|\ i \in Z; j \in Z^+\}.$$

When subsidiary initial conditions of the form

$$u_m^{(0)} = f_m,\ m \in Z, \tag{5.66}$$

and

$$u_m^{(1)} = g_m,\ m \in Z, \tag{5.67}$$

are given, a bivariate sequence $\{u_m^{(t)}\}$ can be calculated successively in the following unique manner:

$$u_0^{(2)}; u_{-1}^{(2)}, u_1^{(2)}; u_0^{(3)}, u_{-2}^{(2)}, u_{-1}^{(3)}, u_2^{(2)}, u_1^{(3)}; u_0^{(4)}, u_{-3}^{(2)}, \ldots$$

Such a unique sequence together with the sequences $\{u_m^{(0)}\}$ and $\{u_m^{(1)}\}$ constitutes a solution of the initial value problem (5.65)–(5.66) and is denoted by $\{u_m^{(t)}|\ m \in Z, t \in N\}$ or $\{u_m^{(t)}\}$.

Even though solutions of (5.65) can be calculated successively, it is of interest to find an explicit form for all of them, that is, its general solution. To obtain these solutions, we will follow closely the concepts developed in the last section. First of all, let us note that equation (5.65) can be written as

$$\{u_m\}^{(t+1)} = a\{u_{m-1}\}^{(t)} + b\{u_m\}^{(t)} + c\{u_{m+1}\}^{(t)} + d\{u_m\}^{(t-1)} + \{p_m\}^{(t)}$$

where $\{u_m\}$ stands for the sequence $\{..., u_{-1}, u_0, u_1, ...\}$, that is,

$$u^{(t+1)} = (a\hat{h}^{[1]} + b\hat{h}^{[0]} + c\hat{h}^{[-1]}) * u^{(t)} + du^{(t-1)} + p^{(t)},\ t \in Z^+,$$

where $u^{(t)} = \{u_m\}^{(t)}$, and $p^{(t)} = \{p_m^{(t)}\}$.

As we have seen in the last section, the concept of a Green's solution is important. Here we have similar concepts. A solution of the homogeneous equation

$$u_m^{(t+1)} = au_{m-1}^{(t)} + bu_m^{(t)} + cu_{m+1}^{(t)} + du_m^{(t-1)},\ (m,t) \in \Omega \tag{5.68}$$

will be called its $(\hat{h}^{[0]}, 0)$-Green's function or the *principal* Green's function if this solution satisfies the initial conditions $u^{(0)} = \hat{h}^{[0]}$ and $u^{(1)} = 0$. A solution of (5.68) is called its $(0, \hat{h}^{[0]})$-Green's function if this solution satisfies the initial conditions $u^{(0)} = 0$ and $u^{(1)} = \hat{h}^{[0]}$. These two solutions exist and are unique and will be denoted by $\tilde{G} = \{\tilde{G}_m^{(t)}\}$ and $K = \{K_m^{(t)}\}$ respectively.

In some special cases the Green's functions can easily be obtained. In the general case, an explicit form of these functions seems to be difficult. We will, however, develop a formal method for deriving the explicit forms of the Green's functions. Since

$$\tilde{G}^{(t+1)} = \delta * \tilde{G}^{(t)} + d\tilde{G}^{(t-1)},\ t \in Z^+, \tag{5.69}$$

where

$$\delta = a\hat{h}^{[1]} + b\hat{h}^{[0]} + c\hat{h}^{[-1]},$$

we suspect that $\tilde{G}^{(t)}$ is of the form λ^t, where $\lambda \in l^Z$. Substituting $\tilde{G}^{(t)}$ by λ^t in (5.69), we obtain a "characteristic" equation

$$\lambda^2 - \delta * \lambda - d\hat{h}^{[0]} = 0.$$

Proceeding formally, we would "solve" for the roots of this equation and obtain formal roots of the form

$$\lambda_\pm = \frac{\delta \pm \sqrt{\delta^2 + 4d\hat{h}^{[0]}}}{2}.$$

Note that the condition

$$\delta^2 + 4d\hat{h}^{[0]} = 0$$

is equivalent to

$$a^2\hat{h}^{[2]} + 2ab\hat{h}^{[1]} + (b^2 + 2ac + 4d)\hat{h}^{[0]} + 2bc\hat{h}^{[-1]} + c^2\hat{h}^{[-2]} = 0,$$

that is,

$$a = c = 0 \text{ and } b^2 + 4d = 0. \tag{5.70}$$

Since we have assumed that $|a| + |c| \neq 0$, the roots λ_+ and λ_- are distinct. The corresponding solution $\tilde{G}^{(t)}$ is of the form

$$\tilde{G}^{(t)} = A * \lambda_+^t + B * \lambda_-^t.$$

In view of the initial conditions $\tilde{G}^{(0)} = \hat{h}^{[0]}$ and $\tilde{G}^{(1)} = 0$, we see further that

$$\tilde{G}^{(t)} = \frac{d\lambda_+^{t-1} - d\lambda_-^{t-1}}{\lambda_+ - \lambda_-},\ t = 2, 3, \dots .$$

The above quotient is only formal, however, if we write $\lambda_+ = u + v$ and $\lambda_- = u - v$, then by means of the binomial theorem,

$$\tilde{G}^{(t)} = d \sum_{k=0}^{\lfloor (t-2)/2 \rfloor} C_{2k+1}^{(t-1)} u^{t-2-2k} (v^2)^k,\ t = 2, 3, \dots .$$

Note that

$$u = \frac{\delta}{2}, v^2 = (\lambda_+ - \lambda_-)^2 = \frac{1}{4}(\delta^2 + 4d\hat{h}^{[0]}),$$

thus we have

$$\tilde{G}^{(t)} = \frac{d}{2^{t-2}} \sum_{k=0}^{\lfloor (t-2)/2 \rfloor} C_{2k+1}^{(t-1)} \delta^{t-2-2k} (\delta^2 + 4d\hat{h}^{[0]})^k,\ t \geq 2. \tag{5.71}$$

By means of direct verification, we may show that the explicit forms obtained above are indeed the $(\hat{h}^{[0]}, 0)$-Green's solution of (5.69). We record this fact in the following result.

Theorem 73 *The unique solution $\{\tilde{G}^{(t)}\}_{t=0}^{\infty}$ of the equation (5.69) subject to the initial conditions $\tilde{G}^{(0)} = \hat{h}^{[0]}$ and $\tilde{G}^{[1]} = 0$ is given by (5.71).*

Once Theorem 73 is checked by direct verification, we may find the explicit form of the solutions of (5.68) through binomial expansions of the terms in (5.71). Indeed,

$$\begin{aligned}
\tilde{G}^{(t)} &= \frac{d}{2^{t-2}} \sum_{k=0}^{\lfloor (t-2)/2 \rfloor} C_{2k+1}^{(t-1)} \delta^{t-2-2k} \sum_{i=0}^{k} C_i^{(k)} \delta^{2k-2i} (4d\hat{h}^{[0]})^i \\
&= \frac{d}{2^{t-2}} \sum_{k=0}^{\lfloor (t-2)/2 \rfloor} \sum_{i=0}^{k} C_{2k+1}^{(t-1)} C_i^{(k)} 2^{2i} d^i \delta^{t-2-2i} \\
&= \frac{d}{2^{t-2}} \sum_{k=0}^{\lfloor (t-2)/2 \rfloor} \sum_{i=0}^{k} C_{2k+1}^{(t-1)} C_i^{(k)} 2^{2i} d^i \\
&\quad \times \sum_{j=0}^{t-2-2i} C_j^{(t-2-2i)} (a\hat{h}^{[1]} + c\hat{h}^{[-1]})^j (b\hat{h}^{[0]})^{t-2-2i-j} \\
&= \sum_{k=0}^{\lfloor (t-2)/2 \rfloor} \sum_{i=0}^{k} \sum_{j=0}^{t-2-2i} C_{2k+1}^{(t-1)} C_i^{(k)} C_j^{(t-2-2i)} \left(\frac{1}{2}\right)^{t-2-2i} d^{i+1} b^{t-2-2i-j} \\
&\quad \times \sum_{m=-j}^{j} C_{(j-m)/2}^{(j)} a^{(j+m)/2} c^{(j-m)/2} \hat{h}^{[m]}.
\end{aligned}$$

By changing the order of summation in the following manner:

$$\begin{aligned}
\sum_{k=0}^{\lfloor (t-2)/2 \rfloor} \sum_{i=0}^{k} \sum_{j=0}^{t-2-2i} \sum_{m=-j}^{j} &= \sum_{k=0}^{\lfloor (t-2)/2 \rfloor} \sum_{i=0}^{k} \sum_{m=-(t-2-2i)}^{t-2-2i} \sum_{j=|m|}^{t-2-2i} \\
&= \sum_{m=-t+2}^{t-2} \sum_{i=0}^{\lfloor (t-2-|m|)/2 \rfloor} \sum_{k=i}^{\lfloor (t-2)/2 \rfloor} \sum_{j=|m|}^{t-2-2i},
\end{aligned}$$

we obtain further that

$$\begin{aligned}
\tilde{G}^{(t)} &= \sum_{m=-t+2}^{t-2} \sum_{i=0}^{\lfloor (t-2-|m|)/2 \rfloor} \sum_{k=i}^{\lfloor (t-2)/2 \rfloor} \sum_{j=|m|}^{t-2-2i} C_{2k+1}^{(t-1)} C_i^{(k)} C_j^{(t-2-2i)} \\
&\quad \times C_{(j-m)/2}^{(j)} \left(\frac{1}{2}\right)^{t-2-2i} a^{(j+m)/2} b^{t-2-2i-j} c^{(j-m)/2} d^{i+1} \hat{h}^{[m]}.
\end{aligned}$$

As a consequence, we see that when $-t+2 \le m \le t-2$,

$$\begin{aligned}
\tilde{G}_m^{(t)} &= \sum_{k=0}^{\lfloor (t-2-|m|)/2 \rfloor} \sum_{k=i}^{\lfloor (t-2)/2 \rfloor} \sum_{j=|m|}^{t-2-2i} C_{2k+1}^{(t-1)} C_i^{(k)} C_j^{(t-2-2i)} \\
&\quad \times C_{(j-m)/2}^{(j)} \left(\frac{1}{2}\right)^{t-2-2i} a^{(j+m)/2} b^{t-2-2i-j} c^{(j-m)/2} d^{i+1}, \qquad (5.72)
\end{aligned}$$

and $\tilde{G}_m^{(t)} = 0$ otherwise.

As an example, consider the equation

$$u_m^{(t+1)} = u_{m-1}^{(t)} + u_{m+1}^{(t)} - u_m^{(t-1)},\ (t,m) \in \Omega,$$

where we have taken $a = c = 1, b = 0$ and $d = -1$ in (5.68). The corresponding $(\hat{h}^{[0]}, 0)$-Green's function is given by

$$\tilde{G}_m^{(t)} = \sum_{i=0}^{\lfloor (t-2-|m|)/2 \rfloor} \sum_{k=i}^{\lfloor (t-2)/2 \rfloor} C_{2k+1}^{(t-1)} C_i^{(k)} C_{(t-2-2i-m)/2}^{(t-2-2i)} \frac{(-1)^{i+1}}{2^{t-2-2i}}$$

when $-t+2 \le m \le t-2$, and $\tilde{G}_m^{(t)} = 0$ otherwise. This result can be strengthened. Indeed, consider the equation (5.68) where $a \neq 0, c \neq 0, b = 0$ and $d = -ac$, then the corresponding $(\hat{h}^{[0]}, 0)$-Green's function is given by

$$\tilde{G}_m^{(t)} = \sum_{i=0}^{\lfloor (t-2-|m|)/2 \rfloor} \sum_{k=i}^{\lfloor (t-2)/2 \rfloor} C_{2k+1}^{(t-1)} C_i^{(k)} C_{(t-2-2i-m)/2}^{(t-2-2i)} \frac{(-1)^{i+1}}{2^{t-2-2i}} a^{(t+m)/2} c^{(t-m)/2}$$

when $-t+2 \le m \le t-2$, and $\tilde{G}_m^{(t)} = 0$ otherwise. Note also that $\tilde{G}_m^{(t)} = 0$ when $t-m$ is odd.

By similar formal arguments and then direct verifications, we may also arrive at the following result for the $(0, \hat{h}^{[0]})$-Green's function.

Theorem 74 *The unique solution $\{K^{(t)}\}_{t=0}^{\infty}$ of the equation (5.69) subject to the initial conditions $K^{(0)} = 0$ and $K^{[1]} = \hat{h}^{[0]}$ is given by*

$$K^{(t)} = \frac{1}{2^{t-1}} \sum_{k=0}^{\lfloor (t-1)/2 \rfloor} C_{2k+1}^{(t)} \delta^{t-1-2k} \left(\delta^2 + 4d\hat{h}^{[0]}\right)^k,\ t \in N.$$

Furthermore, as we have explained when deriving the explicit representation of the principal Green's function, we may also obtain

$$\begin{aligned} K^{(t)} &= \sum_{m=-t+1}^{t-1} \sum_{i=0}^{\lfloor (t-1-|m|)/2 \rfloor} \sum_{k=i}^{\lfloor (t-1)/2 \rfloor} \sum_{j=|m|}^{t-1-2i} C_{2k+1}^{(t)} C_i^{(k)} C_j^{(t-1-2i)} C_{(j-m)/2}^{(j)} \\ &\quad \times \left(\frac{1}{2}\right)^{t-1-2i} a^{(j+m)/2} b^{t-1-2i-j} c^{(j-m)/2} d^i \hat{h}^{[m]}, \end{aligned}$$

for $t \in N$. Thus,

$$\begin{aligned} K_m^{(t)} &= \sum_{i=0}^{\lfloor (t-1-|m|)/2 \rfloor} \sum_{k=i}^{\lfloor (t-1)/2 \rfloor} \sum_{j=|m|}^{t-1-2i} C_{2k+1}^{(t)} C_i^{(k)} C_j^{(t-1-2i)} \\ &\quad \times C_{(j-m)/2}^{(j)} \left(\frac{1}{2}\right)^{t-1-2i} a^{(j+m)/2} b^{t-1-2i-j} c^{(j-m)/2} d^i, \end{aligned} \tag{5.73}$$

for $|m| \le t-1$ and $t = 2, 3, \ldots$.

As an example, consider the equation

$$u_m^{(t+1)} = au_{m-1}^{(t)} + cu_{m+1}^{(t)} - acu_m^{(t-1)},\ (m,t) \in \Omega; a \neq 0.$$

The corresponding $K_m^{(t)}$ is given by

$$K_m^{(t)} = \sum_{i=0}^{\lfloor (t-1-|m|)/2 \rfloor} \sum_{k=i}^{\lfloor (t-1)/2 \rfloor} C_{2k+1}^{(t)} C_i^{(k)} C_{(t-1-2i-m)/2}^{(t-1-2i)} \frac{(-1)^i}{2^{t-1-2i}} a^{(t+m)/2} c^{(t-m)/2}$$

when $-t+1 \leq m \leq t-1$, and $K_m^{(t)} = 0$ otherwise. Note also that $K_m^{(t)} = 0$ when $t-m$ is even.

There are several properties of the Green's functions which can be seen from the above derivations. First of all, we have

$$\sum_{m=-\infty}^{\infty} \tilde{G}_m^{(t)} = \frac{d}{2^{t-2}} \sum_{k=0}^{\lfloor (t-2)/2 \rfloor} C_{2k+1}^{(t-1)} (a+b+c)^{t-2-2k} [(a+b+c)^2 + 4d]^k \tag{5.74}$$

and

$$\sum_{m=-\infty}^{\infty} K_m^{(t)} = \frac{1}{2^{t-1}} \sum_{k=0}^{\lfloor (t-1)/2 \rfloor} C_{2k+1}^{(t)} (a+b+c)^{t-1-2k} [(a+b+c)^2 + 4d]^k \tag{5.75}$$

for $t = 2, 3, 4, \ldots$. We need to explain how (5.74) is obtained: we view the right hand side of (5.71) as a rational function

$$\frac{d}{2^{t-2}} \sum_{k=0}^{\lfloor (t-2)/2 \rfloor} C_{2k+1}^{(t-1)} (ax+b+cx^{-1})^{t-2-2k} ((ax+b+cx^{-1})^2 + 4d)^k,$$

by identifying $\hat{h}^{[i]}$ with x^i, then to obtain the sum of the coefficients of the above expressions after expansion amounts to replacing x with 1 in the above expression. The equality (5.75) is obtained by similar considerations. So are the following inequalities:

$$\begin{aligned} \sum_{m=-\infty}^{\infty} \left| \tilde{G}_m^{(t)} \right| &\leq \frac{|d|}{2^{t-2}} \sum_{k=0}^{\lfloor (t-2)/2 \rfloor} C_{2k+1}^{(t-1)} (|a|+|b|+|c|)^{t-2-2k} \\ &\quad \times [(|a|+|b|+|c|)^2 + 4|d|]^k \\ &= \frac{|d| \left[(\sigma+\tau)^{t-1} - (\sigma-\tau)^{t-1} \right]}{2\tau} \end{aligned} \tag{5.76}$$

for $t \in Z^+$,

$$\begin{aligned} \sum_{m=-\infty}^{\infty} \left| K_m^{(t)} \right| &\leq \frac{1}{2^{t-1}} \sum_{k=0}^{\lfloor (t-1)/2 \rfloor} C_{2k+1}^{(t)} (|a|+|b|+|c|)^{t-1-2k} \\ &\quad \times [(|a|+|b|+|c|)^2 + 4|d|]^k \\ &= \frac{(\sigma+\tau)^t - (\sigma-\tau)^t}{2\tau} \end{aligned} \tag{5.77}$$

for $t \in N$, and

$$\sum_{i=0}^{t-2} \sum_{k=-t+2+i}^{t-2-i} \left| K_k^{(t-1-i)} \right| \leq \sum_{i=0}^{t-2} \frac{(\sigma+\tau)^{t-1-i} - (\sigma-\tau)^{t-1-i}}{2\tau} \tag{5.78}$$

where

$$\sigma = \frac{|a| + |b| + |c|}{2}, \ \tau = \sqrt{\sigma^2 + |d|}. \tag{5.79}$$

Recall that equation (5.65) can be written as

$$u^{(t+1)} = \delta * u^{(t)} + du^{(t-1)} + p^{(t)}, \ t \in Z^+, \tag{5.80}$$

where $p^{(t)} = \{p_m^{(t)}\}_{m \in Z}$ and δ has been defined as $a\hat{h}^{[1]} + b\hat{h}^{[0]} + c\hat{h}^{[-1]}$. Given initial conditions $u^{(0)} = f$ and $u^{(1)} = g$, we assert that the corresponding unique solution is given by

$$u^{(t)} = f * \tilde{G}^{(t)} + g * K^{(t)} + \sum_{i=0}^{t-2} K^{(t-1-i)} * p^{(i+1)} \tag{5.81}$$

for $t \in N$. This assertion is formally obtained by means of the method of variation of parameters and we can verify directly that it is a solution of (5.80). Indeed,

$$f * \tilde{G}^{(0)} + g * K^{(0)} = f * \hat{h}^{[0]} = f,$$

and

$$f * \tilde{G}^{(1)} + g * K^{(1)} = g * \hat{h}^{[0]} = g.$$

Furthermore,

$$\begin{aligned}
& \delta * \left(f * \tilde{G}^{(t)} + g * K^{(t)} + \sum_{i=0}^{t-2} K^{(t-1-i)} * p^{(i+1)} \right) \\
& +d \left(f * \tilde{G}^{(t-1)} + g * K^{(t-1)} + \sum_{i=0}^{t-3} K^{(t-2-i)} * p^{(i+1)} \right) + p^{(t)} \\
= \ & f * \left(\delta * \tilde{G}^{(t)} + d\tilde{G}^{(t-1)} \right) + g * \left(\delta * K^{(t)} + dK^{(t-1)} \right) \\
& + \delta * \sum_{i=0}^{t-2} K^{(t-1-i)} * p^{(i+1)} + d \sum_{i=0}^{t-3} K^{(t-2-i)} * p^{(i+1)} + p^{(t)} \\
= \ & f * \tilde{G}^{(t+1)} + g * K^{(t+1)} + \sum_{i=0}^{t-1} K^{(t-1-i)} * p^{(i+1)}
\end{aligned}$$

as required. The following result is now clear.

Theorem 75 *The unique solution of (5.65) subject to the conditions (5.66) and (5.67) is given by*

$$u_m^{(t)} = \sum_{k=-t+2}^{t-2} \tilde{G}_k^{(t)} f_{m-k} + \sum_{k=-t+1}^{t-1} K_k^{(t)} g_{m-k} + \sum_{i=0}^{t-2} \sum_{k=-t+2+i}^{t-2-i} K_k^{(t-1-i)} p_{m-k}^{(i+1)}$$

for $m \in Z$ and $t \in N$.

There are several immediate consequences of the above theorem. First of all, when f, g are symmetric sequences, then under the additional condition that $a = c$, the corresponding solution $\{u_m^{(t)}\}$ of (5.68) is also symmetric in the sense that

$$u_m^{(t)} = u_{-m}^{(t)},\ (m,t) \in \Omega.$$

Indeed, in view of Theorem 75,

$$\begin{aligned} u_m^{(t)} &= \sum_{k=-t+2}^{t-2} \tilde{G}_k^{(t)} f_{m-k} + \sum_{k=-t+1}^{t-1} K_k^{(t)} g_{m-k} \\ &= \sum_{k=-t+2}^{t-2} \tilde{G}_{-k}^{(t)} f_{-m+k} + \sum_{k=-t+1}^{t-1} K_{-k}^{(t)} g_{-m+k} \\ &= \sum_{k=-t+2}^{t-2} \tilde{G}_k^{(t)} f_{-m-k} + \sum_{k=-t+1}^{t-1} K_k^{(t)} g_{-m-k} \\ &= u_{-m}^{(t)}, \end{aligned}$$

as required. By means of similar arguments, we may verify that when f, g are odd sequences and $a = c$, then the corresponding solution $\{u_m^{(t)}\}$ of (5.68) is odd in the sense that

$$u_m^{(t)} = -u_{-m}^{(t)},\ (m,t) \in \Omega.$$

As another consequence, if the sequence f and g are periodic sequences such that $f_m = f_{m+L}$ and $g_m = g_{m+L}$ for $m \in Z$, then the corresponding solution $\{u_m^{(t)}\}$ of (5.68) satisfies

$$\begin{aligned} u_m^{(t)} &= \sum_{k=-t+2}^{t-2} \tilde{G}_k^{(t)} f_{m-k} + \sum_{k=-t+1}^{t-1} K_k^{(t)} g_{m-k} \\ &= \sum_{k=-t+2}^{t-2} \tilde{G}_k^{(t)} f_{m+L-k} + \sum_{k=-t+1}^{t-1} K_k^{(t)} g_{m+L-k} \\ &= u_{m+L}^{(t)} \end{aligned}$$

for all $m \in Z$. That is, the solution $\{u_m^{(t)}\}$ is periodic in m.

5.7 Method of Linear Systems

As seen in the introductory section of this chapter, a time independent partial difference equation can sometimes be treated as a system of linear equations. In this section, we will expound on this idea further and consider the discrete Poisson equation over a nonempty finite domain Ω :

$$\Xi x_{ij} \equiv \Delta_1^2 x_{i-1,j} + \Delta_2^2 x_{i,j-1} = -p_{ij},\ (i,j) \in \Omega, \tag{5.82}$$

subject to the Dirichlet condition

$$x_{ij} = 0,\ (i,j) \in \partial\Omega. \tag{5.83}$$

In view of Theorem 53, a unique solution of the boundary problem (5.82)–(5.83) exists. Let $\delta_{ij}^{(u,v)}$ be the Dirac delta function defined by

$$\delta_{ij}^{(u,v)} = \begin{cases} 1 & (i,j) = (u,v) \\ 0 & (i,j) \neq (u,v) \end{cases} . \tag{5.84}$$

For each fixed point (u, v) in Ω, the boundary value problem (5.82)–(5.83) with p_{ij} replaced by $\delta_{ij}^{(u,v)}$ has a corresponding solution $G^{(u,v)} = \{G_{ij}^{(u,v)}\}_{(i,j)\in\Omega+\partial\Omega}$. This solution $G^{(u,v)}$ will be called the Green's function associated with the boundary value problem (5.82)–(5.83).

There are a number of important properties of the Green's function. First of all, we show the symmetry property of our Green's function: $G_{st}^{(u,v)} = G_{uv}^{(s,t)}$ for $(u,v),(s,t) \in \Omega$. Indeed, note that in view of the discrete Green's identity in Theorem 2, for any $(u,v),(s,t) \in \Omega$,

$$\begin{aligned} 0 &= \sum_{(i,j)\in\Omega} \left\{G_{ij}^{(u,v)}\Xi G_{ij}^{(s,t)} - G_{ij}^{(s,t)}\Xi G_{ij}^{(u,v)}\right\} \\ &= \sum_{(i,j)\in\Omega} \left\{-G_{ij}^{(u,v)}\delta_{ij}^{(s,t)} + G_{ij}^{(s,t)}\delta_{ij}^{(u,v)}\right\} \\ &= -G_{st}^{(u,v)} + G_{u,v}^{(s,t)}, \end{aligned}$$

as required.

Next, by means of the discrete Green's identity again,

$$\begin{aligned} 0 &= \sum_{(u,v)\in\Omega} G_{uv}^{(i,j)}\Xi x_{uv} + \sum_{(u,v)\in\Omega} G_{uv}^{(i,j)}p_{uv} \\ &= \sum_{(u,v)\in\Omega} G_{ij}^{(u,v)}\Xi x_{uv} + \sum_{(u,v)\in\Omega} G_{uv}^{(i,j)}p_{uv} \\ &= \sum_{(u,v)\in\Omega} x_{uv}\Xi G_{ij}^{(u,v)} + \sum_{(u,v)\in\Omega} G_{uv}^{(i,j)}p_{uv} \\ &= -\sum_{(u,v)\in\Omega} x_{uv}\delta_{ij}^{(u,v)} + \sum_{(u,v)\in\Omega} G_{ij}^{(u,v)}p_{uv} \\ &= -x_{ij} + \sum_{(u,v)\in\Omega} G_{ij}^{(u,v)}p_{uv}, \end{aligned}$$

which implies

$$x_{ij} = \sum_{(u,v)\in\Omega} G_{ij}^{(u,v)}p_{uv},\ (i,j) \in \Omega. \tag{5.85}$$

Theorem 76 *Let Ω be a nonempty finite domain in Z^2. If $\{x_{ij}\}_{(i,j)\in\Omega+\partial\Omega}$ is a solution of (5.82)–(5.83), then $\{x_{ij}\}_{(i,j)\in\Omega}$ is a solution of (5.85). Conversely, if $\{y_{ij}\}_{(i,j)\in\Omega}$ is a solution of (5.85), then the augmented bivariate sequence $x = \{x_{ij}\}_{(i,j)\in\Omega+\partial\Omega}$ defined by $x_{ij} = y_{ij}$ for $(i,j) \in \Omega$ and $x_{ij} = 0$ for $(i,j) \in \partial\Omega$ is a solution of (5.82)–(5.83).*

Next, by means of the maximum principle stated in Theorem 52, it is readily seen that $G_{ij}^{(u,v)} > 0$ for $(i,j) \in \Omega$. The same maximum principle can help us further.

Theorem 77 *For any $(u,v) \in \Omega$, the Green's function associated with the boundary value problem (5.82)–(5.83) satisfies*

$$\max_{(i,j)\in\Omega\backslash\{(u,v)\}} G_{ij}^{(u,v)} < G_{uv}^{(u,v)}.$$

In particular,

$$\max_{(u,v),(i,j)\in\Omega} G_{ij}^{(u,v)} = \max_{(u,v)\in\Omega} G_{u,v}^{(u,v)}.$$

Indeed, let Λ be an arbitrary component of $\Omega\backslash\{(u,v)\}$ containing a neighbor of (u,v). Since $G_{ij}^{(u,v)} > 0$ for $(i,j) \in \Lambda$ and $G_{ij}^{(u,v)} = 0$ for $(i,j) \in \partial\Lambda \cdot \partial\Omega$, thus $G^{(u,v)}$ cannot be constant over $\Lambda + \partial\Lambda$. Note that $\Xi G_{ij}^{(u,v)} = 0$ for $(i,j) \in \Lambda$, thus by means of our maximum principle, we see that the maximum of $\left\{G_{ij}^{(u,v)}\right\}_{(i,j)\in\Lambda+\partial\Lambda}$ occurs on $\partial\Lambda$. This shows that

$$\max_{(i,j)\in\Lambda} G_{ij}^{(u,v)} < \max_{(i,j)\in\partial\Lambda} G_{ij}^{(u,v)} \leq \max_{(i,j)\in\partial\Omega+\{(u,v)\}} G_{ij}^{(u,v)} = \max_{(i,j)\in\{(u,v)\}} G_{ij}^{(u,v)},$$

as desired.

Theorem 78 *Let Ω be a finite nonempty domain. Let $(a,b) \in \partial\Omega$ and let $\tilde{\Omega} = \Omega + \{(a,b)\}$. Let $G^{(u,v)}$ be the Green's function associated with the boundary value problem (5.82)–(5.83), and let $\tilde{G} = \left\{\tilde{G}^{(u,v)}\right\}$ be the Green's function associated with the boundary value problem*

$$\Xi G_{ij}^{(u,v)} = -\delta_{ij}^{(u,v)},\ (i,j) \in \tilde{\Omega},$$
$$\tilde{G}_{ij}^{(u,v)} = 0,\ (i,j) \in \partial\tilde{\Omega}.$$

Then $G_{ij}^{(u,v)} < \tilde{G}_{ij}^{(u,v)}$ for $(i,j) \in \Omega$.

Indeed, let $H_{ij} = G_{ij}^{(u,v)} - \tilde{G}_{ij}^{(u,v)}$ for $(i,j) \in \Omega + \partial\Omega$. Then $\Xi H_{ij} = 0$ for $(i,j) \in \Omega$ and $H_{ij} = -\tilde{G}_{ij}^{(u,v)}$ for $(i,j) \in \partial\Omega$. Since $\tilde{G}_{ab}^{(u,v)} > 0$ and since $\tilde{G}_{ij}^{(u,v)} = 0$ for $(i,j) \in \partial\Omega \cdot \partial\tilde{\Omega}$, $\{H_{ij}\}$ cannot be a constant function over $\Omega + \partial\Omega$. By Theorem 52, we have $H_{ij} < 0$ for $(i,j) \in \Omega$.

We now derive the explicit forms of the Green's function when Ω is a chain or a cycle. Let Ω be a chain of the general form $\{(i_1,j_1),...,(i_n,j_n)\}$ such that (i_s,j_s) and (i_t,j_t) are neighbors if, and only if, $|s-t| = 1$. Then the corresponding Green's function can be expressed as

$$G^{(i_s,j_s)} = \left\{G_{i_t,j_t}^{(i_s,j_s)}\right\}_{t=1}^{n}.$$

For the sake of convenience, we will write

$$G(t|s) = G_{i_t,j_t}^{(i_s,j_s)},\ s,t = 1,...,n.$$

Before finding an explicit formula for $G(t|s)$, we first consider an example.

Example 64 *Let* $\Omega = \{(1,0),(2,0),(3,0)\}$. *Then*

$$\partial\Omega = \{(0,0),(0,1),(0,2),(0,3),(4,0),(3,-1),(2,-1),1,-1)\}.$$

Let $(1,0)$ *be labelled as* $1,(2,0)$ *as* 2 *and* $(3,0)$ *as* 3. *To find* $G(t|1)$ *is the same as solving the linear system*

$$\begin{aligned} -4G(1|1)+G(2|1) &= -1,\\ G(1|1)-4G(2|1)+G(3|1) &= 0,\\ G(2|1)-4G(3|1) &= 0. \end{aligned}$$

Similarly, to find $G(t|s)$ *is the same as finding the inverse of the matrix*

$$\begin{bmatrix} -4 & 1 & 0 \\ 1 & -4 & 1 \\ 0 & 1 & -4 \end{bmatrix}.$$

It turns out that to find inverses of tridiagonal matrices (or band matrices), one way is to find solutions of an associated ordinary difference equation. We will not carry ourselves too far from the main theme, but the details can be found in Cheng and Hsieh [19]. Here we will let $X_{-1} = 0, X_1 = 1$ and X_k be defined by the difference equation

$$X_k = 4X_{k-1} - X_{k-2},\ k = 2,3,\dots .$$

Solving this ordinary difference equation (e.g. by the method of operators), we see that

$$X_k = \frac{1}{2\sqrt{3}}\left(\gamma^{k+1} - \frac{1}{\gamma^{k+1}}\right),\ k \in N, \tag{5.86}$$

where $\gamma = 2+\sqrt{3}$. By means of the above explicit formula, it is easy to check that $X_k > 0$ and $\Delta X_k > 0$ for $k \in N$, as well as

$$G(t|s) = \begin{cases} X_{s-1}X_{n-t}X_n^{-1} & 1 \le s \le t \le n \\ X_{n-s}X_{t-1}X_n^{-1} & 1 \le t \le s \le n \end{cases}.$$

Next, let Ω be a cycle. We may assume that the points in Ω have been ordered as $(i_1,j_1),\dots,(i_n,j_n)$ such that (i_1,j_1) has exactly two neighbors (i_n,j_n) and $(i_2,j_2),(i_2,j_2)$ has exactly two neighbors (i_1,j_1) and (i_3,j_3), etc. As in the previous case, we write the corresponding Green's function $G(t|s)$. To find an explicit formula, let $Y_{-1},Y_0,\dots,Y_n$ be defined by

$$Y_k = \begin{cases} X_k & -1 \le k \le n-1 \\ 4X_{n-1} - 2X_{n-2} - 2 & k = n \end{cases},$$

where X_k has been defined by (5.86). Since $X_k > 1$ for $k \ge 2$, and since $\Delta X_k > 0$ for $k \ge 0$,

$$Y_n = 2(X_{n-1} - X_{n-2}) + 2(X_{n-1} - 1) > 0.$$

It is now easily verified that

$$G(t|s) = \frac{(Y_{|s-t|-1} + Y_{n-1-|t-s|})}{Y_n},\ s,t = 1,2,\dots,n.$$

We remark that when Ω is a rectangular domain, an explicit formula for the associated Green's function has also been given [119]. More specifically, let

$$\Omega = \{1, 2, ..., m\} \times \{1, 2, ..., n\}.$$

Then

$$G_{ij}^{(a,b)} = \frac{2}{m+2}\sum_{r=1}^{m} \sin\frac{ar\pi}{m+1}\sin\frac{ir\pi}{m+1}\frac{\sinh(j\beta_r)}{\sinh(\beta_r)}\frac{\sinh((n+1-j)\beta_r)}{\sinh((n+1)\beta_r)}$$

for $j \le b$, and

$$G_{ij}^{(a,b)} = \frac{2}{m+2}\sum_{r=1}^{m} \sin\frac{ar\pi}{m+1}\sin\frac{ir\pi}{m+1}\frac{\sinh(b\beta_r)}{\sinh(\beta_r)}\frac{\sinh((n+1-j)\beta_r)}{\sinh((n+1)\beta_r}$$

for $j \ge b$, where $\beta_1, ..., \beta_m$ are the roots of the equation

$$\cos\frac{r\pi}{m+1} + \cosh(\beta_r) = 2.$$

5.8 Notes and Remarks

Examples 48 and 49 are due to Lin and Cheng [109]. Example 54 is due to Jeske [94].

Besides those in Section 5.4, there are several additional examples by Jentsch [86], [89] which illustrate the use of operators for obtaining explicit solutions of partial difference equations. The conclusion of Example 58 is also obtained by Corduneanu [49], but his method is formal.

Theorem 65 is due to Cheng and Lu [39].

The section on the method of convolution is based on Cheng and Lin [31] and Cheng and Lu [39].

The last section is based on Cheng *et al.* [20]. The linear systems that represent boundary value problems involving linear partial difference equations defined on finite domains are usually of the form $Au = v$. The matrix A derived is usually sparse. By proper labelling of the lattice points, sometimes, the matrix A may turn out to be a blockwise tridiagonal matrix, or a matrix with structural regularity. For additional information, the references Cornock [51], Maybee [118] may be consulted.

There does not seem to be any collection of explicit solutions for partial difference equations. Some additional examples can be found in Ablowitz and Ladik [1], Glantz and Reissner [71], Gregor [75], Keberle and Montet [97], Long [107], Lynch *et al.* [116], Merzrath [120], Mugler [127], Mugler and Scott [128], Razpet [142], Wang [159], Yu [168], etc.

Once a solution is known to a difference equation, it is desirable to generate additional solutions from it. One method which is well known in ordinary difference equations is the method of variation of parameters. For homogeneous partial difference equations with constant coefficients, related results can be found in Duffin [61], Duffin and Shelly [64], and Hundhausen [84].

Chapter 6

Stability

6.1 Stability Concepts

There are many concepts which are related to the stability of the solutions of partial difference equations. Roughly speaking, we will be concerned with the "sizes" of solutions in terms of the nonstructural variables (see Section 5.1). We will also be concerned with the "changes in sizes" of solutions when the nonstructural variables are changed.

When the changes in sizes of solutions are "continuous" with respect to changes in sizes of the nonstructural variables, we have a "stable" situation. Otherwise, if the changes are "abrupt", we have a "sensitive" situation. Recent interests in the latter case are reflected in the rapid development of bifurcation and chaos theories.

There are a number of measures for the sizes of finite or infinite sequences. First of all, we have defined the p-norms for finite or infinite sequences in Chapter 2. The concept of a subexponential sequence is also useful. A sequence $\{u_k\}_{k=0}^{\infty}$ is said to be exponentially bounded or subexponential if

$$|u_k| \le M\alpha^k,\ k \ge 0$$

for some positive constants M and α. In case $\alpha \in (0,1)$, we may also say that $\{u_k\}$ is subexponentially decaying. A doubly subexponential sequence $\{y_i\}_{i=-\infty}^{\infty}$ is one which satisfies

$$|y_i| \le M\frac{\alpha^i + \alpha^{-i}}{2},\ i \in Z$$

where M and α are positive constants, or equivalently, if there are positive constants N, σ such that

$$|y_i| \le N\sigma^{|i|},\ i \in Z.$$

For bivariate sequences, there are many possibilities. We mention the major ones here, while others will be mentioned later. First of all, a bivariate sequence $u = \left\{u_m^{(n)}\right\}_{m,n\in N}$ such as the solutions of our time dependent discrete heat equations for semi-infinite rods may be viewed as a sequence $\left\{u^{(0)}, u^{(1)}, ...\right\}$ of sequences in l^N. Therefore, it is said to be "spatially" subexponential if each $u^{(n)}$ is subexponential and is said to be "time" convergent if $\lim_{n\to\infty} u^{(n)} = w$ for some w in l^N. Other combinations of stability concepts are also possible. Similarly, $v = \left\{v_m^{(n)}\right\}_{(m,n)\in Z\times N}$ may be viewed as a sequence $\left\{v^{(0)}, v^{(1)}, ...\right\}$ of sequences in l^Z, therefore, spatially doubly subexponential sequences and time bounded sequences may be needed.

A bivariate sequence $u = \{u_{ij}\}_{i,j\in N}$ may arise as solutions of time independent equations. The associated concepts of stability are then different. For instance, a bivariate sequence $\{u_{ij}\}_{i,j\in N}$ is said to be subexponential if

$$|u_{ij}| \le \Gamma\beta^i\gamma^j, \ (i,j) \in \Omega$$

for some positive numbers Γ, β and γ. Another concept is also useful. A real bivariate sequence $u = \{u_{ij}\}_{ij\in N}$ is said to be weakly subexponentially decaying if

$$|u_{ij}| \le M\alpha^{\min(i,j)}, \ i,j \in N,$$

for some $M \ge 0$ and $\alpha \in (0,1)$. Note that u_{ij} may not tend to zero as $j \to \infty$ when its independent variable i remains finite.

There are a number of methods which can be used for obtaining stability criteria. We will explain by giving examples how these methods can be applied to various partial difference equations.

6.2 Equations Over Cylinders

Recall that a cylinder in the lattice plane is of the form

$$\left\{(i,j) \in Z^2 | \, i = 1,2,...,n; j \in N\right\},$$

which arises when heat diffusion is considered for finite rods. A number of methods are known for obtaining stability criteria for partial difference equations defined on cylinders.

6.2.1 Method of General Solutions

When general solutions can be found, it is sometimes easy to establish the sizes of solutions. As an example, consider the partial difference equation (5.34)

$$\Delta_2 u_{j,k} = a\Delta_1^2 u_{j-1,k+1} + b\Delta_1^2 u_{j-1,k}, \ j = 1,...,J; \ k \in N,$$

subject to the conditions

$$u_{0,k} = 0 = u_{J+1,k} = 0, \ k \in N,$$

and

$$u_{j,0} = \psi_j, \ j = 0,1,...,J+1.$$

Recall that the general solution is given by

$$u_{jk} = \sum_{m=1}^{J} c_k \sigma_{(m)}^k \sin\frac{m\pi j}{J+1}, \ j = 0,1,...,J+1; \ k \in N,$$

where

$$\sigma_{(m)} = \frac{1 + \rho_{(m)} b}{1 - \rho_{(m)} a},\ m = 1, 2, ..., J.$$

$$\rho_{(m)} = -4 \sin^2 \frac{m\pi}{2(J+1)},\ m = 1, 2, ..., J,$$

$$c_m = \frac{2}{J+1} \sum_{j=1}^{J} \psi_j \sin \frac{m\pi j}{J+1},\ m = 1, ..., J,$$

and

$$\psi_j = \sum_{m=1}^{J} c_m \sin \frac{m\pi j}{J+1},\ j = 1, ..., J.$$

We may show directly that

$$\sum_{j=1}^{J} \psi_j^2 = \frac{J+1}{2} \sum_{m=1}^{J} c_m^2$$

and

$$\sum_{j=1}^{J} u_{jk}^2 = \frac{J+1}{2} \sum_{m=1}^{J} \sigma_{(m)}^{2k} c_m^2.$$

Thus when

$$\sigma_{(m)}^2 \leq 1,\ m = 1, 2, ...J,$$

we have the stability result

$$\sum_{j=1}^{J} u_{jk}^2 \leq \sum_{j=1}^{J} \psi_j^2,\ k \in Z^+.$$

Conversely, when one of the numbers $\sigma_{(1)}^2, ..., \sigma_{(J-1)}^2$, say $\sigma_{(t)}^2$, is strictly greater than 1, then by picking $c_t = 1$ and $c_m = 0$ for $m \neq t$, we see that $u_{1k}^2 + ... + u_{Jk}^2$ will diverge to positive infinity as $k \to \infty$.

6.2.2 Method of Maximum Principles

Consider the discrete implicit heat equation

$$\Delta_2 u_i^{(j-1)} = r \Delta_1^2 u_{i-1}^{(j)},\ i = 1, ..., I;\ j \in Z^+,$$

subject to the conditions $r > 0$,

$$u_0^{(j)} = u_{I+1}^{(j)} = 0,\ j \in N,$$

and

$$u_i^{(0)} = \psi_i,\ i = 1, ..., I.$$

Since for each $n \geq 0$, we see from the maximum principle stated in Theorem 55 that

$$u_i^{(n+1)} \leq \max\left\{u_0^{(n+1)}, u_{I+1}^{(n+1)}, \max_{1\leq k\leq I}\{u_k^{(n)}\}\right\} = \max_{1\leq k\leq I}\{u_k^{(n)}\},$$

for $i = 1, ..., I$, and by symmetry considerations, we see that

$$\max_{1\leq k\leq I}\left|u_k^{(n+1)}\right| \leq \max_{1\leq k\leq I}\left|u_k^{(n)}\right| \leq ... \leq \max_{1\leq k\leq I}\left|u_k^{(0)}\right| = \max_{1\leq i\leq I}|\psi_i|.$$

In particular, if $\max_{1\leq i\leq I}|\psi_i| \leq M$, then $\left|u_k^{(n+1)}\right| \leq M$ for $1 \leq k \leq I$ and $n \in N$.

6.2.3 Method of Energies

Consider the partial difference equation

$$\Delta_2 u_j^{(n)} = a\Delta_1^2 u_{j-1}^{(n)},\ j = 1, ..., J;\ n \in N, \tag{6.1}$$

subject to the conditions $a > 0$,

$$u_0^{(n)} = 0 = u_{J+1}^{(n)},\ n \in N, \tag{6.2}$$

and

$$u_j^{(0)} = \psi_j,\ j = 1, ..., J. \tag{6.3}$$

If we multiply (6.1) by $u_j^{(n+1)} + u_j^{(n)}$ and then sum the resulting equation from $j = 1$ to $j = J$, we obtain

$$\sum_{j=1}^{J}\left(u_j^{(n+1)}\right)^2 - \sum_{j=1}^{J}\left(u_j^{(n)}\right)^2 = -a\left[\sum_{j=0}^{J}\left\{\left(\Delta_1 u_j^{(n)}\right)^2 + \Delta_1 u_j^{(n+1)}\Delta_1 u_j^{(n)}\right\}\right]$$

If we define

$$E_n = \sum_{j=1}^{J}\left(u_j^{(n)}\right)^2 - \frac{a}{2}\sum_{j=0}^{J}\left(\Delta_1 u_j^{(n)}\right)^2,\ n \in N,$$

then

$$E_n \leq \sum_{j=1}^{J}\left(u_j^{(n)}\right)^2,$$

and

$$E_{n+1} - E_n = -\frac{a}{2}\sum_{j=0}^{J}\left(\Delta_1 u_j^{(n+1)} + \Delta_1 u_j^{(n)}\right)^2 \leq 0.$$

We conclude therefore that E_n is a monotonic decreasing function of n.

Since

$$\sum_{j=0}^{J}\left(\Delta u_j^{(n)}\right)^2 \leq 2\sum_{j=0}^{J}\left[\left(u_{j+1}^{(n)}\right)^2 + \left(u_j^{(n)}\right)^2\right] = 4\sum_{j=1}^{J}\left(u_j^{(n)}\right)^2,$$

so from the definition of E_n, we see that

$$E_n \geq (1-2a)\sum_{j=1}^{J}\left(u_j^{(n)}\right)^2.$$

If $0 < a < 1/2$, then

$$(1-2a)\sum_{j=1}^{J}\left(u_j^{(n)}\right)^2 \leq E_n \leq E_{n-1} \leq \ldots \leq E_0 = \sum_{j=1}^{J}\left(u_j^{(0)}\right)^2 = \sum_{j=1}^{J}\psi_j^2$$

and stability results.

The quantity E_n is actually a function $E(x_0,\ldots,x_{J+1})$ with $x_0,\ldots,x_{J+1}$ substituted by $u_0^{(n)}, u_1^{(n)}, \ldots, u_{J+1}^{(n)}$ respectively. When such a function is "monotonic decreasing along solutions" of our dynamical models, then it is called a Lyapunov function. Such a statement is not precise but we will not give a rigorous definition. As is well known, Lyapunov functions are usually (but not always) motivated by considering "energies" associated with the difference equations at hand. Therefore, the term "energy method" is sometimes used to refer to such an approach.

6.2.4 Method of Functional Inequalities

Next we consider the following equation

$$\Delta_2 u_i^{(j)} = a_j \Delta_1^2 u_{i-1}^{(j)} + b_j u_i^{(j)} + c_j u_i^{(j-\sigma)},\ 1 \leq i \leq n,\ j \in N, \tag{6.4}$$

where σ is a nonnegative integer, and $\{a_j\}_{j=0}^{\infty}$, $\{b_j\}_{j=0}^{\infty}$ as well as $\{c_j\}_{j=0}^{\infty}$ are real sequences. Under the boundary conditions

$$u_0^{(j)} = 0 = u_{n+1}^{(j)},\ j \geq -\sigma, \tag{6.5}$$

and initial conditions

$$u_i^{(j)} = \psi_i^{(j)},\quad 1 \leq i \leq n,\ -\sigma \leq j \leq 0, \tag{6.6}$$

it is easily seen that an existence and uniqueness theorem holds for solutions of this equation.

We now prove the following stability criterion [165]: *Suppose $a_j > 0$ for $j \geq 0$. Suppose further that there are nonnegative numbers δ and η such that*

$$\left|1 + 2b_j + |c_j| + 8a_j^2 + 4(b_j - 2a_j)^2\right| \leq \delta < 1,\ j \in N$$

and

$$4c_j^2 + |c_j| \leq \eta < 1 - \delta,\ j \in N.$$

Then there exist $M > 0$ and $\xi > 1$ such that the solution $\{u_{ij}\}$ of (6.4)-(6.6) will satisfy

$$\sum_{i=1}^{n}\left(u_i^{(j)}\right)^2 \leq M\xi^{-j} \max_{-\sigma \leq s \leq 0}\sum_{i=1}^{n}\left(u_i^{(s)}\right)^2,\ j \in N. \tag{6.7}$$

Two preparatory results will be needed. First, let $\{u_i^{(j)}\}$ be a solution of the boundary problem (6.4)–(6.6), then

$$\Delta_2 u_i^{(j)} = (b_j - 2a_j)\, u_i^{(j)} + a_j u_{i+1}^{(j)} + a_j u_{i-1}^{(j)} + c_j u_i^{(j-\sigma)}.$$

Thus

$$\begin{aligned}
&\sum_{i=1}^{n}\left(\Delta_2 u_i^{(j)}\right)^2\\
\le\ &\sum_{i=1}^{n} 4\left\{(b_j-2a_j)^2\left(u_i^{(j)}\right)^2 + a_j^2\left(u_{i+1}^{(j)}\right)^2 + a_j^2\left(u_{i-1}^{(j)}\right)^2 + c_j^2\left(u_i^{(j-\sigma)}\right)^2\right\}\\
\le\ &4\,(b_j-2a_j)^2\sum_{i=1}^{n}\left(u_i^{(j)}\right)^2 + 4a_j^2\sum_{i=1}^{n}\left(u_i^{(j)}\right)^2\\
&+4a_j^2\sum_{i=1}^{n}\left(u_i^{(j)}\right)^2 + 4c_j^2\sum_{i=1}^{n}\left(u_i^{(j-\sigma)}\right)^2,
\end{aligned}$$

or

$$\sum_{i=1}^{n}\left(\Delta_2 u_i^{(j)}\right)^2 \le 4\left((b_j-2a_j)^2 + 2a_j^2\right)\sum_{i=1}^{n}\left(u_i^{(j)}\right)^2 + 4c_j^2\sum_{i=1}^{n}\left(u_i^{(j-\sigma)}\right)^2. \tag{6.8}$$

Next, in view of (6.4), we also have

$$u_i^{(j+1)} = a_j\Delta_1^2 u_{i-1}^{(j)} + (1+b_j)u_i^{(j)} + c_j u_i^{(j-\sigma)},$$

thus

$$u_i^{(j)}u_i^{(j+1)} = a_j u_i^{(j)}\Delta_1^2 u_{i-1}^{(j)} + (1+b_j)\left(u_i^{(j)}\right)^2 + c_j u_i^{(j)}u_i^{(j-\sigma)}.$$

Since

$$\left(\Delta_2 u_i^{(j)}\right)^2 = \left(u_i^{(j+1)} - u_i^{(j)}\right)^2 = \left(u_i^{(j+1)}\right)^2 - 2u_i^{(j)}u_i^{(j+1)} + \left(u_i^{(j)}\right)^2,$$

therefore, we obtain

$$\begin{aligned}
&\left(u_i^{(j+1)}\right)^2\\
=\ &\left(\Delta_2 u_i^{(j)}\right)^2 + 2u_i^{(j)}u_i^{(j+1)} - \left(u_i^{(j)}\right)^2\\
=\ &\left(\Delta_2 u_i^{(j)}\right)^2 + 2a_j u_i^{(j)}\Delta_1^2 u_{i-1}^{(j)} + (1+2b_j)\left(u_i^{(j)}\right)^2 + 2c_j u_i^{(j)}u_i^{(j-\sigma)}.
\end{aligned} \tag{6.9}$$

Let $\{u_i^{(j)}\}$ be the solution of (6.4)–(6.6). Then in view of (6.9), we have

$$\begin{aligned}
\sum_{i=1}^{n}\left(u_i^{(j+1)}\right)^2 =\ &\sum_{i=1}^{n}\left(\Delta_2 u_i^{(j)}\right)^2 + 2a_j\sum_{i=1}^{n}u_i^{(j)}\Delta_1^2 u_{i-1}^{(j)}\\
&+(1+2b_j)\sum_{i=1}^{n}\left(u_i^{(j)}\right)^2 + 2c_j\sum_{i=1}^{n}u_i^{(j)}u_i^{(j-\sigma)}.
\end{aligned} \tag{6.10}$$

Furthermore, in view of (6.5), we have

$$\begin{aligned}\sum_{i=1}^{n} u_i^{(j)}\Delta_1^2 u_{i-1}^{(j)} &= u_{n+1}^{(j)}\Delta_1 u_n^{(j)} - u_1^{(j)}\Delta_1 u_0^{(j)} - \sum_{i=1}^{n}\left(\Delta_1 u_i^{(j)}\right)^2 \\ &= -\sum_{i=0}^{n}\left(\Delta_1 u_i^{(j)}\right)^2. \qquad (6.11)\end{aligned}$$

Thus, substituting (6.8) and (6.11) into (6.10), we obtain

$$\begin{aligned}&\sum_{i=1}^{n}\left(u_i^{(j+1)}\right)^2 \\ \le\ & 4\left(2a_j^2+(b_j-2a_j)^2\right)\sum_{i=1}^{n}\left(u_i^{(j)}\right)^2 + 4c_j^2\sum_{i=1}^{n}\left(u_i^{(j-\sigma)}\right)^2 - 2a_j\sum_{i=0}^{n}\left(\Delta_1 u_i^{(j)}\right)^2 \\ & +(1+2b_j)\sum_{i=1}^{n}\left(u_i^{(j)}\right)^2 + |c_j|\sum_{i=1}^{n}\left(u_i^{(j)}\right)^2 + \left(u_i^{(j-\sigma)}\right)^2 \\ =\ & \left\{1+2b_j+|c_j|+8a_j^2+4(b_j-2a_j)^2\right\}\sum_{i=1}^{n}\left(u_i^{(j)}\right)^2 \\ & +\left(|c_j|+4c_j^2\right)\sum_{i=1}^{n}\left(u_i^{(j-\sigma)}\right)^2 - 2a_j\sum_{i=0}^{n}\left(\Delta_1 u_i^{(j)}\right)^2 \\ \le\ & \delta\sum_{i=1}^{n}u_{ij}^2 + \eta\sum_{i=1}^{n}\left(u_i^{(j-\sigma)}\right)^2 - 2a_j\sum_{i=0}^{n}\left(\Delta_1 u_i^{(j)}\right)^2 \qquad (6.12)\end{aligned}$$

for $j \ge 0$. In view of the Gronwall type inequality in Theorem 19, we then obtain

$$\begin{aligned}&\sum_{i=1}^{n}\left(u_i^{(j)}\right)^2 \\ \le\ & \delta^j\sum_{i=1}^{n}\left(u_i^{(0)}\right)^2 + \sum_{t=0}^{j-1}\eta\sum_{i=1}^{n}\left(u_i^{(t-\sigma)}\right)^2\delta^{j-t-1} - 2\sum_{t=0}^{j-1}a_t\sum_{i=0}^{n}\left(\Delta_1 u_i^{(t)}\right)^2\delta^{j-t-1}\end{aligned}$$

for $j \ge 0$, which implies

$$\sum_{i=1}^{n}\left(u_i^{(j)}\right)^2 \le \sum_{i=1}^{n}\left(u_i^{(0)}\right)^2\delta^j + \sum_{t=0}^{j-1}\eta\sum_{i=1}^{n}\left(u_i^{(t-\sigma)}\right)^2\delta^{j-t-1},\ j \ge 0. \qquad (6.13)$$

Note that since $\eta < 1-\delta$, the function

$$f(x) = 1 - x^{\sigma+1}\left(\frac{\eta}{1-\delta x}\right)$$

will satisfy $f(1) > 0$. Thus by continuity, there is a number $1 < \xi < 1/\delta$ such that $f(\xi) > 0$. By multiplying both sides of (6.13) by ξ^j, we obtain

$$\sum_{i=1}^{n}\left(u_i^{(j)}\right)^2\xi^j \le \sum_{i=1}^{n}\left(u_i^{(0)}\right)^2(\delta\xi)^j + \sum_{t=0}^{j-1}\eta(\delta\xi)^{j-t-1}\xi^{\sigma+1}\sum_{i=1}^{n}\left(u_i^{(t-\sigma)}\right)^2\xi^{t-\sigma}$$

$$\leq \sum_{i=1}^{n}\left(u_i^{(0)}\right)^2+\sum_{t=0}^{j-1}\eta\xi^{\sigma+1}(\delta\xi)^{j-t-1}\sup_{-\sigma\leq s\leq t}\sum_{i=1}^{n}\left(u_i^{(s)}\right)^2\xi^s$$
$$\leq \sum_{i=1}^{n}\left(u_i^{(0)}\right)^2+\frac{\eta\xi^{\sigma+1}}{1-\delta\xi}\sup_{-\sigma\leq s\leq j}\sum_{i=1}^{n}\left(u_i^{(s)}\right)^2\xi^s.$$

In view of (2.29), we see that

$$\begin{aligned}\sum_{i=1}^{n}\left(u_i^{(j)}\right)^2\xi^j &\leq \sup_{-\sigma\leq s\leq j}\sum_{i=1}^{n}\left(u_i^{(s)}\right)^2\xi^s\\ &\leq \frac{1}{f(\xi)}\sum_{i=1}^{n}\left(u_i^{(0)}\right)^2+\frac{1}{f(\xi)}\sup_{-\sigma\leq s\leq 0}\sum_{i=1}^{n}\left(u_i^{(s)}\right)^2\xi^s\\ &\leq \frac{2}{f(\xi)}\sup_{-\sigma\leq s\leq 0}\sum_{i=1}^{n}\left(u_i^{(s)}\right)^2,\ j\in N,\end{aligned}$$

which implies

$$\sum_{i=1}^{n}\left(u_i^{(j)}\right)^2\leq\frac{2}{f(\xi)}\xi^{-j}\sup_{-\sigma\leq s\leq 0}\sum_{i=1}^{n}\left(u_i^{(s)}\right)^2,\ j\in N,$$

as required.

6.2.5 Spectral Methods

Here we will be concerned with the stability problem of the more general discrete heat equation with perturbation

$$u_i^{(j+1)}=au_{i-1}^{(j)}+bu_i^{(j)}+cu_{i+1}^{(j)}+g_i^{(j)}+F\left(j,u_i^{(j)}\right),\ i=1,...,n;\ j\in N, \quad (6.14)$$

where $a,b,c\in R$, $g=\{g_i^{(j)}\}$ is a real function defined for $i=1,2,...,n$ and $j\in N$, and F is a real function defined on R. We will also assume that side conditions

$$u_0^{(j)}=\phi_j,\ j\in N, \quad (6.15)$$

$$u_{n+1}^{(j)}=\psi_j,\ j\in N, \quad (6.16)$$

$$u_i^{(0)}=w_i,\ i=1,2,...,n \quad (6.17)$$

are imposed, where $w=\text{col}(w_1,...,w_n)$ is a real vector, and $\phi=\{\phi_j\}_{j=0}^{\infty}$ and $\psi=\{\psi_j\}_{j=0}^{\infty}$ are real sequences. Let

$$\Omega=\{(i,j)|\ i=0,1,...,n+1;\ j\in N\}.$$

A solution of (6.14)–(6.17) is a discrete function $u=\{u_i^{(j)}\}_{(i,j)\in\Omega}$ which satisfies the functional relation (6.14) and also the side conditions (6.15)–(6.17). It is easily seen that an existence and uniqueness theorem holds for solutions of the problem (6.14)–(6.17).

By designating $\text{col}(u_1^{(j)}, u_2^{(j)}, ..., u_n^{(j)})$ as the real vector $u^{(j)}$, we see that the sequence $\{u^{(j)}\}_{j\in N}$ will satisfy the vector equation

$$u^{(j+1)} = Au^{(j)} + f^{(j)} + F(j, u^{(j)}), \quad j \in N, \tag{6.18}$$

and the initial condition

$$u^{(0)} = w \tag{6.19}$$

where

$$A = \begin{bmatrix} b & c & 0 & \dots & 0 \\ a & b & c & \dots & 0 \\ & \cdot & \cdot & \cdot & \\ 0 & \dots & a & b & c \\ 0 & \dots & 0 & a & b \end{bmatrix}, \tag{6.20}$$

$$f^{(j)} = \text{col}(g_1^{(j)}, ..., g_n^{(j)}) + \text{col}(a\phi_j, 0, ..., 0, c\psi_j), \tag{6.21}$$

and

$$F(k, x) = \text{col}(F(k, x_1), ..., F(k, x_n)), \ x = \text{col}(x_1, ..., x_n). \tag{6.22}$$

Conversely, if $\{u^{(j)}\}_{j\in N}$ is a solution of (6.18)–(6.19), then by augmenting each $u^{(j)} = \text{col}(u_1^{(j)}, ..., u_n^{(j)})$ with the terms ϕ_j and ψ_j to form $\left\{\phi_j, u_1^{(j)}, ..., u_n^{(j)}, \psi_j\right\}$, we see that the resulting family forms a solution of (6.14)–(6.17).

For any real n vector $x = \text{col}(x_1, ..., x_n)$ and n by n real matrix $B = [b_{ij}]$, the following norms will be used:

$$\|x\|_1 = \sum_{i=1}^{n} |x_i|,$$

and

$$\|B\| = \max_{1\le j\le n} \sum_{i=1}^{n} |b_{ij}|.$$

It is well known that $\|Bx\|_1 \le \|B\| \, \|x\|_1$. Let $\{\lambda_i\}_{i=1}^n$ be the eigenvalues of the matrix B. The spectral radius $\max_{1\le i\le n} |\lambda_i|$ will be denoted by $\rho(B)$. It is well known that $\lim_{n\to\infty} B^n = 0$ if, and only if, $\rho(B) < 1$. A more precise statement can also be made (see e.g. [50]): Let $B = [b_{ij}]_{n\times n}$ be a real matrix and $\rho(B)$ be its spectral radius. Then there exists a constant $\Gamma \ge 1$, such that

$$\|B^i\| \le \Gamma(\rho(B))^i, \ i \in N.$$

An important case of the difference problem (6.14)–(6.17) arises when $\phi \equiv 0, \psi \equiv 0, g \equiv 0$ and $F \equiv 0$. As noted before, this case is equivalent to the initial value problem

$$u^{(j+1)} = Au^{(j)}, \tag{6.23}$$

$$u^{(0)} = w.$$

Theorem 79 *Suppose $ac > 0$. Then all solutions of (6.23) are bounded if, and only if,*

$$|b| + 2\sqrt{ac}\cos\frac{\pi}{n+1} \le 1, \tag{6.24}$$

and all solutions of (6.23) tend to zero if, and only if

$$|b| + 2\sqrt{ac}\cos\frac{\pi}{n+1} < 1. \tag{6.25}$$

Indeed, in view of Theorem 16, when $ac > 0$, the spectral radius $\rho(A)$ of A is equal to

$$\max_{1\le k\le n}\left|b + 2\sqrt{ac}\cos\frac{k\pi}{n+1}\right| = \max\{|\lambda_1(A)|, |\lambda_n(A)|\} = |b| + 2\sqrt{ac}\cos\frac{\pi}{n+1}.$$

Thus, the solution $\{u^{(j)}\}_{j\in N}$ of (6.19)–(6.23) satisfies

$$\left\|u^{(j+1)}\right\|_1 \le \|A\| \left\|u^{(j)}\right\|_1 \le \dots \le \left\|A^j\right\| \|w\|_1 \le \Gamma\left(\rho(A)\right)^j \|w\|_1 \le \Gamma \|w\|_1, \quad j \in N,$$

for some $\Gamma \ge 1$. This shows that all solutions are bounded. Conversely, suppose $\rho(A) = |\lambda_1(A)| > 1$, let

$$w = \text{col}\left(\sin\frac{\pi}{n+1}, \left(\frac{a}{c}\right)^{1/2}\sin\frac{2\pi}{n+1}, \dots, \left(\frac{a}{c}\right)^{(n-1)/2}\sin\frac{n\pi}{n+1}\right),$$

which is the eigenvector of A corresponding to the eigenvalue $\lambda_1(A)$ (see Theorem 16). Then the solution $\{u^{(j)}\}_{j\in N}$ of (6.23) corresponding to the initial condition $u^{(0)} = w$ satisfies

$$u^{(j+1)} = Au^{(j)} = \dots = A^j u^{(0)} = (\lambda_1(A))^{j+1} w,$$

so that

$$\left\|u^{(j+1)}\right\|_1 = |\lambda_1(A)|^{j+1} \|w\|_1 \to \infty.$$

The other case where $\rho(A) = |\lambda_n(A)|$ is similarly proved. Next, if $a = 0$ or $c = 0$, then $\rho(A) = |b|$. By means of arguments similar to those just given, we see that $|b| \le 1$ ($|b| < 1$) is a necessary and sufficient condition for all solutions of (6.23) to be bounded (respectively to converge to zero).

We remark that under the condition that $ac > 0$, in view of the inequality $2\sqrt{|xy|} \le |x| + |y|$, we may replace (6.25) by the easily recognized condition

$$|a| + |b| + |c| < 1,$$

and conclude that every solution of (6.23) converges to zero.

Theorem 80 *Suppose $ac < 0$. Then every solution of equation (6.23) is bounded if, and only if,*

$$b^2 - 4ac\cos^2\frac{\pi}{n+1} \le 1, \tag{6.26}$$

and every solution tends to zero if, and only if,

$$b^2 - 4ac\cos^2\frac{\pi}{n+1} < 1. \tag{6.27}$$

The proof is similar to that of Theorem 79, except for the fact that the spectral radius of A is now given by

$$|\lambda_1(A)| = \sqrt{b^2 + \left(2\sqrt{|ac|}\cos\frac{\pi}{n+1}\right)^2} = \sqrt{b^2 - 4ac\cos^2\frac{\pi}{n+1}}. \tag{6.28}$$

We now derive stability criteria for (6.18). First consider the case where $F \equiv 0$. By inductive arguments, it is easily seen that the unique solution $\{u^{(j)}\}_{j\in N}$ of (6.18) subject to $u^{(0)} = w$ now satisfies

$$u^{(j+1)} = A^{j+1}w + \sum_{k=0}^{j} A^{j-k} f^{(k)},\ j \in N. \tag{6.29}$$

Thus,

$$\left\|u^{(j+1)}\right\| \le \Gamma\rho^{j+1}\|w\| + \Gamma\sum_{k=0}^{j}\rho^{j-k}\left\|f^{(k)}\right\|,\ j \in N, \tag{6.30}$$

where Γ is some positive number greater than or equal to 1, and $\rho = \rho(A)$ is the spectral radius of A (recall that $\rho(A) = |b| + 2\sqrt{ac}\cos(\pi/(n+1))$ if $ac \ge 0$ and $\rho(A) = \sqrt{b^2 - 4ac\cos^2(\pi/(n+1))}$ if $ac < 0$.). If $\rho \le 1$, then

$$\left\|u^{(j+1)}\right\| \le \Gamma\|w\| + \Gamma\sum_{k=0}^{j}\left\|f^{(k)}\right\|,\ j \in N.$$

The following result, in view of (6.21), is now clear.

Theorem 81 *Suppose $F \equiv 0$, and $\rho(A) \le 1$. Suppose further that*

$$\sum_{k=0}^{\infty}\left\|\mathrm{col}(g_1^{(k)},...,g_n^{(k)})\right\| < \infty,\ \sum_{k=0}^{\infty}|\phi_k| < \infty \textit{ and } \sum_{k=0}^{\infty}|\psi_j| < \infty.$$

Then every solution of (6.14) is bounded.

Theorem 82 *Suppose that $\phi = 0, \psi = 0$ and $g = 0$, and that*

$$\lim_{|x|\to 0}\left|\frac{F(k,x)}{x}\right| = 0 \textit{ uniformly in } k. \tag{6.31}$$

Suppose further that $|a| + |b| + |c| < 1$. Then there is a positive number δ such that whenever $\|\mathrm{col}(w_1,...,w_n)\| < \delta$, the solution of (6.14) subject to the initial condition $u^0 = \mathrm{col}(w_1,...,w_n)$ will converge to 0.

Proof. First note that $\|A\| = |a| + |b| + |c|$. Since $|a| + |b| + |c| < 1$, there is some positive number ε such that

$$|a| + |b| + |c| + \varepsilon < 1.$$

In view of (6.31), there is a positive number δ such that

$$\|F(k,x)\|_1 \le \varepsilon\|x\|_1,\ k \in N,$$

whenever $\|x\|_1 < \delta$. Therefore, if $\|w\|_1 < \delta$, then

$$\left\|u^{(1)}\right\|_1 \le \|A\| \, \|w\|_1 + \|F(w)\|_1 \le \|A\| \, \|w\|_1 + \varepsilon \|w\|_1 < (\|A\| + \varepsilon)\delta,$$

and by induction, we see that

$$\left\|u^{(j+1)}\right\|_1 < (\|A\| + \varepsilon)^{j+1}\delta, \; j \in N.$$

The proof is complete. ■

We now turn to the general case. Again, by inductive arguments, it is easily seen that the unique solution $\{u^{(j)}\}_{j=0}^{\infty}$ of (6.18) subject to $u^{(0)} = w$ now satisfies

$$u^{(j+1)} = A^{j+1}w + \sum_{k=0}^{j} A^{j-k} f^{(k)} + \sum_{k=0}^{j} A^{j-k} F(k, u^{(k)}), \; j \in N. \tag{6.32}$$

Thus,

$$\left\|u^{(j+1)}\right\|_1 \le \Gamma\rho^{j+1} \|w\|_1 + \Gamma \sum_{k=0}^{j} \rho^{j-k} \left\|f^{(k)}\right\|_1 + \Gamma \sum_{k=0}^{j} \rho^{j-k} \left\|F(k, u^{(k)})\right\|_1, \tag{6.33}$$

for $j \in N$, where $\rho = \rho(A)$ is the spectral radius of A, and Γ is some positive number greater than or equal to 1. Suppose that

$$\|F(k,u)\|_1 \le h(k) \|u\|_1, \tag{6.34}$$

then we see further that

$$\frac{\|u^{(j+1)}\|_1}{\rho^{j+1}} \le \Gamma \|w\|_1 + \frac{\Gamma}{\rho} \sum_{k=0}^{j} \rho^{-k} \left\|f^{(k)}\right\|_1 + \frac{\Gamma}{\rho} \sum_{k=0}^{j} h(k) \frac{\|u^{(k)}\|_1}{\rho^k}, \; j \in N.$$

By means of Theorem 19, we see that

$$\frac{\|u^{(j+1)}\|_1}{\rho^{j+1}} \le \left\{ \Gamma \|w\|_1 + \frac{\Gamma}{\rho} \sum_{k=0}^{j} \rho^{-k} \left\|f^{(k)}\right\|_1 \right\} \exp\left\{ \frac{\Gamma}{\rho} \sum_{k=0}^{j} h(k) \right\}, \; j \in N.$$

Several consequences can now be concluded. For instance, we have the following.

Theorem 83 *Suppose $|F(k,x)| \le h(k)\,|x|$ for all $x \in R$ and*

$$\sum_{k=0}^{\infty} h(k) < \infty.$$

Suppose further that the spectral radius ρ of A is less than 1, and

$$\sum_{k=0}^{\infty} \rho^{-k} \left\| \mathrm{col}(g_1^{(k)} + a\phi_k, g_2^{(k)} ..., g_{n-1}^{(k)}, g_n^{(k)} + c\psi_k) \right\|_1 < \infty.$$

Then every solution of (6.14) converges to 0.

6.2.6 Method of Separable Solutions

Consider the three-level discrete reaction-diffusion equation of the form

$$u_i^{(t+1)} - u_i^{(t)} = p\left(u_{i-1}^{(t)} - 2u_i^{(t)} + u_{i+1}^{(t)}\right) + qu_i^{(t-\sigma)}, \ i = 1, ..., n; t \in N, \tag{6.35}$$

where σ is a positive integer and p, q are real numbers. Dirichlet boundary conditions of the form

$$u_0^{(t)} = 0 = u_{n+1}^{(t)}, \ t \in N, \tag{6.36}$$

will be imposed. Given an arbitrary set of initial values $u_i^{(t)}$, $-\sigma \le t \le 0, 1 \le i \le n$, we can successively calculate $u_1^{(1)}, ..., u_n^{(1)}; u_1^{(2)}, ..., u_n^{(2)}; ...$ in a unique manner. Such a bivariate sequence $u = \{u_i^{(t)} \mid i = 1, ..., n; t = -\sigma, -\sigma+1, ...\}$ is called a solution of (6.35)–(6.36). By designating col $\left(u_1^{(t)}, u_2^{(t)}, ..., u_n^{(t)}\right)$ as the vector $u^{(t)}$, we see that a solution of (6.35)–(6.36) can also be regarded as a vector sequence $\{u^{(t)}\}_{t=-\sigma}^{\infty}$. Furthermore, such a sequence satisfies the delay vector recurrence relation

$$u^{(t+1)} - u^{(t)} = pAu^{(t)} + qu^{(t-\sigma)}, \ t \in N, \tag{6.37}$$

where

$$A = \begin{pmatrix} -2 & 1 & 0 & \dots & 0 \\ 1 & -2 & 1 & \dots & 0 \\ & & \cdot & \cdot & \cdot \\ 0 & \dots & 1 & -2 & 1 \\ 0 & \dots & 0 & 1 & -2 \end{pmatrix}_{n\times n}.$$

We will obtain a necessary and sufficient condition for all solutions of (6.37) to be zero convergent. First recall from Theorem 16 in Chapter 2 that the eigenvalue problem

$$Av = \lambda v$$

has eigenvalues

$$\lambda_j = -2 + 2\cos\frac{j\pi}{n+1}, \ j = 1, ..., n, \tag{6.38}$$

with corresponding orthogonal eigenvectors

$$\psi_j = \text{col}\left(\sin\frac{j\pi}{n+1}, \sin\frac{2j\pi}{n+1}, ..., \sin\frac{nj\pi}{n+1}\right), \ j = 1, ..., n.$$

Next, let us denote the transpose of a vector v by v', then the inner product of two vectors u and v is given by $v'u$.

Suppose (6.37) has a zero convergent solution $\{u^{(t)}\}$. Then taking the inner product of (6.37) with the vector ψ_j, we see that

$$\begin{aligned} \psi_j'u^{(t+1)} - \psi_j'u^{(t)} &= p\psi_j'Au^{(t)} + q\psi_j'u^{(t-\sigma)} \\ &= p(A\psi_j)'u^{(t)} + q\psi_j'u^{(t-\sigma)} \\ &= p\lambda_j\psi_j'u^{(t)} + q\psi_j'u^{(t-\sigma)} \end{aligned}$$

for $t \geq -\sigma$. In other words, the scalar sequence $\{x_t\}_{t=-\sigma}^{\infty} \equiv \{\psi_j' u^{(t)}\}_{t=-\sigma}^{\infty}$ is a zero convergent solution of the scalar difference equation

$$x_{t+1} - x_t = p\lambda_j x_t + q x_{t-\sigma},\ t \in N. \tag{6.39}$$

Conversely, note that if $\{x_t\}_{t=-\sigma}^{\infty}$ is a solution of (6.39), then the "separable" sequence $\{u^{(t)}\}_{t=-\sigma}^{\infty}$ defined by

$$u^{(t)} = x_t \psi_j,\ t \geq -\sigma,$$

is a solution of (6.37). Indeed, it is clear that $u^{(t)} = x_t\psi_j \to 0$ as $t \to \infty$. Furthermore, in view of (6.39), we see that

$$\begin{aligned} & u^{(t+1)} - u^{(t)} \\ = \;& x_{t+1}\psi_j - x_t\psi_j = p\lambda_j x_t \psi_j + q x_{t-\sigma}\psi_j \\ = \;& p x_t(\lambda_j\psi_j) + q x_{t-\sigma}\psi_j = pA(x_t\psi_j) + q x_{t-\sigma}\psi_j \\ = \;& pAu^{(t)} + qu^{(t)} \end{aligned}$$

as required. In particular, when $\{x_t\}$ is zero convergent of (6.39), then $\{u^{(t)}\}$ is a zero convergent solution of (6.37). We have thus shown that (6.37) has a zero convergent solution if and only if (6.39) has a zero convergent solution.

We will go one step further and show that all solutions of (6.37) are zero convergent if, and only if, for every $j = 1, 2, ..., n$, all solutions of (6.39) are zero convergent. First of all, if $\{x_t\}$ is a solution of (6.39) which does not converge to zero, then clearly the solution $\{x_t\psi_j\}$ is a solution of (6.37) which does not converge to zero. Next, we show that every solution $\{u^{(t)}\}$ of (6.37) is of the form

$$u^{(t)} = \sum_{j=1}^{n} x_t^{[j]}\psi_j, t = -\sigma, -\sigma+1, .., \tag{6.40}$$

where each $\left\{x_t^{[j]}\right\}$ is some solution of (6.39) to be determined. Indeed, we have already shown that $\left\{x_t^{[j]}\psi_j\right\}, 1 \leq j \leq n$, are solutions of (6.37) and hence a linear combination of these solutions is also a solution of (6.37). Furthermore, since $\{\psi_1, ..., \psi_n\}$ is a basis of R^n, there exist unique vectors

$$\left(c_{-\sigma}^{[1]}, c_{-\sigma}^{[2]}, ..., c_{-\sigma}^{[n]}\right), \left(c_{-\sigma+1}^{[1]}, ...c_{-\sigma+1}^{[n]}\right), \;...\;, \left(c_0^{[1]}, ..., c_0^{[n]}\right),$$

such that

$$u^{(-\sigma)} = \sum_{j=1}^{n} c_{-\sigma}^{[j]}\psi_j, \;...\;, u^{(0)} = \sum_{j=1}^{n} c_0^{[j]}\psi_j$$

respectively. If we now let $\left\{x_t^{[j]}\right\}$ be the unique solution of (6.39) determined by the initial conditions

$$x_{-\sigma}^{[j]} = c_{-\sigma}^{[j]},\ x_{-\sigma+1}^{[j]} = c_{-\sigma+1}^{[j]},\ ...\ , x_0^{[j]} = c_0^{[j]},$$

then by uniqueness, $u^{(t)}$ must equal to $\sum_{j=1}^{n} x_t^{[j]}\psi_j$ for $t \in Z^+$.

Now it is easily seen from (6.40) that when all the solutions $\left\{x_t^{[j]}\right\}, 1 \leq j \leq n$, are zero convergent, then $\{u^{(t)}\}$ is also zero convergent.

Theorem 84 *Every solution of (6.37) converges to zero if, and only if, for each $j = 1, ..., n$, every solution of (6.39) converges to zero, where $\lambda_1, ..., \lambda_n$ are defined by (6.38).*

We remark that trivial modifications of the above arguments show that every solution of (6.37) is bounded if, and only if, for each $j = 1, 2, ..., n$, every solution of (6.39) is bounded.

The usefulness of the above theorem lies in the fact that we have reduced the problem of asymptotic stability of a partial difference equation into one involving an ordinary difference equation, and that much is known about the ordinary difference equations. Indeed, it is well known that every solution of the difference equation

$$x_{n+1} = ax_n + bx_{n-\sigma},\ n \in N, \tag{6.41}$$

converges to zero if, and only if, all the roots of its characteristic equation

$$z^{n+1} = az^n + bz^{n-\sigma}$$

have magnitudes less than 1. For instance, when $\sigma = 1$, then (see e.g. [102]) all the roots of the corresponding equation have magnitude less than one if, and only if,

$$-1 < b < 1 - a \text{ and } b < 1 + a.$$

Therefore, if we assume in addition that $n = 3$, then every solution of (6.37) converges to zero if, and only if,

$$-1 < q < -p\lambda_j \text{ and } q < 2 + p\lambda_j,\ 1 \le j \le 3,$$

or, equivalently,

$$-1 < q < \max\left\{2 + (-2 - \sqrt{2})p, (2 - \sqrt{2})p, (2 + \sqrt{2})p\right\}.$$

For a general positive integer σ, necessary and sufficient condition in terms of the coefficients a and b for all roots of (6.41) to have magnitude less than one is also known (see Kuruklis [102]). Hence necessary and sufficient conditions for the asymptotic stability of (6.37) in terms of the coefficients p and q as well as the dimensional constant n can be obtained in similar ways as illustrated above.

The same technique can be used again to deal with (vector) equations such as

$$u^{(t+1)} - u^{(t)} = pAu^{(t-\tau)} + \sum_{i=1}^{m} q_i u^{(t-\sigma_i)},\ t \in N,$$

$$u^{(t+1)} - u^{(t)} = pBu^{(t)} + qu^{(t-\sigma)},\ t \in N,$$

etc., where B is a matrix possessing real eigenvalues. No new principles, however, are involved.

6.3 Equations Over Half Planes

We have discussed several partial difference equations defined over the upper half lattice plane. Since their general solutions are relatively simple, it is understandable that stability criteria can be derived by means of these solutions. For comparison purposes, we will also explain how induction can be used to obtain stability criteria.

6.3.1 Method of Exact Solutions for Two-Level Equations

In this section, we will derive several stability criteria for the solutions of the discrete heat equation (5.49):

$$u_m^{(n+1)} = au_{m-1}^{(n)} + bu_m^{(n)} + cu_{m+1}^{(n)}, \ m \in Z, n \in N, \tag{6.42}$$

subject to the initial condition (5.50):

$$u_m^{(0)} = f_m, \ m \in Z. \tag{6.43}$$

First of all, recall that the value $u_0^{(n)}$ is completely determined by the values $f_{-n}, f_{-n+1}, ..., f_n$; and similarly, u_m^n by f_k, $m-n \le k \le m+n$. For this reason, we say that the integers $m-n, ..., m+n$ are the domain of dependence for the value $u_m^{(n)}$.

There are many concepts which are related to the stability of our equation (6.42). In particular, the trivial solution of (6.42) is said to be stable if there exists a positive constant Γ such that for every initial sequence $f = \{f_m\}_{m=-\infty}^{\infty}$, the corresponding solution of (6.42)–(6.43) will satisfy

$$\left|u_m^{(n)}\right| \le \Gamma \|f\|_m^{(n)}, \ (m,n) \in Z \times N,$$

where

$$\|f\|_m^{(n)} = \max\{|f_k| \mid m-n \le k \le m+n\}.$$

Clearly, if $\sup_{-\infty<k<\infty} |f_k| \le \delta$ then $\|f\|_m^{(n)} \le \Gamma\delta$ for all m and n. Another important concept of stability can also be defined as follows. The trivial solution of (6.42), is said to be exponentially asymptotically stable if there is a real number $\xi \in (0,1)$ and a positive number Γ such that for every initial sequence $f = \{f_m\}_{m=-\infty}^{\infty}$, the corresponding solution will satisfy

$$\left|u_m^{(n)}\right| \le \Gamma\xi^n \|f\|_m^{(n)}, \ (m,n) \in Z \times N,$$

Recall from Section 6 of Chapter 5 that

$$u^{(n)} = G^{(n)} * f,$$

where $u^{(n)} = \{u_m^{(n)}\}_{m\in Z}$, $G^{(n)} = \left(a\hat{h}^{[1]} + b\hat{h}^{[0]} + c\hat{h}^{[-1]}\right)^n$ and $f = \{f_m\}_{m\in Z}$.

Theorem 85 *Let* $u = \left\{u_i^{(j)}\right\}$ *be a solution of the initial value problem (6.42)–(6.43). Then*

$$\left|u_m^{(n)}\right| \le (|a| + |b| + |c|)^n \|f\|_{mn}, \ (m,n) \in Z \times N, \tag{6.44}$$

where $\|f\|_m^{(n)} = \max_{m-n\le k\le m+n} |f_k|$.

Proof. In view of

$$u_m^{(n)} = \sum_{k=-n}^{n} G_k^{(n)} f_{m-k}, \ (m,n) \in Z \times N,$$

and Theorem 5.67,

$$\begin{aligned}
\left|u_m^{(n)}\right| &\le \sum_{j=-n}^{n} |G_j^m| \, |f_{m-j}| \le \|f\|_{mn} \sum_{j=-n}^{n} |G_j^m| \\
&\le \|f\|_{mn} \sum_{j=-n}^{n} \sum_{i=|j|}^{n} \delta^{(i+j)} C_i^m C_{(i-|j|)/2}^i |a|^{(i+j)/2} |b|^{n-i} |c|^{(i-j)/2} \\
&= \|f\|_{mn} \left(|a| + |b| + |c|\right)^n,
\end{aligned}$$

for $(m,n) \in Z \times N$. The proof is complete. ■

As an immediate consequence, we see that if $|a| + |b| + |c| \le 1$, then

$$\left|u_m^{(n)}\right| \le \|f\|_m^{(n)}, \quad (m,n) \in Z \times N, \tag{6.45}$$

that is, the trivial solution of (6.42) is stable.

Theorem 86 *If $|a| + |b| + |c| \le 1$, then the trivial solution of (6.42) is stable. The converse also holds either when $abc = 0$, or, when $abc \neq 0$ and $ac > 0$.*

Proof. Suppose $|a| + |b| + |c| > 1$, $b \ge 0$ and $a, c > 0$. Let $f = \{d\}_{m \in Z}$ where $d > 0$. Then by (5.60), the corresponding solution of (6.42) is given by

$$u_m^{(n)} = \sum_{k=-n}^{n} G_k^{(n)} d = (a+b+c)^n d = (a+b+c)^n \|f\|_m^{(n)}.$$

Since $a+b+c > 1$, for any $\Gamma > 0$, there exists an integer n such that $(a+b+c)^n > \Gamma$. Thus

$$u_m^{(n)} > \Gamma \|f\|_m^{(n)},$$

which shows that the trivial solution of (6.42) is not stable. Next suppose $|a| + |b| + |c| > 1$, $a > 0$, $b < 0$ as well as $c > 0$. Let $f = \{(-1)^m d\}_{m=-\infty}^{\infty}$ where $d > 0$. Then in view of (5.61), the corresponding solution of (6.42) is given by

$$u_m^{(n)} = \sum_{k=-n}^{n} G_k^{(n)} (-1)^{m-k} d = (-1)^{m+n} d (a-b+c)^n.$$

Again, for any $\Gamma > 0$, there exists an integer n such that $(a-b+c)^n > \Gamma$. Thus

$$\left|u_m^{(n)}\right| > \Gamma \|f\|_{mn},$$

which shows that the trivial solution of (5.49) is not stable. Next, suppose $|a| + |b| + |c| > 1$, $b = 0$ and $ac < 0$. Let $f = \{(-1)^{m/2} \delta^{(m)} d\}_{m \in Z}$ where $d > 0$, i.e. $f_m = 0$ when i is odd and $f_m = \pm d$ when m is even. Then it is easily checked that $u_m^{(n)}$ is either equal to zero or to $\pm(a-c)^n d$. The same conclusion can thus be drawn as seen in the previous two cases. Next, suppose $c = 0$ and $ab \neq 0$. We may take $f = \{1\}_{m \in Z}$ and arrive at the same conclusion. The other cases are similarly proved. The proof is complete. ■

We remark that when $a = b = r > 0$ and $b = 1 - 2r$, the condition $|a| + |b| + |c| \le 1$ is equivalent to $r \le 1/2$.

Theorem 87 *If $|a|+|b|+|c|<1$, then the trivial solution of (6.42) is exponentially asymptotically stable. The converse also holds either when $abc=0$, or, $abc \neq 0$ and $ac>0$.*

The proof is similar to that of Theorem 86 and is thus omitted.

Our final result in this section provides a stability criterion for subexponential solutions of (6.42).

Theorem 88 *Suppose $f=\{f_m\}_{m\in Z}$ is subexponential. Then the solution $u=\left\{u_m^{(n)}\right\}$ of (6.42)–(6.43) is also subexponential.*

Proof. Suppose

$$|f_m| \leq M\alpha^m, \quad m \in Z$$

for some positive numbers M and α. In view of Theorem 5.67, we have

$$\begin{aligned}
\left|u_m^{(n)}\right| &\leq \sum_{j=-n}^{n} \left|G_j^{(n)}\right| |f_{m-j}| \leq M\alpha^m \sum_{j=-n}^{n} \left|G_j^{(n)}\right| \alpha^{-j} \\
&\leq M\alpha^m \sum_{j=-n}^{n} \sum_{i=|j|}^{n} \delta^{(i+j)} C_i^n C^i_{(i-|j|)/2} \left|\frac{a}{\alpha}\right|^{(i+j)/2} |b|^{n-i} |c\alpha|^{(i-j)/2} \\
&= M\alpha^m \left(\frac{|a|}{\alpha}+|b|+|c|\,\alpha\right)^n .
\end{aligned}$$

The proof is complete. ■

Similar arguments also lead to a stability criterion for the doubly subexponential solution of (6.42): if f is doubly subexponential, then the solution of (6.42)–(6.43) is also doubly subexponential.

6.3.2 Method of Exact Solutions for Three-Level Equations

We will, in this section, derive several stability criteria for three-level partial difference scheme of the form (5.65)

$$u_m^{(t+1)} = au_{m-1}^{(t)} + bu_m^{(t)} + cu_{m+1}^{(t)} + du_m^{(t-1)} + p_m^{(t)}, \ m \in Z, t \in Z^+,$$

where $|a|+|c| \neq 0$, subject to conditions (5.66)

$$u_m^{(0)} = f_m, \ m \in Z,$$

and (5.67)

$$u_m^{(1)} = g_m, \ m \in Z.$$

First, in view of Theorem 75, we see that the solution of equation (5.65) subject to (5.66) and (5.67) satisfies

$$\left|u_m^{(t)}\right| \leq \sum_{k=-t+2}^{t-2} \left|\tilde{G}_k^{(t)}\right| |f_{m-k}| + \sum_{k=-t+1}^{t-1} \left|K_k^{(t)}\right| |g_{m-k}|$$

$$+\sum_{i=0}^{t-2}\sum_{k=-t+2-i}^{t-2-i}\left|K_k^{(t-1-i)}\right|\left|p_k^{(i+1)}\right|$$

$$\leq \max_{m-t+2\leq j\leq m+t-2}|f_j|\sum_{k=-t+2}^{t-2}\left|\tilde{G}_k^{(t)}\right| + \max_{m-t+1\leq j\leq m+t-1}|g_j|\sum_{k=-t+1}^{t-1}\left|K_k^{(t)}\right|$$

$$+\max_{0\leq s\leq t-2,-t+2-s\leq j\leq t-2-s}\left|p_j^{(s+1)}\right|\sum_{i=0}^{t-2}\sum_{k=-t+2-i}^{t-2-i}\left|K_k^{(t-1-i)}\right|$$

for $m \in Z$ and $t \in N$. In view of (5.76), (5.77) and (5.78), we see that

$$\begin{aligned}\left|u_m^{(t)}\right| &\leq \frac{|d|\left[(\sigma+\tau)^{t-1}-(\sigma-\tau)^{t-1}\right]}{2\tau}\max_{m-t+2\leq j\leq m+t-2}|f_j| \\ &+\frac{(\sigma+\tau)^t-(\sigma-\tau)^t}{2\tau}\max_{m-t+1\leq j\leq m+t-1}|g_j| \\ &+\left(\max_{0\leq s\leq t-2,-t+2-s\leq j\leq t-2-s}\left|p_j^{(s+1)}\right|\right)\sum_{k=1}^{t-1}\frac{(\sigma+\tau)^k-(\sigma-\tau)^k}{2\tau} \quad (6.46)\end{aligned}$$

for $t = 2, 3, \ldots$, where σ and τ have been defined by (5.79). In particular, when $d = 0$ and $\{p_m^{(t)}\} \equiv 0$, we have $\sigma = \tau$ and

$$\left|u_m^{(t)}\right| \leq (|a|+|b|+|c|)^{t-1}\max_{m-t+1\leq j\leq m+t-1}|g_j|,$$

which coincide with Theorem 85.

As an example, consider the DuFort Frankel scheme

$$u_m^{(t+1)} = \frac{2\gamma}{2\gamma+1}\left(u_{m-1}^{(t)}+u_{m+1}^{(t)}\right)+\frac{1-2\gamma}{2\gamma+1}u_m^{(t-1)},\ \gamma>0,\ (m,t)\in Z\times Z^+.$$

Note that $|a|+|c| \neq 0$ and $\sigma = 2\gamma/(2\gamma+1)$. Furthermore,

$$\tau = \sqrt{\frac{4\gamma^2}{(2\gamma+1)^2}+\frac{|1-2\gamma|}{2\gamma+1}},$$

which is equal to

$$\tau = \frac{1}{2\gamma+1}$$

when $0 < \gamma \leq 1/2$, and

$$\tau = \frac{\sqrt{8\gamma^2-1}}{2\gamma+1}$$

when $\gamma > 1/2$. Thus

$$\sigma+\tau = 1,\ \sigma-\tau = \frac{2\gamma-1}{2\gamma+1}$$

when $0 < \gamma \leq 1/2$, and

$$\sigma+\tau = \frac{2\gamma+\sqrt{8\gamma^2-1}}{2\gamma+1},\ \sigma-\tau = \frac{2\gamma-\sqrt{8\gamma^2-1}}{2\gamma+1}$$

when $\gamma > 1/2$. In view of (6.46), we see that when the initial sequence $\{f_m\}$ and $\{g_m\}$ are bounded by F and $\tilde{G}$ respectively, we have

$$\left|u_m^{(t)}\right| \leq \frac{1-2\gamma}{2}\left\{1-\left(\frac{2\gamma-1}{2\gamma+1}\right)^{t-1}\right\}F + \frac{1+2\gamma}{2}\left\{1-\left(\frac{2\gamma-1}{2\gamma+1}\right)^{t}\right\}\tilde{G}$$

for $t \geq 2$ and $0 < \gamma \leq 1/2$, and

$$\begin{aligned}\left|u_m^{(t)}\right| &\leq \frac{(2\gamma-1)F}{2\sqrt{8\gamma^2-1}}\left(\frac{2\gamma+\sqrt{8\gamma^2-1}}{2\gamma+1}\right)^{t-1}\left\{1-\left(\frac{2\gamma-\sqrt{8\gamma^2-1}}{2\gamma+\sqrt{8\gamma^2-1}}\right)^{t-1}\right\}\\ &+\frac{(2\gamma+1)\tilde{G}}{2\sqrt{8\gamma^2-1}}\left(\frac{2\gamma+\sqrt{8\gamma^2-1}}{2\gamma+1}\right)^{t}\left\{1-\left(\frac{2\gamma-\sqrt{8\gamma^2-1}}{2\gamma+\sqrt{8\gamma^2-1}}\right)^{t}\right\}\end{aligned}$$

for $t \geq 2$ and $\gamma > 1/2$. Clearly, when $0 < \gamma \leq 1/2$, $\left|u_m^{(t)}\right| \leq \max\{F, \tilde{G}\}$ for $m \in Z$ and $t \in N$. This implies that the DuFort Frankel scheme is stable when $0 < \gamma \leq 1/2$. The same assertion, however, cannot be made for $\gamma > 1/2$ yet, since the corresponding $\sigma + \tau > 1$. Fortunately, we can approach this problem via another route. First, let us recall that $\sigma = 2\gamma/(2\gamma+1)$. Note that $\gamma > 1/2$ if, and only if, $1/2 < \sigma < 1$. In view of (5.74) and (5.75), note also that

$$\begin{aligned}&\left|\sum_{m=-\infty}^{\infty}\tilde{G}_m^{(t)}\right|\\ &= \left|\frac{1-2\sigma}{2^{t-2}}\sum_{k=0}^{[(t-2)/2]}C_{2k+1}^{(t-1)}(2\sigma)^{t-2-2k}\left[4\sigma^2+4(1-2\sigma)\right]^k\right|\\ &= \left|(1-2\sigma)\sum_{k=0}^{[(t-2)/2]}C_{2k+1}^{(t-1)}\sigma^{t-2-2k}(1-\sigma)^{2k}\right|\\ &\leq \frac{2\sigma-1}{1-\sigma}\left|\sum_{k=0}^{t-1}C_k^{(t-1)}\sigma^{t-1-k}(1-\sigma)^k\right|\\ &\leq \frac{2\sigma-1}{1-\sigma},\end{aligned}$$

and

$$\begin{aligned}&\left|\sum_{m=-\infty}^{\infty}K_m^{(t)}\right|\\ &= \left|\frac{1}{2^{t-1}}\sum_{k=0}^{[(t-1)/2]}C_{2k+1}^{(t)}(2\sigma)^{t-1-2k}\left[4\sigma^2+4(1-2\sigma)\right]^k\right|\\ &= \left|\sum_{k=0}^{[(t-1)/2]}C_{2k+1}^{(t)}\sigma^{t-1-2k}(\sigma-1)^{2k}\right|\end{aligned}$$

$$\leq \left| \sum_{k=0}^{t} C_k^{(t)} \sigma^{t-k} (1-\sigma)^{k-1} \right|$$
$$\leq \frac{1}{1-\sigma}.$$

Now we assert that for $1/2 < \sigma < 1$, the symmetric Green's function $\tilde{G}_{-m}^{(t)} = \tilde{G}_m^{(t)} \leq 0$. Indeed, our assertion is true when $t-m$ is odd since the corresponding $\tilde{G}_m^{(t)}$ is 0. To see that our assertion also holds for the case where $t-m$ is even, first observe that

$$\begin{aligned} \tilde{G}_{t-2}^{(t)} &= \sigma \tilde{G}_{t-3}^{(t-1)} + \sigma \tilde{G}_{t-1}^{(t-1)} + (1-2\sigma)\tilde{G}_{t-2}^{(t-2)} \\ &= \sigma \tilde{G}_{t-3}^{(t-1)} = \sigma^2 \tilde{G}_{t-4}^{(t-2)} = \dots \\ &= \sigma^{t-2} \tilde{G}_0^{(2)} \\ &= \sigma^{t-2}(1-2\sigma) \\ &< 0, \end{aligned} \tag{6.47}$$

where the second equality holds since $\tilde{G}_m^{(t)} = 0$ for $-t+2 \leq m \leq t-2$. We assert that $\tilde{G}_m^{(t)} \leq \sigma \tilde{G}_{m-1}^{(t-1)}$ when $0 \leq m \leq t-2$. Indeed, the cases where $t=2$ and $t=3$ are easily verified. Assume by induction that our assertion holds for $t = 2, 3, \dots, j$. Then

$$\begin{aligned} \tilde{G}_m^{(j+1)} &= \sigma \tilde{G}_{m-1}^{(j)} + \sigma \tilde{G}_{m+1}^{(j)} + (1-2\sigma)\tilde{G}_m^{(j-1)} \\ &\leq \sigma \tilde{G}_{m-1}^{(j)} + \sigma^2 \tilde{G}_m^{(j-1)} + (1-2\sigma)\tilde{G}_m^{(j-1)} \\ &\leq \sigma \tilde{G}_{m-1}^{(j)} + \sigma \tilde{G}_{j-2}^{(j)} + (1-2\sigma)\tilde{G}_m^{(j-1)} \\ &= \sigma \tilde{G}_{m-1}^{(j)} + (1-\sigma)^2 \tilde{G}_m^{(j-1)} \\ &\leq \tilde{G}_{m-1}^{(j)} \end{aligned}$$

as required. Similarly, we may show that for $1/2 < \sigma < 1$, the symmetric Green's function $K_m^{(t)} = K_{-m}^{(t)} \geq 0$. Thus

$$\begin{aligned} \left| u_m^{(t)} \right| &\leq \left| \sum_{k=-\infty}^{\infty} \tilde{G}_k^{(t)} f_{m-k} \right| + \left| \sum_{k=-\infty}^{\infty} K_k^{(t)} g_{m-k} \right| \\ &\leq F \frac{2\sigma - 1}{1-\sigma} + \tilde{G} \frac{1}{1-\sigma} \\ &\leq \frac{2\sigma}{1-\sigma} \max\{F, \tilde{G}\} \end{aligned}$$

for $t = 2, 3, \dots$ and $m \in Z$. This shows that the DuFort Frankel scheme is stable (for an alternate approach by means of the theory of a complex variable, see pages 128–130 of [67].).

As an interesting remark, note that (6.47) suggests that the principal Green's function $\{\tilde{G}_m^{(t)}\}$ for the equation

$$u_m^{(t+1)} = \sigma u_{m-1}^{(t)} + \sigma u_{m+1}^{(t)} + (1-2\sigma) u_m^{(t)}, \quad \sigma \in R, (m,t) \in \Omega$$

satisfies

$$\tilde{G}_{t-2}^{(t)} = \sigma^{t-2}\tilde{G}_0^{(2)} = \sigma^{t-2}(1-2\sigma).$$

Thus $\lim_{t\to\infty}\left|\tilde{G}_{t-2}^{(t)}\right| = \infty$, which shows that the corresponding partial difference equation is not stable.

Exact solutions can also be used to show instability. Let us consider the Richardson partial difference scheme again:

$$u_m^{(t+1)} = \gamma u_m^{(t)} - 2\gamma u_m^{(t)} + \gamma u_{m+1}^{(t)} + u_m^{(t-1)},\ \gamma > 0,\ (m,t) \in \Omega.$$

Under the initial conditions

$$u_m^{(0)} = \alpha_0\beta^m,\ m \in Z,$$

and

$$u_m^{(1)} = \alpha_1\beta^m,\ m \in Z,$$

where $|\alpha_0| + |\alpha_1| \neq 0$ and β is some nonzero number, from Theorem 65, we know that the corresponding solution is of the form

$$u_m^{(t)} = \alpha_t\beta^m,$$

where α_t is defined by

$$\alpha_{t+1} = \lambda\alpha_t + \alpha_{t-1}, \lambda = \frac{\gamma}{\beta} - 2\gamma + \gamma\beta$$

for $t \geq 2$. An explicit form of $\{\alpha_t\}$ can be obtained by first solving the characteristic equation

$$\eta^2 - \lambda\eta - 1 = 0$$

to yield

$$\eta_\pm = \frac{\lambda}{2} \pm \sqrt{\left(\frac{\lambda}{2}\right)^2 + 1}$$

and then solving for the constants in $\alpha_t = E\eta_+^t + F\eta_-^t$ from the initial conditions. After these calculations, we obtain

$$u_m^{(t)} = \beta^m\left\{\frac{-\alpha_0\eta_- + \alpha_1}{\eta_+ - \eta_-}\eta_+^t + \frac{\alpha_0\eta_+ - \alpha_1}{\eta_+ - \eta_-}\eta_-^t\right\}, m \in Z, t \in N.$$

Note that when $\beta > 0$,

$$\lambda = \gamma\left(\frac{1}{\beta} + \beta\right) - 2\gamma \geq 2\gamma - 2\gamma = 0,$$

where equality holds only if $\beta = 1$. Thus $\eta_+ \geq 1$ and equality holds only if $\beta = 1$. Similarly, when $\beta < 0$, we have $\eta_- \leq 1$ and equality holds only if $\beta = -1$. In particular, the solution $\{u_m^{(t)}\}$ will diverge to infinity as $t \to \infty$ or $m \to \infty$ if $\beta > 1$. This shows that the Richardson scheme is unstable as has already be shown on pages 125–127 of [67]. In general, we have the following result, the proof of which is clear from the above development: Suppose $\beta \neq 0$ and the equation

$$\eta^2 - \left(\frac{a}{\beta} + b + c\beta\right)\eta - d\eta = 0$$

has a root η of magnitude strictly greater than 1, then the solution $\{u_m^{(t)}\}$ of (5.68) subject to the conditions (5.44) and (5.45) will satisfy

$$\lim_{t\to\infty} \left|u_m^{(t)}\right| = \infty$$

for each $m \in Z$.

6.3.3 Method of Induction for Three-Level Equations

Consider the equation

$$u_i^{(j+1)} = au_{i-1}^{(j)} + bu_i^{(j)} + cu_{i+1}^{(j)} + du_i^{(j-1)} + p_i^{(j)},\ i \in Z, j \in N, \tag{6.48}$$

where we assume that $d \neq 0$ and $|a| + |b| + |c| \neq 0$ to avoid degenerate cases. Auxiliary initial conditions of the form

$$u_i^{(0)} = \phi_i,\ i \in Z, \tag{6.49}$$

and

$$u_i^{(-1)} = \psi_i,\ i \in Z, \tag{6.50}$$

are imposed. Note that a solution of (6.48) is a bivariate sequence $u = \{u_i^{(j)}\}$ defined for $i \in Z$ and $j = \{-1\} \cup N$ and can be calculated successively in the following unique manner: $u_0^{(1)}$; $u_{-1}^{(1)}, u_1^{(1)}, u_0^{(2)}$; $u_{-2}^{(1)}, u_2^{(1)}, u_{-1}^{(2)}, u_1^{(2)}, u_0^{(3)}$; ... . Thus the value $u_i^{(1)}$ is determined by the values $\phi_{i-1}, \phi_i, \phi_{i+1}, \psi_i$ and $p_i^{(0)}$. By inductive arguments, it is not difficult to see that for $j \geq 1$, the value $u_i^{(j)}$ depends on $\phi_{i-j}, \phi_{i-j+1}, ..., \phi_{i+j}, \psi_{i-j+1}, .., \psi_{i+j-1}$ and

$$p_i^{(j-1)}; p_{i-1}^{(j-2)}, p_i^{(j-2)}, p_{i+1}^{(j-2)}; ..., p_{i-j+1}^{(0)}, ..., p_{i+j-1}^{(0)}.$$

For this reason, we will employ the following notations:

$$\|(\phi,\psi)\|_{ij} = \max\{|\phi_{i-j}|, |\phi_{i-j+1}|, ..., |\phi_{i+j}|; |\psi_{i-j+1}|, ..., |\psi_{i+j-1}|\},$$

and

$$\|p\|_s^{(t)} = \max\left\{\left|p_i^{(j)}\right| \mid s-t+j \leq i \leq s+t-j, 0 \leq j \leq t\right\}.$$

Theorem 89 *If* $0 < |a| + |b| + |c| + |d| \leq 1$, *then the unique solution* $u = \{u_i^{(j)}\}$ *of (6.48)–(6.50) will satisfy*

$$\left|u_i^{(j)}\right| \leq \|(\phi,\psi)\|_{ij}\, \xi^{\lfloor (j+1)/2 \rfloor} + \|p\|_i^{(j-1)} \left(1 + \xi + ... + \xi^{j-1}\right),$$

for $j \in Z^+$ *and* $i \in Z$, *where* $\xi = |a| + |b| + |c| + |d|$, *and* $\lfloor t \rfloor$ *denotes the greatest integer less than or equal to* t.

Proof. First of all, we have

$$\begin{aligned}\left|u_i^{(1)}\right| &\le |a|\,|\phi_{i-1}| + |b|\,|\phi_i| + |c|\,|\phi_{i+1}| + |d|\,|\psi_i| + \left|p_i^{(0)}\right| \\ &\le \xi\,\|(\phi,\psi)\|_{i1} + \|p\|_i^{(0)}\end{aligned}$$

for all i. Similarly, we have

$$\begin{aligned}\left|u_i^{(2)}\right| &\le |a|\left|u_{i-1}^{(1)}\right| + |b|\left|u_i^{(1)}\right| + |c|\left|u_{i+1}^{(1)}\right| + |d|\left|u_i^{(0)}\right| + \left|p_i^{(1)}\right| \\ &\le |a|\left(\xi\,\|(\phi,\psi)\|_{i-1,1} + \|p\|_{i-1}^{(0)}\right) + |b|\left(\xi\,\|(\phi,\psi)\|_{i1} + \|p\|_i^{(0)}\right) \\ &\quad + |c|\left(\xi\,\|(\phi,\psi)\|_{i+1,1} + \|p\|_{i+1}^{(0)}\right) + |d|\,|\phi_i| + \left|p_i^{(1)}\right| \\ &\le \xi\,(|a|+|b|+|c|)\,\|(\phi,\psi)\|_{i2} + |d|\,|\phi_i| \\ &\quad + (|a|+|b|+|c|)\,\|p\|_i^{(1)} + \left|p_i^{(1)}\right| \\ &\le \xi\,\|(\phi,\psi)\|_{i2} + (|a|+|b|+|c|+|d|+1)\,\|p\|_i^{(1)} \\ &\le \xi\,\|(\phi,\psi)\|_{i2} + (1+\xi)\,\|p\|_i^{(1)}\end{aligned}$$

for all i. Assume by induction that

$$\left|u_i^{(j)}\right| \le \|(\phi,\psi)\|_{ij}\,\xi^{\lfloor (j+1)/2\rfloor} + \|p\|_i^{(j-1)}\left(1+\xi+\xi^2+\dots+\xi^{j-1}\right)$$

holds for all $i \in Z$ and $j = 1, 2, \dots, k$. Then

$$\left|u_i^{(k+1)}\right|$$

$$\begin{aligned}&\le |a|\left|u_{i-1}^{(k)}\right| + |b|\left|u_i^{(k)}\right| + |c|\left|u_{i+1}^{(k)}\right| + |d|\left|u_i^{(k-1)}\right| + \left|p_i^{(k)}\right| \\ &\le |a|\,\|(\phi,\psi)\|_{i-1,k}\,\xi^{\lfloor (k+1)/2\rfloor} + |b|\,\|(\phi,\psi)\|_{ik}\,\xi^{\lfloor (k+1)/2\rfloor} \\ &\quad + |c|\,\|(\phi,\psi)\|_{i+1,k}\,\xi^{\lfloor (k+1)/2\rfloor} + |d|\,\|(\phi,\psi)\|_{i,k-1}\,\xi^{\lfloor k/2\rfloor} \\ &\quad + \left(|a|\,\|p\|_{i-1}^{(k-1)} + |b|\,\|p\|_i^{(k-1)} + |c|\,\|p\|_{i+1}^{(k-1)}\right)\left(1+\xi+\xi^2+\dots+\xi^{k-1}\right) \\ &\quad + |d|\,\|p\|_i^{(k-2)}\left(1+\xi+\xi^2+\dots+\xi^{k-2}\right) + \left|p_i^{(k)}\right|\end{aligned}$$

for all i. Since

$$\xi^{\lfloor (k+1)/2\rfloor}(|a|+|b|+|c|) + \xi^{\lfloor k/2\rfloor}\,|d| \le \xi^{\lfloor (k+2)/2\rfloor},$$

we finally see that

$$\left|u_i^{(k+1)}\right| \le \|(\phi,\psi)\|_{i,k+1}\,\xi^{\lfloor (k+1)/2\rfloor} + \|p\|_i^{(k)}\left(1+\xi+\dots+\xi^k\right)$$

for all i. The proof is complete. ■

As an immediate corollary, we see that when either one of the following sets of conditions hold: (i) $\{p_i^{(j)}\}$ is identically zero, $|a|+|b|+|c|+|d| \le 1$, and $\{\phi_i\}$ as well as $\{\psi_i\}$ are bounded; or (i) $|a|+|b|+|c|+|d|<1$, and $\{\phi_i\}$, $\{\psi_i\}$ as well as $\{p_i^{(j)}\}$ are bounded, then the corresponding solution of (6.48)–(6.50) is also bounded.

As another corollary, we see that if $\{p_i^{(j)}\}$ is identically zero, $|a|+|b|+|c|+|d|<1$, and $\{\phi_i\}$ as well as $\{\psi_i\}$ are bounded, then the corresponding solution of (6.50) will tend to zero uniformly in i as j tends to infinity.

We remark that in the above result, the bivariate sequence $p=\{p_i^{(j)}\}$ is quite general. In case $p=\{p_i^{(j)}\}$ is a subexponential bivariate sequence satisfying the condition

$$\left|p_i^{(j)}\right| \le M\beta^j,\ j\in N; i\in Z, \tag{6.51}$$

where M and β are some positive numbers, a more specific result can be given.

Theorem 90 *Suppose* $0<|a|+|b|+|c|+|d|\le 1$ *and* $p=\{p_i^{(j)}\}$ *satisfies condition (6.51). Then the unique solution of (6.48)–(6.50) satisfies*

$$\left|u_i^{(j)}\right| \le \|(\phi,\psi)\|_{ij}\,\xi^{\lfloor (j+1)/2\rfloor} + M\omega_j$$

for $j\in Z^+$ *and* $i\in Z$, *where* $\xi=|a|+|b|+|c|+|d|$, *and the sequence* $\{\omega_j\}_{j=1}^{\infty}$ *is defined by* $\omega_1=1, \omega_2=|a|+|b|+|c|+\beta$, *and*

$$\omega_{n+1}=(|a|+|b|+|c|)w_n+|d|\,w_{n-1}+\beta^n,\ n=2,3,\dots. \tag{6.52}$$

Proof. First of all, we have

$$\left|u_i^{(1)}\right| \le \xi\,\|(\phi,\psi)\|_{i1}+\|p\|_i^{(0)} \le \xi\,\|(\phi,\psi)\|_{i1}+M\omega_1$$

for all i. Similarly, we have

$$\begin{aligned}
\left|u_i^{(2)}\right| &\le |a|\left|u_{i-1}^{(1)}\right|+|b|\left|u_i^{(1)}\right|+|c|\left|u_{i+1}^{(1)}\right|+|d|\left|u_i^{(0)}\right|+\left|p_i^{(1)}\right| \\
&\le |a|\left(\xi\,\|(\phi,\psi)\|_{i-1,1}+M\right)+|b|\left(\xi\,\|(\phi,\psi)\|_{i1}+M\right) \\
&\quad +|c|\left(\xi\,\|(\phi,\psi)\|_{i+1,1}+M\right)+|d|\,|\phi_i|+M\beta \\
&\le \xi\left(|a|+|b|+|c|\right)\|(\phi,\psi)\|_{i2}+|d|\,|\phi_i| \\
&\quad +\left(|a|+|b|+|c|\right)M+M\beta \\
&\le \xi\,\|(\phi,\psi)\|_{i2}+\left(|a|+|b|+|c|+\beta\right)M \\
&\le \xi\,\|(\phi,\psi)\|_{i2}+M\omega_2
\end{aligned}$$

for all i. Next,

$$\begin{aligned}
\left|u_i^{(3)}\right| &\le |a|\left|u_{i-1}^{(2)}\right|+|b|\left|u_i^{(2)}\right|+|c|\left|u_{i+1}^{(2)}\right|+|d|\left|u_i^{(1)}\right|+\left|p_i^{(2)}\right| \\
&\le \xi^2\,\|(\phi,\psi)\|_{i3}+(|a|+|b|+|c|)M\omega_2+|d|\,M\omega_1+M\beta^2 \\
&= \xi^2\,\|(\phi,\psi)\|_{i3}+M\omega_3.
\end{aligned}$$

The rest of the proof may now be completed by induction. ■

The condition (6.51) implies that for each $j\in N$, $F_j=\sup\left\{\left|p_i^{(j)}\right| | i\in Z\right\}$ is finite. The question then arises as to whether the finiteness of each F_j already implies a result similar to Theorem 90. Indeed, we have the following result, the proof of which is similar to that of Theorem 90: Suppose $0 < |a| + |b| + |c| + |d| \le 1$ and $F_j < \infty$ for $j \in N$. Then the unique solution of (6.48)–(6.50) satisfies

$$\left|u_i^{(j)}\right| \le \|(\phi, \psi)\|_{ij} \, \xi^{\lfloor (j+1)/2 \rfloor} + \eta_j$$

for $j \in Z^+$ and $i \in Z$, where $\xi = |a| + |b| + |c| + |d|$, and the sequence $\{\eta_j\}_{j=1}^{\infty}$ is defined by

$$\eta_1 = F_0, \ \eta_2 = (|a| + |b| + |c|)F_0 + F_1, \tag{6.53}$$

and

$$\eta_{n+1} = (|a| + |b| + |c|)\eta_n + |d| \, \eta_{n-1} + F_n, \ n = 2, 3, \dots . \tag{6.54}$$

In view of this result, the asymptotic behavior of the solutions of (6.48) depends on the asymptotic behavior of the solution of (6.54) defined by (6.53). The explicit solution of (6.53)–(6.54) can, however, be calculated by means of the method of variation of parameters. Indeed, for the more general difference equation

$$y_{n+2} + f_n y_{n+1} + g_n y_n = h_n, \ n \in Z^+, \tag{6.55}$$

where $g_n \neq 0$ for $n \in Z^+$, a particular solution can be found by means of this method and is given by

$$-u_n \sum_{k=1}^{n-1} \frac{h_k v_{k+1}}{u_{k+1}v_{k+2} - v_{k+1}u_{k+2}} + v_n \sum_{k=1}^{n-1} \frac{h_k u_{k+1}}{u_{k+1}v_{k+2} - v_{k+1}u_{k+2}}, \tag{6.56}$$

where $\{u_k\}_{k=1}^{\infty}$ and $\{v_k\}_{k=1}^{\infty}$ are independent solutions of the homogeneous equation

$$y_{n+2} + f_n y_{n+1} + g_n y_n = 0, \ n \in Z^+.$$

Note that the homogeneous equation

$$\eta_{n+2} - (|a| + |b| + |c|)\eta_{n+1} - |d| \, \eta_n = 0, \ n \in Z^+,$$

has the solutions

$$u_n = \lambda_+^n, \ v_n = \lambda_-^n, \ n \in Z^+,$$

where λ_+ and λ_- are the solutions of the characteristic equation

$$\lambda^2 - (|a| + |b| + |c|)\lambda - |d| = 0,$$

i.e.

$$\lambda_\pm = \frac{(|a| + |b| + |c|) \pm \sqrt{(|a| + |b| + |c|)^2 + 4|d|}}{2}.$$

The solutions $\{u_n\}$ and $\{v_n\}$ are independent since their Casoratian is

$$\{u_n v_{n+1} - v_n u_{n+1}\}_{n=1}^{\infty} = \left\{\lambda_+^n \lambda_-^n (\lambda_- - \lambda_+)\right\}_{n=1}^{\infty}$$

and $|a| + |b| + |c| > 0, |d| > 0$ (recall that they are part of our assumptions on equation (6.48)). Thus, in view of (6.56), we see that

$$\eta_n = C\lambda_+^n + D\lambda_-^n - \frac{\lambda_+^n}{\lambda_- - \lambda_+} \sum_{k=1}^{n-1} \frac{F_{k+1}}{\lambda_+^{k+1}} + \frac{\lambda_-^n}{\lambda_- - \lambda_+} \sum_{k=1}^{n-1} \frac{F_{k+1}}{\lambda_-^{k+1}}, \ n \in Z^+,$$

is the general solution of (6.54), where C and D are arbitrary constants. Finally, substituting (6.53) into this general solution yields

$$C = \frac{(\lambda_- - \omega)F_0 - F_1}{\lambda_+(\lambda_- - \lambda_+)},\ D = \frac{(\omega - \lambda_+)F_0 + F_1}{\lambda_-(\lambda_- - \lambda_+)},$$

where $\omega = |a| + |b| + |c|$. We can now state the following result.

Theorem 91 *Suppose* $0 < |a| + |b| + |c| + |d| \leq 1$ *and* $F_j < \infty$ *for* $j \in N$. *Then the unique solution of (6.48)–(6.50) satisfies*

$$\left|u_i^{(j)}\right| \leq \|(\phi, \psi)\|_{ij}\, \xi^{\lfloor (j+1)/2 \rfloor} + \frac{(\lambda_- - \omega)F_0 - F_1}{\lambda_+(\lambda_- - \lambda_+)}\lambda_+^j +$$

$$\frac{(\omega - \lambda_+)F_0 + F_1}{\lambda_-(\lambda_- - \lambda_+)}\lambda_-^j - \frac{\lambda_+^j}{\lambda_- - \lambda_+}\sum_{k=1}^{j-1}\frac{F_{k+1}}{\lambda_+^{k+1}} + \frac{\lambda_-^j}{\lambda_- - \lambda_+}\sum_{k=1}^{j-1}\frac{F_{k+1}}{\lambda_-^{k+1}},$$

for $j \in Z^+$ *and* $i \in Z$, *where* $\omega = |a| + |b| + |c|$ *and* $\xi = \omega + |d|$.

Note that the conditions $\omega = |a|+|b|+|c| > 0$, $|d| > 0$ and $\omega+|d| \leq 1$ are sufficient for $|\lambda_\pm| \leq 1$. Indeed, since $|d| \leq 1-\omega$, we see that $\omega^2+4|d| \leq \omega^2+4(1-\omega) = (\omega-2)^2$, so that

$$\omega - 2 \leq \sqrt{\omega^2 + 4|d|} \leq 2 - \omega,$$

and

$$-1 \leq \frac{\omega \pm \sqrt{\omega^2 + 4|d|}}{2} \leq 1$$

as required. As a consequence, if in addition to the assumptions of Theorem 91, we assume further that $\{\phi_i\}$ and $\{\psi_i\}$ are bounded, and that the convolution products

$$\left\{\sum_{k=1}^{j-1} F_{k+1}\lambda_+^{j-k-1}\right\}_{j=2}^{\infty},\quad \left\{\sum_{k=1}^{j-1} F_{k+1}\lambda_-^{j-k-1}\right\}_{j=2}^{\infty}$$

are bounded sequences, then the corresponding solution of (6.48)–(6.50) is bounded.

As a final remark, our method of induction does not make use of any explicit solutions and is applicable to nonlinear equations such as

$$u_i^{(j+1)} = au_{i-1}^{(j)} + b\left(u_i^{(j)}\right)^2 + cu_{i+1}^{(j)} + du_i^{(j-1)}.$$

6.4 Equations Over Quadrants

As for equations defined on half planes, the method of exact solutions and the method of induction are useful tools for obtaining stability criteria. Emphasis will be placed on the latter method, however.

6.4.1 Method of Exact Solutions for a Two-Level Equation

Recall that the general solution of (5.23)

$$u_m^{(n+1)} - bu_{m+1}^{(n)} - cu_m^{(n)} = 0, \ m, n \in N, \tag{6.57}$$

subject to of the condition

$$u_j^{(0)} = f_j, \ j \in N, \tag{6.58}$$

has been found in Example 56. The trivial solution of (6.57) is said to be stable with respect to the above initial condition if there exists a positive constant Γ such that for every $f = \{f_j\}_{j=0}^{\infty}$, the solution of (6.57)–(6.58) will satisfy

$$\left|u_j^{(t)}\right| \le \Gamma \|f\|'_{jt}, \ j, t \in N, \tag{6.59}$$

where

$$\|\psi\|'_{jt} = \max_{j \le k \le j+t} |\psi_k|.$$

The trivial solution of (6.57) is said to be exponentially asymptotically stable with respect to the initial condition (6.58) if there is a real number $\xi \in (0,1)$ and a positive number Λ such that for every $\psi = \{\psi_k\}_{k=0}^{\infty}$, the solution of (6.57)–(6.58) satisfies

$$\left|u_j^{(t)}\right| \le \Lambda \xi^t \|\psi\|'_{jt}, \ j, t \in N. \tag{6.60}$$

We will derive explicit necessary and sufficient conditions in terms of the coefficients b and c for the trivial solution of (6.57) to be stable or asymptotically stable.

Theorem 92 *The trivial solution of (6.57)–(6.58) is stable if, and only if,* $|b|+|c| \le 1$.

Proof. Suppose $|b| + |c| \le 1$. Then in view of (5.25),

$$\left|u_j^{(t)}\right| \le \sum_{k=0}^{t} C_k^{(t)} |b|^{t-k} |c|^k |\psi_{j+t-k}|$$

$$\le \|\psi\|'_{jt} \sum_{k=0}^{j} C_k^{(t)} |b|^{t-k} |c|^k = (|b| + |c|)^t \|\psi\|'_{jt} \le \|\psi\|'_{jt}$$

as required. Conversely, suppose to the contrary that $|b| + |c| > 1$ and there is a positive constant Γ such that (6.59) holds for every $\psi = \{\psi_j\}_{j=0}^{\infty}$. There are several cases to consider. First of all, suppose $b, c > 0$. Let $\psi = \{d\}_{k=0}^{\infty}$ where $d > 0$. Then in view of (5.25),

$$u_j^{(t)} = (-1)^t \sum_{k=0}^{t} C_k^{(t)} b^{t-k} c^k d = (-1)^t (b+c)^t \|\psi\|'_{jt},$$

which implies

$$\left|u_j^{(t)}\right| = (b+c)^t \|\psi\|'_{jt}, \ j, i \in N.$$

Since $(b+c) > 1$, there is an integer T such that $(b+c)^t > \Gamma$ for $t \geq T$. Therefore,

$$\left|u_j^{(t)}\right| > \Gamma \|\psi\|'_{jt}, \; t \geq T, \tag{6.61}$$

which is a contradiction. Next, consider the case $b > 0$ and $c < 0$. Let $\Psi = \{(-1)^j d\}_{j=0}^{\infty}$ where $d > 0$. Then in view of (5.25),

$$u_j^{(t)} = (-1)^t \sum_{k=0}^{t} C_k^{(t)} b^{t-k} c^k (-1)^{j+t-k} d = (-1)^j (b-c)^t \|\psi\|'_{jt}.$$

Again (6.61) holds by the same arguments. The other cases can be proved by similar arguments. The proof is complete. ■

We remark that by choosing a positive divergent sequence $\psi = \{\psi_j\}_{j=0}^{\infty}$, it is easily seen from (6.61) that $\left|u_j^{(t)}\right|$ will also diverge to $+\infty$ as $t \to \infty$. This kind of instability may be important in numerical computations by means of finite difference schemes.

Theorem 93 *The trivial solution of (6.57)-(6.58) is exponentially asymptotically stable if, and only if,* $|b| + |c| < 1$.

Proof. Suppose $|b| + |c| < 1$. Let $\xi = |b| + |c|$. In view of (5.25),

$$\left|u_j^{(t)}\right| = \sum_{k=0}^{t} C_k^{(t)} |b|^{t-k} |c|^k |\psi_{j+t-k}| = (b+c)^t \|\psi\|'_{jt} \leq \xi^t \|\psi\|'_{jt}$$

as required. Conversely, suppose to the contrary that $|b| + |c| \geq 1$ and there is a positive constant Λ and $\xi \in (0,1)$ such that (6.60) holds for every $\psi = \{\psi_k\}_{k=0}^{\infty}$. As in the proof of previous theorem, there are several cases to consider. We will only consider the case $b, c > 0$. Let $\psi = \{d\}_{k=0}^{\infty}$ where $d > 0$. Then in view of (5.25),

$$\left|u_j^{(t)}\right| = (a+c)^t \|\psi\|'_{jt}, \; j, t \in N.$$

Since $(a+c) \geq 1$, there is an integer T such that $(a+c)^t > \Lambda \xi^t$ for $t \geq T$. Therefore,

$$\left|u_j^{(t)}\right| > \Lambda \xi^t \|\psi\|'_{jt}, \; t \geq T,$$

which is a contradiction. The proof is complete. ■

6.4.2 Method of Induction for a Two-Level Nonhomogeneous Equation

In this section, we will consider the nonhomogeneous discrete heat equation

$$u_j^{(t+1)} + b u_{j+1}^{(t)} + c u_j^{(t)} = f_j^{(t)}, \; j, t \in N, \tag{6.62}$$

where a, b and c are real numbers and $f = \{f_j^{(t)}\}_{j,t=0}^{\infty}$ is a real bivariate sequence.

A solution of (6.62) is a real bivariate sequence $u = \{u_j^{(t)}\}_{j,t=0}^{\infty}$ which satisfies (6.62). First of all, recall the subsets

$$Q_k = \{(j,t) | \; j,t \in N, \; j+t = k\}, \; k \in N,$$

and the ordering $\preccurlyeq$ introduced for the nonnegative quadrant $N \times N$ in Section 1 of Chapter 2. By means of this ordering , we have a precise method for traversing the values of a solution of (6.62): for any $(j,t),(m,n) \in N \times N$, $(j,t) \preccurlyeq (m,n)$ if either (i) $(j,t) \in Q_k$ and $(m,n) \in Q_l$ and $k < l$, or (ii) (j,t), $(m,n) \in Q_k$ and $j > m$. Since (6.62) can be written in the form

$$u_j^{(t+1)} = -bu_{j+1}^{(t)} - cu_j^{(t)} + f_j^{(t)}, \; j,t \in N, \tag{6.63}$$

it is clear that if $u_j^{(0)}$ is given for each $j \in N$,

$$u_j^{(0)} = \psi_j, \; j \in N, \tag{6.64}$$

then we can calculate $u_0^{(1)}$; $u_1^{(1)}, u_0^{(2)}$; $u_2^{(1)}, u_1^{(2)}, u_0^{(3)}$; ... successively in a unique manner. Indeed, $u(Q_0)$ is ψ_0. Assume by induction that $u(Q_k)$ has been determined for each $k \leq K$. We show that $u(Q_{K+1})$ can also be calculated. To see this, note first that $u_{K+1}^{(0)} = \psi_{K+1}$. Assume by induction that $(m,n) \in Q_{K+1}$ and $u_j^{(t)}$ has been determined for each $(j,t) \in Q_{K+1}$ which satisfies $(j,t) \preccurlyeq (m,n)$ and $(j,t) \neq (m,n)$. Then since $(m-1,n+1) \preccurlyeq (m,n)$ and $(m-1,n) \in Q_K$, thus $u_n^{(m)}$ can be calculated from

$$u_n^{(m)} = -bu_{n+1}^{(m-1)} - cu_n^{(m-1)} + f_n^{(m-1)} \tag{6.65}$$

as required.

We remark that since $u(Q_{k+1})$ can be determined by $u(Q_k)$, which in turn is determined by $u(Q_{k-1})$, etc., it is clear that $u(Q_{k+1})$ is determined by the initial values $\{\psi_0, ..., \psi_{k+1}\}$ and the forcing terms $\{f_j^{(t)} | \; 0 \leq j+t \leq k\}$. For this reason, we will employ the following two notations in this section:

$$\|\psi\|_j^{(t)} = \max\{|\psi_k| \; | \; 0 \leq k \leq j+t\}, \; j,t \in N,$$

and

$$\|f\|_j^{(t)} = \begin{cases} \max\{\left|f_k^{(s)}\right| \; | \; 0 \leq s,k \leq j+t\} & j,t \in N \\ 0 & \text{otherwise} \end{cases}.$$

Theorem 94 *Suppose $|b| + |c| < 1$. Then there exists a number ξ, which is zero if $|b| + |c| = 0$ and belongs to $(0,1)$ otherwise, such that the solution $\{u_j^{(t)}\}$ of (6.62) determined by the initial condition (6.64) will also satisfy*

$$\left|u_j^{(t)}\right| \leq \|\psi\|_j^{(t)} \xi^t + \frac{\|f\|_j^{(t-1)}}{1-\xi}, \; j,t \in N. \tag{6.66}$$

Proof. Take $\xi = |b| + |c|$. Then $\xi \in [0,1)$ by our assumptions. We assert that

$$\left|u_j^{(t)}\right| \leq \|f\|_j^{(t-1)} (1 + \xi + ... + \xi^{t-1}) + \|\psi\|_j^{(t)} \xi^t \tag{6.67}$$

for $j, t \in N$. Note first that, in view of (6.64),

$$\left|u_0^{(0)}\right| = |\psi_0| = \|\psi\|_{00},$$

$$\left|u_1^{(0)}\right| = |\psi_1| \le \|\psi\|_{01},$$

and in view of (6.62),

$$\left|u_0^{(1)}\right| \le |b| \left|u_1^{(0)}\right| + |c| \left|u_0^{(0)}\right| + \left|f_0^{(0)}\right| \le \|\psi\|_0^{(1)} \xi + \|f\|_0^{(0)}.$$

We have thus shown that (6.67) holds for $(j,t) \in Q_0$ and $(j,t) \in Q_1$. Assume by induction that (6.67) holds for $(j,t) \in Q_k$ and $k \le K$; we assert that (6.67) also holds for $(j,t) \in Q_{K+1}$. Indeed,

$$\left|u_{K+1}^{(0)}\right| = |\psi_{K+1}| \le \|\psi\|_{K+1}^{(0)}.$$

Assume by induction that $(m,n) \in Q_{K+1}$ and (6.67) holds for each $(j,t) \in Q_{K+1}$ which satisfies $(j,t) \preceq (m,n)$ and $(j,t) \ne (m,n)$. Then in view of (6.65),

$$\begin{aligned}
\left|u_n^{(m)}\right| &\le |b| \left|u_{n+1}^{(m-1)}\right| + |c| \left|u_n^{(m-1)}\right| + \left|f_n^{(m-1)}\right| \\
&\le (|b| + |c|) \left\{ \|f\|_n^{(m-1)} (1 + \xi + \dots + \xi^{m-2}) + \|\psi\|_n^{(m)} \xi^{m-1} \right\} + \|f\|_n^{(m-1)} \\
&= \|f\|_n^{(m-1)} (1 + \xi + \dots + \xi^{m-1}) + \|\psi\|_n^{(m)} \xi^m,
\end{aligned}$$

as required. Finally,

$$\left|u_j^{(t)}\right| \le \|f\|_j^{(t-1)} (1 + \xi + \dots) + \|\psi\|_j^{(t)} \xi^t = \frac{\|f\|_j^{(t-1)}}{1-\xi} + \|\psi\|_j^{(t)} \xi^t, \quad j,t \in N.$$

■

As a consequence, suppose $|b| + |c| < 1$, $\sup_{j \ge 0} |\psi_j| = \|\psi\| < \infty$ and

$$\sup_{j,t \ge 0} \left|f_j^{(t)}\right| = \|f\| < \infty;$$

then there exists $\xi \in [0,1)$ such that the solution $\{u_j^{(t)}\}$ of (6.62) determined by the initial condition (6.64) will also satisfy

$$\left|u_j^{(t)}\right| \le \|\psi\| \xi^t + \frac{\|f\|}{1-\xi}, \; j,t \in N.$$

Roughly speaking, this says that when the system parameters b and c have small magnitudes, then under bounded input and bounded initial conditions, the corresponding output is also bounded. Furthermore, if the input f is identically zero, then

$$\left|u_j^{(t)}\right| \le \|\psi\| \xi^t, \; j,t \in N,$$

that is, $\{u_j^{(t)}\}$ is uniformly subexponentially decaying in t.

By means of symmetric arguments, we may also show the following dual statement of Theorem 94.

Theorem 95 *Suppose $|b| > |c| + 1$. Then there exists $\xi \in (0,1)$ such that the solution $\{u_j^{(t)}\}$ of (6.62) determined by the initial condition*

$$u_0^{(t)} = \phi_t,\ t \in N, \tag{6.68}$$

will also satisfy

$$\left|u_j^{(t)}\right| \le \|\phi\|_j^{(t)} \xi^j + \frac{\|f\|_j^{(t-1)}}{|b|(1-\xi)},\ j,t \in N.$$

Theorem 96 *Suppose $|b| + |c| < 1$ and f is subexponentially decaying, i.e.,*

$$\left|f_j^{(t)}\right| \le M\alpha^t\beta^j, \quad \alpha,\beta \in (0,1);\ j,t \in N;\ M \ge 0.$$

Then there exists $\xi \in (\alpha, 1)$ such that the solution of (6.62) determined by the initial condition (6.64) is also subexponentially decaying and satisfies

$$\left|u_j^{(t)}\right| \le \frac{M}{\xi-\alpha}\xi^t\beta^j + \|\psi\|_j^{(t)}\xi^t,\ j,t \in N.$$

Proof. Take $\xi = \max\{|b| + |c|, (\alpha+1)/2\}$. Then $\xi \in (\alpha,1)$ by assumption. We assert that

$$\left|u_j^{(t)}\right| \le M\beta^j\xi^{t-1}\left(1 + \frac{\alpha}{\xi} + \dots + \left(\frac{\alpha}{\xi}\right)^{t-1}\right) + \|\psi\|_j^{(t)}\xi^t \tag{6.69}$$

for $j,t \in N$. Note first that, in view of (6.64),

$$\left|u_0^{(0)}\right| \le \|\psi\|_0^{(0)}, \quad \left|u_1^{(0)}\right| \le \|\psi\|_1^{(0)},$$

and in view of (6.62),

$$\left|u_0^{(1)}\right| \le \|\psi\|_0^{(1)}\xi + \left|f_0^{(0)}\right| \le \|\psi\|_0^{(1)}\xi + M.$$

We have thus shown that (6.69) holds for $(j,t) \in Q_0$ and for $(j,t) \in Q_1$. Assume by induction that (6.69) holds for $(j,t) \in Q_k$ and $k \le K$; we assert that (6.69) also holds for $(j,t) \in Q_{K+1}$. Indeed,

$$\left|u_{K+1}^{(0)}\right| = |\psi_{K+1}| \le \|\psi\|_{K+1}^{(0)}.$$

Assume by induction that $(m,n) \in Q_{K+1}$ and (6.69) holds for $(j,t) \in Q_{K+1}$ which satisfies $(j,t) \preccurlyeq (m,n)$ and $(j,t) \neq (m,n)$. Then in view of (6.65), we have

$$\begin{aligned}\left|u_n^{(m)}\right| \quad &\le \quad |b|\left\{M\beta^{n+1}\xi^{m-2}\left[1+\left(\frac{\alpha}{\xi}\right)+\dots\left(\frac{\alpha}{\xi}\right)^{m-2}\right] + \|\psi\|_n^{(m)}\xi^{m-1}\right\} \\ &\quad +|c|\left\{M\beta^n\xi^{m-2}\left[1+\left(\frac{\alpha}{\xi}\right)+\dots+\left(\frac{\alpha}{\xi}\right)^{m-2}\right] + \|\psi\|_{mn}\xi^{m-1}\right\} \\ &\quad +M\alpha^{m-1}\beta^n\end{aligned}$$

$$\begin{aligned}
&= M\beta^n\xi^{m-2}\left[1+\left(\frac{\alpha}{\xi}\right)+\ldots+\left(\frac{\alpha}{\xi}\right)^{m-2}\right](\beta|b|+|c|) \\
&\quad +\|\psi\|_n^{(m)}\xi^{m-1}(|b|+|c|)+M\alpha^{m-1}\beta^n \\
&\le M\beta^n\xi^{m-1}\left[1+\left(\frac{\alpha}{\xi}\right)+\ldots+\left(\frac{\alpha}{\xi}\right)^{m-2}\right]+\|\psi\|_n^{(m)}\xi^m+M\alpha^{m-1}\beta^n \\
&\le M\beta^n\xi^{m-1}\left[1+\left(\frac{\alpha}{\xi}\right)+\ldots+\left(\frac{\alpha}{\xi}\right)^{m-1}\right]+\|\psi\|_n^{(m)}\xi^m,
\end{aligned}$$

as required. Finally,

$$\begin{aligned}
\left|u_j^{(t)}\right| &\le M\beta^j\xi^{t-1}\left[1+\left(\frac{\alpha}{\xi}\right)+\ldots\right]+\|\psi\|_j^{(t)}\xi^t \\
&\le \frac{M}{\xi-\alpha}\xi^t\beta^j+\|\psi\|_j^{(t)}\xi^t, \quad j,t\in N.
\end{aligned}$$

■

As a consequence, suppose $|b|+|c|<1$, $\sup_{j\ge 0}|\psi_j| = \|\psi\| < \infty$, and f is subexponentially decaying as in the above theorem. Then there exists $\xi \in (\alpha,1)$ such that the solution of (6.62) determined by the initial condition (6.64) is also subexponentially decaying and satisfies

$$\left|u_j^{(t)}\right| \le \frac{M}{\xi-\alpha}\xi^t\beta^j+\|\psi\|\xi^t, \quad j,t\in N.$$

Roughly speaking, this says that when the system parameters b and c have small magnitudes, then under subexponentially decaying input and bounded initial conditions, the corresponding output is also subexponentially decaying.

By means of symmetric arguments, we may also show the following dual statement of Theorem 96.

Theorem 97 *Suppose $|b|>|c|+1$, and f is subexponentially decaying, i.e.,*

$$\left|f_j^{(t)}\right| \le M\alpha^t\beta^j, \quad \alpha,\beta\in(0,1);\ j,t\in N;\ M\ge 0.$$

Then there exists $\xi\in(\beta,1)$ such that the solution of (6.62) determined by the initial condition (6.68) is also subexponentially decaying and satisfies

$$\left|u_j^{(t)}\right| \le \frac{M}{|b|(\xi-\beta)}\alpha^t\xi^j+\|\phi\|_j^{(t)}\xi^j,\ j,t\in N,$$

where

$$\|\phi\|_j^{(t)} = \max\{|\phi_k| \mid 0\le k\le j+t\}.$$

It is well known that every solution of a linear ordinary difference equation with constant coefficients is subexponential. This fact is important since it implies that small errors introduced in the initial data may accumulate only in an exponential fashion when the solution of the difference equation is calculated in a recursive manner. However, not all solutions of partial difference equations of the form (6.62) are subexponential. Indeed Example 48 provides a counter-example. However, as we will see, if the sequences $\{f_j^{(t)}\}$ and $\{u_j^{(0)}\}$ are subexponential, then every solution of (6.62) is also subexponential.

Theorem 98 *Let $\{f_j^{(t)}\}$ and $\{\psi_j\}$ be subexponential such that*

$$\left|f_j^{(t)}\right| \leq M_1\alpha_1^t\beta_1^j,\ j,t \in N;\ M_1 \geq 0,$$

and

$$|\psi_j| \leq M_2\beta_2^j,\ j \in N;\ M_2 \geq 0$$

respectively. Then every solution $\{u_j^{(t)}\}$ of (6.62) subject to the initial condition (6.64) is also subexponential.

Proof. We assume that $N \times N$ is linearly ordered by the same ordering defined at the beginning. Let $M = \max\{M_1, M_2\}$, $\beta = \max\{\beta_1, \beta_2\}$ and $\alpha = |b|\beta + |c| + 1$. We assert that

$$\left|u_j^{(t)}\right| \leq M\alpha^t\beta^j \tag{6.70}$$

holds for $j, t \in N$. Note first that

$$\left|u_0^{(0)}\right| = |\psi_0| \leq M_2 \leq M,$$

$$\left|u_1^{(0)}\right| = |\psi_1| \leq M_2\beta_2 \leq M\beta,$$

and

$$\begin{aligned}\left|u_0^{(1)}\right| &\leq |b|\left|u_1^{(0)}\right| + |c|\left|u_0^{(0)}\right| + \left|f_0^{(0)}\right| \\ &\leq |b|\,M_2\beta_2 + |c|\,M_2 + M_1 \leq M\{|b|\,\beta_2 + |c| + 1\} \\ &\leq M\alpha.\end{aligned}$$

We have thus shown that (6.70) holds for $(j,t) \in Q_0$ and for $(j,t) \in Q_1$. Assume by induction that (6.70) holds for $(j,t) \in Q_k$ and $k \leq K$; we assert that (6.70) also holds for $(j,t) \in Q_{K+1}$. Indeed,

$$\left|u_{K+1}^{(0)}\right| = |\psi_{K+1}| \leq M_2\beta_2^{K+1} \leq M\beta^{K+1}.$$

Assume by induction that $(m,n) \in Q_{K+1}$ and (6.70) holds for $(j,t) \in Q_{K+1}$ which satisfies $(j,t) \preccurlyeq (m,n)$ and $(j,t) \neq (m,n)$. Then in view of (6.65), we have

$$\begin{aligned}\left|u_n^{(m)}\right| &\leq |b|\,M\alpha^{m-1}\beta^{n+1} + |c|\,M\alpha^{m-1}\beta^n + M_1\alpha_1^{m-1}\beta_1^n \\ &\leq M\alpha^{m-1}\beta^n\,\{|b|\,\beta + |c| + 1\} = M\alpha^m\beta^n,\end{aligned}$$

as required. ∎

By means of symmetric arguments, we may also show that if $\{f_j^{(t)}\}$ and $\{u_0^{(t)}\}$ are subexponential, then every solution of (6.62) subject to the initial condition (6.68) is also subexponential.

6.4.3 Method of Induction for a Four-Point Equation

In this section, we will consider the nonhomogeneous partial difference equation

$$u_{i+1,j+1} + au_{i+1,j} + bu_{i,j+1} + cu_{ij} = f_{ij},\ i,j \in N, \tag{6.71}$$

where a, b and c are real numbers and $f = \{f_{ij}\}_{i,j\in Z}$ is a real function.

Since (6.71) can be written as

$$u_{i+1,j+1} = -au_{i+1,j} - bu_{i,j+1} - cu_{ij} + f_{ij},\ i,j \in N,$$

it is clear that if the conditions

$$u_{0j} = \psi_j,\ j \in N, \tag{6.72}$$

$$u_{i0} = \phi_i,\ i \in N. \tag{6.73}$$

and the compatibility condition

$$\psi_0 = \phi_0, \tag{6.74}$$

are imposed, we can calculate u_{11}; u_{12}, u_{21}; u_{13}, u_{22}, u_{31}; ...successively in a unique manner.

In the proof of our next result, we will need the ordering $\preccurlyeq$ introduced in Section 1 of Chapter 2.

Theorem 99 *Suppose $\{f_{ij}\}$, $\{\phi_i\}_{i\in N}$ and $\{\psi_j\}_{j\in N}$ are subexponential sequences such that $\phi_0 = \psi_0$. Then every solution of (6.71) under the conditions (6.72) and (6.73) is also subexponential.*

Proof. Let

$$|\phi_i| \le M_1\alpha_1^i,\ i \in N,$$

$$|\psi_j| \le M_2\beta_2^j,\ j \in N,$$

and

$$|f_{ij}| \le M_3\alpha_3^i\beta_3^j,\ i,j \in N.$$

Let $M = \max\{M_1, M_2, M_3\}$, and let α and β be positive numbers such that $\alpha \ge \max\{\alpha_1, \alpha_3\}$, $\beta \ge \max\{\beta_2, \beta_3\}$ and

$$\frac{|a|}{\beta} + \frac{|b|}{\alpha} + \frac{|c|+1}{\alpha\beta} \le 1.$$

We assert that

$$|u_{ij}| \le M\alpha^i\beta^j \tag{6.75}$$

holds for $i, j \in N$. Note first that

$$|u_{00}| = |\psi_0| \le M,$$

$$|u_{01}| = |\psi_1| \le M_2\beta_2 \le M\beta,$$

and

$$|u_{10}| = |\phi_1| \le M_1\alpha_1 \le M\alpha.$$

We have thus shown that (6.75) holds for $(i,j) \in Q_0$ and for $(i,j) \in Q_1$. Assume by induction that (6.75) holds for $(i,j) \in Q_k$ and $k \le K$; we assert that (6.75) also holds for $(i,j) \in Q_{K+1}$. Indeed,

$$|u_{0,K+1}| = |\psi_{K+1}| \le M_2\beta_2^{K+1} \le M\beta^{K+1},$$

and

$$|u_{K+1,0}| = |\phi_{K+1}| \le M_1\alpha_1^{K+1} \le M\alpha^{K+1}.$$

Furthermore, for $(m,n) \in Q_{K+1}$ which does not equal to either $(0, K+1)$ or $(K+1, 0)$, we see from (6.71) that

$$\begin{aligned} |u_{mn}| &\le |a|\, M\alpha^m\beta^{n-1} + |b|\, M\alpha^{m-1}\beta^n + |c|\, M\alpha^{m-1}\beta^{n-1} + M_3\alpha_3^{m-1}\beta_3^{n-1} \\ &\le M\alpha^m\beta^n \left\{ \frac{|a|}{\beta} + \frac{|b|}{\alpha} + \frac{|c|+1}{\alpha\beta} \right\} \\ &\le M\alpha^m\beta^n. \end{aligned}$$

The proof is complete. ■

We remark that if the assumption of the above theorem is not satisfied, then its conclusion may not hold. Indeed, a counter-example has been provided (see Example 49).

The set $N \times N$ can also be partitioned into equivalence classes $W_0, W_1, \dots,$ defined by

$$W_k = \{(i,j) |\ i,j \in N,\ \max(i,j) = k\},\ k \in N.$$

The equivalence classes can be linearly ordered by their indices. Each W_k can further be broken down into disjoint subsets

$$\{(k,k)\},$$

$$W_k^- = \{(i,j) \in W_k |\ i < j\} = \{(0,k), (1,k), \dots, (k-1,k)\}$$

and

$$W_k^+ = \{(i,j) \in W_k |\ i > j\} = \{(k,0), (k,1), \dots, (k,k-1)\}.$$

The subsets W_k^- and W_k^+ can be linearly ordered by their first and second arguments respectively. Thus Z can be linearly ordered, and a solution of (6.71) subject to (6.72)–(6.74) determined.

We remark that since $u(W_{k+1})$ can be determined by $u(W_k)$, which in turn is determined by $u(W_{k-1})$, etc., it is clear that $u(W_{k+1})$ is determined by the initial values $\{\psi_0, \dots, \psi_{k+1}\}$, $\{\phi_0, \dots, \phi_{k+1}\}$ and the forcing terms $\{f_{ij} |\ 0 \le i,j \le k+1\}$. For this reason, we will employ the following two notations in this section:

$$\|(\psi,\phi)\|_{ij} = \max_{0 \le k \le \max(i,j)} \{|\psi_k|, |\phi_k|\},\ i,j \in N,$$

and

$$\|f\|_{ij}^* = \max\{|f(s,t)|\ |\ 0 \le s \le i, 0, \le t \le j\},\ i,j \in N.$$

Theorem 100 *Suppose $f_{ij} = 0$ for $i,j \in N$ and $|a| + |b| + |c| < 1$. Then there exists a number ξ, which is zero if $|a| + |b| + |c| = 0$, and belongs to $(0,1)$ otherwise, such that the solution $\{u_{ij}\}$ of (6.71) determined by the initial conditions (6.72) and (6.73) and the compatibility condition (6.74) will also satisfy*

$$|u_{ij}| \le \|(\psi,\phi)\|_{ij}\, \xi^{\min(i,j)},\ i,j \in N.$$

Proof. Let $\xi = |a| + |b| + |c|$. Then $0 \le \xi < 1$ by assumption. We assert that

$$|u_{ij}| \le \|(\psi, \phi)\|_{ij} \, \xi^{\min(i,j)} \tag{6.76}$$

holds for $i, j \in N$. Note first that, in view of (6.72) and (6.73),

$$|u_{00}| \le \|(\psi, \phi)\|_{00}, \quad |u_{01}| \le \|(\psi, \phi)\|_{01}, \quad |u_{10}| \le \|(\psi, \phi)\|_{10},$$

and in view of (6.71),

$$|u_{11}| \le (|a| + |b| + |c|) \, \|(\psi, \phi)\|_{10} = \|(\psi, \phi)\|_{11} \, \xi.$$

We have thus shown that (6.76) holds for $(i, j) \in W_0$ and for $(i, j) \in W_1$. Assume by induction that (6.76) holds for $(i, j) \in W_k$ and $k \le K$; we assert that (6.76) also holds for $(i, j) \in W_{K+1} = W_{K+1}^- \cup \{(K+1, K+1)\} \cup W_{K+1}^+$. Consider first the case $(i, j) \in W_{K+1}^-$. We show that this case is true by another induction. Note that

$$|u_{0,K+1}| = |\psi_{K+1}| \le \|(\psi, \phi)\|_{0,K+1}.$$

Assume by induction that

$$|u_{i,K+1}| \le \|(\psi, \phi)\|_{i,K+1} \, \xi^i$$

for all $0 \le i < m \le K$; then in view of (6.71), we see that

$$\begin{aligned}
|u_{m,K+1}| &\le |a| \, |u_{mK}| + |b| \, |u_{m-1,K+1}| + |c| \, |u_{m-1,K}| \\
&\le |a| \, \|(\psi, \phi)\|_{mK} \, \xi^m + |b| \, \|(\psi, \phi)\|_{m-1,K+1} \, \xi^{m-1} \\
&\quad + |c| \, \|(\psi, \phi)\|_{m-1,K} \, \xi^{m-1} \\
&\le \|(\psi, \phi)\|_{m,K+1} \, \xi^{m-1} (|a| + |b| + |c|) \\
&= \|(\psi, \phi)\|_{m,K+1} \, \xi^m = \|(\psi, \phi)\|_{m,K+1} \, \xi^{\min(m,K+1)}
\end{aligned}$$

We have thus shown that (6.76) holds for all $(i, j) \in W_{K+1}^-$. The case $(i, j) \in W_{K+1}^+$ is similarly proved. Finally,

$$\begin{aligned}
|u_{K+1,K+1}| &\le |a| \, |u_{K+1,K}| + |b| \, |u_{K,K+1}| + |c| \, |u_{KK}| \\
&\le |a| \, \|(\psi, \phi)\|_{K+1,K} \, \xi^K + |b| \, \|(\psi, \phi)\|_{K,K+1} \, \xi^K \\
&\quad + |c| \, \|(\psi, \phi)\|_{K,K} \, \xi^K \\
&\le \|(\psi, \phi)\|_{K+1,K+1} \, \xi^{K+1}.
\end{aligned}$$

■

As a consequence, suppose $f_{ij} = 0$ for $i, j \in N$, $|a| + |b| + |c| < 1$, and

$$\max \left\{ \sup_{j \ge 0} |\psi_j|, \sup_{i \ge 0} |\phi_i| \right\} = \|(\psi, \phi)\| < \infty.$$

Then there exists $\xi \in [0, 1)$ such that the solution $\{u_{ij}\}$ of (6.71) determined by the initial conditions (6.72) and (6.73) and the compatibility condition (6.74) will also satisfy

$$|u_{ij}| \le \|(\psi, \phi)\| \, \xi^{\min(i,j)}, \quad i, j \in N.$$

Roughly speaking, this says that when the system parameters a, b, c have small magnitudes, then under bounded initial conditions, the corresponding output is weakly subexponentially decaying.

Next, we consider the case where the input f is a subexponentially decaying function and the initial conditions are identically zero.

Theorem 101 *Suppose $\psi_j = \phi_j = 0$ for $j \in N$. Suppose $|a| + |b| + |c| < 1$, and*

$$|f_{ij}| \le M\sigma^i \tau^j, \ \sigma, \tau \in (0,1); \ M \ge 0; \ i, j \in N.$$

Then there exist $\alpha, \beta \in (0,1)$ and $M_1, M_2 \ge 0$ such that the solution $\{u_{ij}\}$ of (6.71) determined by the conditions (6.72)–(6.74) will also satisfy

$$|u_{ij}| \le \min\left(M_1\alpha^i, M_2\beta^j\right), \ i, j \in N.$$

Proof. Let

$$\alpha = \max\left\{\sigma, \frac{|a| + |b| + |c| + 1}{2}\right\},$$

and

$$\beta = \max\left\{\tau, \frac{|a| + |b| + |c| + 1}{2}\right\}.$$

Then by assumptions, $\alpha, \beta \in (0,1)$, $\sigma \le \alpha$, $\tau \le \beta$, $|b| \le |a| + |b| + |c| < \alpha$ and $|a| \le |a| + |b| + |c| < \beta$. We assert that

$$\begin{aligned} &|u_{ij}| \\ \le\ & \frac{M\alpha^i\beta\left(1+\beta+\ldots+\beta^{j-2}\right)}{(\alpha-|b|)(\beta-|a|)} \\ &+M\alpha^{i-1}\beta^{j-1}\left[1+\frac{|a|}{\beta}+\ldots+\left(\frac{|a|}{\beta}\right)^{j-1}\right]\left[1+\ldots+\left(\frac{|b|}{\alpha}\right)^{i-1}\right] \end{aligned} \tag{6.77}$$

for $i, j \in N$. Note first that, in view of (6.72) and (6.73),

$$|u_{00}| = |u_{01}| = |u_{10}| = 0,$$

and in view of (6.71),

$$|u_{11}| = |f_{00}| \le M.$$

We have thus shown that (6.77) holds and hence that

$$\begin{aligned} |u_{ij}| \le\ & \frac{M\alpha^i\beta\left(1+\beta+\ldots+\beta^{j-2}\right)}{(\alpha-|b|)(\beta-|a|)} \\ &+M\alpha^{i-1}\beta^{j-1}\left[1+\frac{|a|}{\beta}+\ldots+\left(\frac{|a|}{\beta}\right)^{j-1}\right]\left[1+\frac{|b|}{\alpha}+\ldots+\left(\frac{|b|}{\alpha}\right)^{i-1}\right] \\ \le\ & \frac{M\alpha^i\beta\left(1+\beta+\ldots+\beta^{j-2}\right)}{(\alpha-|b|)(\beta-|a|)} + \frac{M\alpha^i\beta^j}{(\alpha-|b|)(\beta-|a|)} \\ \le\ & \frac{M\alpha^i\beta\left(1+\beta+\ldots+\beta^{j-1}\right)}{(\alpha-|b|)(\beta-|a|)} \\ \le\ & \frac{M\beta}{(\alpha-|b|)(\beta-|a|)(1-\beta)}\alpha^i \end{aligned}$$

for $(i,j) \in W_0$ and for $(i,j) \in W_1$. Assume by induction that (6.77) holds for $(i,j) \in W_k$ and $k \le K$; we assert that (6.77) also holds for $(i,j) \in W_{K+1} = W_{K+1}^- \cup \{(K+1,K+1)\} \cup W_{K+1}^+$. Consider first the case $(i,j) \in W_{K+1}^-$. Note that

$$|u_{0,K+1}| = 0.$$

Assume by induction that

$$\begin{aligned} |u_{i,K+1}| \ \le\ & \frac{M\alpha^i\beta\left(1+\beta+\ldots+\beta^{K-1}\right)}{(\alpha-|b|)(\beta-|a|)} \\ & +M\alpha^{i-1}\beta^K\left[1+\frac{|a|}{\beta}+\ldots+\left(\frac{|a|}{\beta}\right)^K\right]\left[1+\frac{|b|}{\alpha}+\ldots+\left(\frac{|b|}{\alpha}\right)^{i-1}\right] \end{aligned}$$

holds for all $0 \le i < m \le K$; then in view of (6.71), we see that

$$\begin{aligned} & |u_{m,K+1}| \\ \le\ & M\alpha^{m-1}\beta^K + |a|\frac{M\alpha^m\beta\left(1+\beta+\ldots+\beta^{K-1}\right)}{(\alpha-|b|)(\beta-|\alpha|)} \\ & +|b|\frac{M\alpha^{m-1}\beta\left(1+\beta+\ldots+\beta^{K-1}\right)}{(\alpha-|b|)(\beta-|\alpha|)} \\ & +|b|\,M\alpha^{m-2}\beta^K\left[1+\frac{|a|}{\beta}+\ldots+\left(\frac{|a|}{\beta}\right)^K\right]\left[1+\frac{|b|}{\alpha}+\ldots+\left(\frac{|b|}{\alpha}\right)^{m-2}\right] \\ & +|c|\frac{M\alpha^{m-1}\beta\left(1+\beta+\ldots+\beta^{K-1}\right)}{(\alpha-|b|)(\beta-|a|)} \\ \le\ & M\alpha^{m-1}\beta^K\left[1+\frac{|a|}{\beta}+\ldots+\left(\frac{|a|}{\beta}\right)^K\right]\left[1+\frac{|b|}{\alpha}+\ldots+\left(\frac{|b|}{\alpha}\right)^{m-1}\right] \\ & +\frac{M\alpha^m\beta\left(1+\beta+\ldots+\beta^{K-1}\right)}{(\alpha-|b|)(\beta-|a|)}. \end{aligned}$$

We have thus shown that (6.77) holds and hence that

$$|u_{ij}| \le \frac{M\beta}{(\alpha-|b|)(\beta-|a|)(1-\beta)}\alpha^i$$

for all $(i,j) \in W_{K+1}^-$. The cases $(i,j) \in W_{K+1}^+$ and $(i,j) = (K+1,K+1)$ are similarly proved. Finally, by means of the same arguments, we may show that

$$|u_{ij}| \le \frac{M\alpha}{(\alpha-|b|)(\beta-|a|)(1-\alpha)}\beta^j,\ i,j \in N.$$

The proof is completed by taking

$$M_1 = \frac{M\beta}{(\alpha-|b|)(\beta-|a|)(1-\beta)},$$

and

$$M_2 = \frac{M\alpha}{(\alpha - |b|)(\beta - |a|)(1-\alpha)}.$$

■

By combining the above two results, we obtain the following stability criterion: Suppose $|a| + |b| + |c| < 1$, and

$$|f_{ij}| \le M\sigma^i\tau^j,\ \sigma, \tau \in (0,1);\ M \ge 0;\ i, j \in N.$$

Then there exists $\xi, \alpha, \beta \in (0,1)$ and $M_1, M_2 \ge 0$ such that the solution $\{u_{ij}\}$ of (6.71) determined by the initial conditions (6.72) and (6.73) and the compatibility condition (6.74) will also satisfy

$$|u_{ij}| \le \|(\psi, \phi)\|_{ij}\, \xi^{\min(i,j)} + \min\left(M_1\alpha^i, M_2\beta^j\right),\ i, j \in N.$$

Next, we will derive a result which implies the standard Lyapunov stability of the trivial solution of (6.71). It involves the following identity:

$$\begin{aligned}
&\sum_{t=0}^{i-1}\sum_{s=0}^{j-1}(-a)^{j-1-s}(-b)^{i-1-t}\left\{u_{t+1,s+1} + au_{t+1,s} + bu_{t,s+1} + cu_{ts}\right\} \\
= \;& u_{ij} - (-a)^j u_{i0} - (-b)^i u_{0j} + (-b)^i(-a)^j u_{00} - (ab-c)(-a)^{j-1}(-b)^{i-1}u_{00} \\
& -(ab-c)\left\{(-b)^{i-1}\sum_{s=1}^{j-1}(-a)^{j-1-s}u_{0s} + (-a)^{j-1}\sum_{t=1}^{i-1}(-b)^{i-1-t}u_{t0}\right\} \\
& -(ab-c)\sum_{s=1}^{j-1}\sum_{t=1}^{i-1}(-a)^{j-1-s}(-b)^{i-1-t}u_{ts},
\end{aligned}$$

which holds when $|a| + |b| > 0$ and $i, j \in Z^+$. To see that this identity holds, note that given a nontrivial number ω and an arbitrary sequence $\{x_k\}_{k=0}^{\infty}$, we have

$$\sum_{\beta=0}^{\gamma-1}\omega^{\gamma-\beta-1}(x_{\beta+1} - \omega x_\beta) = \sum_{\beta=0}^{\gamma-1}\Delta(\omega^{\gamma-\beta}x_\beta) = x_\gamma - \omega^\gamma x_0. \tag{6.78}$$

By means of the above equality, we see that

$$\begin{aligned}
&\sum_{t=0}^{i-1}(-b)^{i-1-t}\left\{u_{t+1,s+1} + au_{t+1,s} + bu_{t,s+1} + cu_{ts}\right\} \\
= \;& u_{i,s+1} + (-b)^i u_{0,s+1} + \sum_{t=0}^{i-1}(-b)^{j-1-t}\left\{au_{t+1,s} + cu_{ts}\right\} \\
= \;& u_{i,s+1} + (-b)^i u_{0,s+1} + au_{is} + \sum_{t=1}^{i-1}(-b)^{i-1-t}(c-ab)u_{ts} + (-b)^{i-1}cu_{0s}.
\end{aligned}$$

Multiplying both sides of the above equality by $(-a)^{j-1-s}$, then summing the resulting equation from $s = 0$ to $s = j-1$, and then invoking the equality (6.78), we will obtain the desired identity.

Theorem 102 *Suppose* $|a| < 1$, $|b| < 1$, $|ab-c| < (1-|a|)(1-|b|)$. *Then the solution* $\{u_{ij}\}$ *of (6.71) determined by the conditions (6.72)–(6.74) will satisfy*

$$|u_{ij}| \le \frac{3(1-|a|)(1-|b|)\,\|(\psi,\phi)\|_{ij} + \|f\|^*_{i-1,j-1}}{1-|a|-|b|+|ab|-|ab-c|}, \ i,j\in N, \tag{6.79}$$

where

$$\|(\psi,\phi)\|_{ij} = \max_{0\le k\le \max(i,j)} \{|\psi_k|, |\phi_k|\}, \ i,j\in N,$$

and

$$\|f\|^*_{ij} = \max\{|f_{st}| \mid 0\le s\le i, 0\le t\le j\}, \ i,j\in N.$$

Proof. We first consider the case that $|a|+|b|>0$. Substituting (6.71), (6.72), (6.73) and (6.74) into the above stated identity, we obtain

$$\begin{aligned} u_{ij} &= (-a)^j\phi_i + (-b)^i\psi_j - (-a)^j(-b)^i\phi_0 \\ &\quad + (ab-c)\sum_{t=0}^{i-1}\sum_{s=0}^{j-1}(-a)^{j-1-s}(-b)^{i-1-t}u_{ts} - \sum_{t=0}^{i-1}\sum_{s=0}^{j-1}(-a)^{j-1-s}(-b)^{i-1-t}f_{ts}. \end{aligned}$$

For the sake of convenience, let

$$\Gamma_{ij} = \max_{0\le t\le i, 0\le s\le j} |u_{ts}|.$$

Then

$$\begin{aligned} |u_{ij}| &\le |a|^j|\phi_i| + |b|^i|\psi_j| + |a|^j|b|^i|\phi_0| \\ &\quad + (ab-c)\Gamma_{ij}\sum_{t=0}^{i-1}\sum_{s=0}^{j-1}|a|^{j-1-s}|b|^{i-1-t} \\ &\quad + \|f\|^*_{i-1,j-1}\sum_{t=0}^{i-1}\sum_{s=0}^{j-1}|a|^{j-1-s}|b|^{i-1-t} \\ &\le 3\|(\psi,\phi)\|_{ij} + \left(|ab-c|\,\Gamma_{ij} + \|f\|^*_{i-1,j-1}\right)\frac{(1-|a|^j)(1-|b|^i)}{(1-|a|)(1-|b|)} \\ &\le 3\|(\psi,\phi)\|_{ij} + \frac{|ab-c|\,\Gamma_{ij} + \|f\|^*_{i-1,j-1}}{(1-|a|)(1-|b|)}. \end{aligned}$$

Thus

$$\Gamma_{ij} \le 3\|(\psi,\phi)\|_{ij} + \frac{|ab-c|\,\Gamma_{ij} + \|f\|^*_{i-1,j-1}}{(1-|a|)(1-|b|)},$$

which implies

$$\frac{1-|a|-|b|+|ab|-|ab-c|}{(1-|a|)(1-|b|)}\Gamma_{ij} \le 3\|(\psi,\phi)\|_{ij} + \frac{\|f\|^*_{i-1,j-1}}{(1-|a|)(1-|b|)}.$$

Since $|ab-c| < (1-|a|)(1-|b|)$ by assumption, we have

$$\frac{1-|a|-|b|+|ab|-|ab-c|}{(1-|a|)(1-|b|)} > 0,$$

thus

$$\Gamma_{ij} \leq \frac{3(1-|a|)(1-|b|)\,\|(\psi,\phi)\|_{ij} + \|f\|^*_{i-1,j-1}}{1-|a|-|b|+|ab|-|ab-c|},$$

which implies (6.79) as required when $|a|+|b|>0$. When $a=b=0$, then $|c|<1$ by assumption and (6.71) reduces to

$$u_{i+1,j+1} + cu_{ij} = f_{ij},\ i,j \in N. \tag{6.80}$$

If $i-j=k\geq 0$, then $u_{k0}=\phi_k$, $u_{k+1,1} = -cu_{k0}+f_{k0}$,

$$\begin{aligned} u_{k+2,2} &= -c(-c\phi_k + f_{k0}) + f_{k+1,1} \\ &= (-c)^2\phi_k + (-cf_{k0} + f_{k+1,1}. \end{aligned}$$

In general, the unique solution of (6.80) subject to the conditions (6.72), (6.73) and (6.74) is given by

$$u_{ij} = (-c)^j\phi_{i-j} + \sum_{t=0}^{j-1}(-c)^{j-1-t}f_{i-j+t,t}$$

when $i\geq j\geq 0$, and

$$u_{ij} = (-c)^i\psi_{j-i} + \sum_{t=0}^{i-1}(-c)^{i-1-t}f_{t,j-i+t}$$

when $j>i\geq 0$. Thus

$$|u_{ij}| \leq \|(\psi,\phi)\|_{ij} + \|f\|^*_{i-1,j-1}\frac{1-|c|^i}{1-|c|} \leq \frac{\|(\psi,\phi)\|_{ij} + \|f\|^*_{i-1,j-1}}{1-|c|}.$$

The proof is complete. ∎

As a consequence, suppose $|a|<1$, $|b|<1$, $|ab-c|<(1-|a|)(1-|b|)$,

$$\sup_{i,j\geq 0}|f_{ij}| = \|f\| < \infty,$$

and

$$\max\left\{\sup_{j\geq 0}|\psi_j|, \sup_{i\geq 0}|\phi_i|\right\} = \|(\psi,\phi)\| < \infty.$$

Then the solution $\{u_{ij}\}$ of (6.71) determined by the conditions (6.72)–(6.74) will satisfy

$$|u_{ij}| \leq \frac{3(1-|a|)(1-|b|)\,\|(\psi,\phi)\| + \|f\|}{1-|a|-|b|+|ab|-|ab-c|},\ i,j\in N.$$

6.4.4 Method of Induction for a Four-Point Delay Equation

We are now dealing with the following equation

$$u_{i,j+1} = -\left\{au_{i+1,j} + cu_{ij} + p_{ij}u_{i-\sigma,j-\tau}\right\},\ (i,j)\in N\times N. \tag{6.81}$$

where a and c are real numbers, σ and τ are nonnegative integers, and $\{p_{ij}\}$ is a real bivariate sequence defined for $(i,j) \in N \times N$. A solution of (6.81) is a real bivariate sequence $u = \{u_{ij}\}$ which is defined for

$$\Omega = \{(i,j) |\ i \geq -\sigma, j \geq -\tau\},$$

and satisfies (6.81) respectively. Furthermore, it is clear that if we are given initial conditions of the form

$$u_{ij} = \phi_{ij},\ (i,j) \in \{(i,j) |\ i \geq -\sigma, -\tau \leq j \leq 0\}, \tag{6.82}$$

$$u_{ij} = \psi_{ij},\ (i,j) \in \{(i,j) |\ -\sigma \leq i \leq -1, j \geq -\tau\}, \tag{6.83}$$

and the compatibility condition

$$\phi_{ij} = \psi_{ij},\ -\sigma \leq i \leq -1,\ \ -\tau \leq j \leq 0, \tag{6.84}$$

then we may successively calculate $u_{10}; u_{11}, u_{02}; u_{21}, u_{12}, u_{30}; \ldots$ in a unique manner. An existence and uniqueness theorem for solutions of the initial value problem (6.81)–(6.84) is thus obtained.

We first note the following identity for a sequence $\{x_k\}_{k=0}^{t+1}$:

$$\sum_{k=0}^{d} \omega^{d-k}\left(x_{k+1} - \omega x_k\right) = \sum_{k=0}^{d} \Delta\left(\omega^{d-(k-1)} x_k\right) = x_{d+1} - \omega^{d+1} x_0, \quad \omega \in R. \tag{6.85}$$

Now let $u = \{u_{ij}\}_{(i,j)\in\Omega}$ be a solution of (6.81). Then in view of (6.81), we see that

$$\sum_{s=0}^{j-1} (-c)^{j-1-s}\left\{u_{i,s+1} + cu_{is} + au_{i+1,s} + p_{is}u_{i-\sigma,s-\tau}\right\} = 0, \quad i \geq 0, j \geq 1$$

Therefore, substituting (6.85) and rewriting, we obtain

$$\begin{aligned} u_{ij} &= (-1)^j \left\{ c^{j-1}\left[cu_{i0} + au_{i+1,0}\right] + \sum_{s=1}^{j-1}(-1)^s ac^{j-1-s} u_{i+1,s} \right. \\ &\left. + \sum_{s=-\tau}^{j-1-\tau} (-1)^{s+\tau} c^{j-1-\tau-s} p_{i,s+\tau} u_{i-\sigma,s} \right\}, \end{aligned}$$

for $i \geq 0, j \geq 1$. In view of the conditions (6.82)–(6.84), we see that when $i \geq 0$ and $1 \leq j \leq 1+\tau$,

$$\begin{aligned} u_{ij} &= (-1)^j \left\{ c^{j-1}\left[c\phi_{i,0} + a\phi_{i+1,0}\right] + \sum_{s=1}^{j-1}(-1)^s ac^{j-1-s} u_{i+1,s} \right. \\ &\left. + \sum_{s=-\tau}^{j-1-\tau} (-1)^{s+\tau} c^{j-1-\tau-s} p_{i,s+\tau} \phi_{i-\sigma,s} \right\}, \end{aligned} \tag{6.86}$$

when $0 \leq i \leq \sigma - 1$ and $j \geq 1$,

$$\begin{aligned} u_{ij} &= (-1)^j \left\{ c^{j-1}\left[c\phi_{i,0} + a\phi_{i+1,0}\right] + \sum_{s=1}^{j-1}(-1)^s ac^{j-1-s} u_{i+1,s} \right. \\ &\left. + \sum_{s=-\tau}^{j-1-\tau} (-1)^{s+\tau} c^{j-1-\tau-s} p_{i,s+\tau} \psi_{i-\sigma,s} \right\}, \end{aligned} \tag{6.87}$$

and when $i \geq \sigma$ as well as $j \geq 2+\tau$,

$$\begin{aligned} u_{ij} &= (-1)^j \left\{ c^{j-1}\left[c\phi_{i,0} + a\phi_{i+1,0}\right] + \sum_{s=1}^{j-1}(-1)^s a c^{j-1-s} u_{i+1,s} \right. \\ &\quad + \sum_{s=-\tau}^{0} (-1)^{s+\tau} c^{j-1-\tau-s} p_{i,s+\tau}\phi_{i-\sigma,s} \\ &\quad \left. + \sum_{s=1}^{j-1-\tau} (-1)^{s+\tau} c^{j-1-\tau-s} p_{i,s+\tau} u_{i-\sigma,s} \right\}. \end{aligned} \tag{6.88}$$

Theorem 103 *Suppose that* $|p_{ij}| \leq p$ *for* $(i,j) \in N \times N$, *that the sequences* $\phi = \{\phi_{ij}\}$ *and* $\psi = \{\psi_{ij}\}$ *are subexponential such that*

$$|\phi_{ij}| \leq M_1 \alpha^i,\ -\tau \leq j \leq 0;\ i \in N$$

and

$$|\psi_{ij}| \leq M_2 \beta^j,\ -\sigma \leq i \leq -1, j \geq -\tau.$$

Then the solution $\{u_{ij}\}$ *of (6.81) subject to the conditions (6.82)-(6.84) is also subexponential so that there exist positive numbers* M, λ, ω *such that* $|u_{ij}| \leq M\lambda^i\omega^j$ *for* $i, j \in N$.

Proof. Let

$$M_3 = \max_{-\sigma \leq t \leq -1, -\tau \leq s \leq 0} |\phi_{ts}|,$$

and

$$M = \max\{M_1, M_2, pM_1, pM_2, pM_3, 2|a|M_1\}.$$

Also let

$$f(x,y) = \frac{2}{3} + \frac{1}{y}\left(1 + \frac{1}{\alpha^\sigma} + \frac{\tau+1}{4\alpha^\sigma}\right) + \frac{1}{xy}\left(\frac{4}{3} + \frac{\tau+1}{4} + \frac{4p}{3|b|}\right).$$

Since $\lim_{x,y\to\infty} f(x,y) = 2/3$, there exist λ_0 and w_0 such that $|f(x,y)| \leq 1$ for $x \geq \lambda_0$ and $y \geq w_0$. Take $\lambda = \max\{\alpha, \lambda_0, 1\}$ and $\omega = \max\{\beta, w_0, 2\lambda, 4|c|, 4\lambda|a|, 4\}$. We will show that

$$|u_{ij}| \leq M\lambda^i\omega^j \tag{6.89}$$

holds for $i \geq 0$ and $j \geq 1$. There are three exhaustive cases to consider: (i) $i \geq 0$, $1 \leq j \leq 1+\tau$; (ii) $0 \leq i \leq \sigma - 1$, $j \geq 1$; and (iii) $i \geq \sigma$, $j \geq 1$. Case (i): Consider first the case $j = 1$ and $0 \leq i \leq \sigma - 1$. In view of (6.86), we have

$$\begin{aligned} |u_{ij}| &\leq |c|M_1\alpha^i + |a|M_1\alpha^{i+1} + pM_3 \\ &\leq \frac{1}{4}M\lambda^i\omega + \frac{1}{2}M\lambda^{i+1} + M \leq M\lambda^i\omega\left(\frac{1}{4} + \frac{1}{2}\cdot\frac{\lambda}{\omega} + \frac{1}{\lambda^i\omega}\right) \\ &\leq M\lambda^i\omega\left(\frac{1}{4} + \frac{1}{2}\cdot\frac{1}{2} + \frac{1}{\omega}\right) \\ &\leq M\lambda^i\omega. \end{aligned}$$

Next, if $j = 1$ and $i \geq \sigma$, we see from (6.86) that

$$\begin{aligned} |u_{ij}| &\leq |c| M_1 \alpha^i + |a| M_1 \alpha^{i+1} + pM_1 \alpha^{i-\sigma} \\ &\leq \frac{1}{4} M \lambda^i \omega + \frac{1}{2} M \lambda^{i+1} + M \lambda^i \frac{1}{\alpha^\sigma} \\ &< M \lambda^i \omega \left(\frac{1}{2} + \frac{1}{\omega \alpha^\sigma} \right) \\ &\leq M \lambda^i \omega. \end{aligned}$$

Assume by induction that

$$|u_{ik}| \leq M \lambda^i \omega^k$$

for $1 \leq k \leq \tau$ and $i \geq 0$. Then when $0 \leq i \leq \sigma - 1$, we see from (6.86) that

$$\begin{aligned} &|u_{i,k+1}| \\ \leq\ & |c|^{k+1} M_1 \alpha^i + |c|^k |a| M_1 \alpha^{i+1} + \sum_{s=-\tau}^{k-\tau} |c|^{k-\tau-s} pM_3 + \sum_{s=1}^{k} |c|^{k-s} |a| M \lambda^{i+1} \omega^s \\ \leq\ & \frac{1}{4} M \lambda^i \omega^{k+1} + \frac{1}{4} \cdot \frac{1}{2} M \omega^k \lambda^{i+1} + pM_3 \sum_{s=-\tau}^{k-\tau} \left(\frac{\omega}{4} \right)^{k-\tau-s} \\ & + M \lambda^{i+1} |a| \sum_{s=1}^{k} \left(\frac{\omega}{4} \right)^{k-s} \omega^s \\ \leq\ & M \lambda^i \omega^{k+1} \left(\frac{1}{4} + \frac{1}{8} \cdot \frac{\lambda}{\omega} \right) + M \omega^k \frac{\tau+1}{4} + M \lambda^{i+1} \omega^k |a| \frac{4}{3} \\ \leq\ & M \lambda^i \omega^{k+1} \left(\frac{31}{48} + \frac{1}{\lambda \omega} \frac{\tau+1}{4} \right) \leq M \lambda^i \omega^{k+1}, \end{aligned}$$

and when $i \geq \sigma$,

$$\begin{aligned} |u_{i,k+1}| \leq\ & |c|^{k+1} M_1 \alpha^i + |c|^k |a| M_1 \alpha^{i+1} \\ & + \sum_{s=-\tau}^{k-\tau} |c|^{k-\tau-s} pM_1 \alpha^{i-\alpha} + \sum_{s=1}^{k} |c|^{k-s} |a| M \lambda^{i+1} \omega^s \\ \leq\ & M \lambda^i \omega^{k+1} \frac{1}{4} + M \lambda^{i+1} \omega^k \frac{1}{8} + pM_1 \alpha^i \frac{1}{\alpha^\sigma} \sum_{s=-\tau}^{k-\tau} \left(\frac{\omega}{4} \right)^{k-\tau-s} \\ & + M \lambda^{i+1} |a| \sum_{s=1}^{k} \left(\frac{\omega}{4} \right)^{k-s} \omega^s \\ \leq\ & M \lambda^i \omega^{k+1} \left(\frac{1}{4} + \frac{1}{8} \cdot \frac{\lambda}{\omega} \right) + M \lambda^i \omega^k \frac{\tau+1}{4\alpha^\sigma} + M |a| \lambda^{i+1} \omega^k \frac{4}{3} \\ \leq\ & M \lambda^i \omega^{k+1} \left(\frac{31}{48} + \frac{1}{\omega} \frac{\tau+1}{4\alpha^\sigma} \right) \leq M \lambda^i \omega^{k+1}. \end{aligned}$$

Case (ii): We have seen that (6.89) holds for $0 \leq i \leq \sigma - 1$ and $1 \leq j \leq 1 + \tau$. Assume by induction that (6.89) holds for $0 \leq i \leq \sigma - 1$ and $1 \leq j \leq k$ where

$k \geq 1 + \tau$. Then in view of (6.87),

$$\begin{aligned}
|u_{i,k+1}| &\leq |c|^{k+1} M_1 \alpha^i + |c|^k |a| M_1 \alpha^{i+1} + \sum_{s=-\tau}^{0} |c|^{k-\tau-s} pM_3 \\
&\quad + \sum_{s=1}^{k-\tau} |c|^{k-\tau-s} pM_2 \beta^s + \sum_{s=1}^{k} |c|^{k-s} |a| M \lambda^{i+1} \omega^s \\
&\leq \frac{1}{4} M \lambda^i \omega^{k+1} + \frac{1}{4} \cdot \frac{1}{2} M \lambda^{i+1} \omega^k + M \sum_{s=-\tau}^{0} \left(\frac{\omega}{4}\right)^{k-\tau-s} \\
&\quad + M \sum_{s=1}^{k-\tau} \left(\frac{\omega}{4}\right)^{k-\tau-s} \omega^s + M \lambda^{i+1} |a| \sum_{s=1}^{k} \left(\frac{\omega}{4}\right)^{k-s} \omega^s \\
&\leq M \lambda^i \omega^{k+1} \left(\frac{1}{4} + \frac{1}{8} \cdot \frac{\lambda}{\omega}\right) + M \omega^k \frac{\tau + 1}{4} + M \omega^k \cdot \frac{4}{3} + M \lambda^{i+1} \omega^k |a| \frac{4}{3} \\
&\leq M \lambda^i \omega^{k+1} \left[\frac{31}{48} + \frac{1}{\lambda \omega} \left(\frac{4}{3} + \frac{\tau + 1}{4}\right)\right] \leq M \lambda^i \omega^{k+1}.
\end{aligned}$$

Case (iii): We have seen that (6.89) holds for $i \geq \sigma$ and $1 \leq j \leq 1 + \tau$. Assume by induction that (6.89) holds for $i \geq \sigma$ and $1 \leq j \leq k$ where $k \geq 1 + \tau$. Then in view of (6.88),

$$\begin{aligned}
|u_{i,k+1}| &\leq |c|^{k+1} M_1 \alpha^i + |c|^k |a| M_1 \alpha^{i+1} + \sum_{s=-\tau}^{0} |c|^{k-\tau-s} pM_1 \alpha^{i-\sigma} \\
&\quad + \sum_{s=1}^{k} |c|^{k-s} |a| M \lambda^{i+1} \omega^s + \sum_{s=1}^{k-\tau} |c|^{k-\tau-s} pM \lambda^{i-\sigma} \omega^s \\
&\leq \frac{1}{4} M \lambda^i \omega^{k+1} + \frac{1}{2} \cdot \frac{1}{4} M \lambda^{i+1} \omega^k + pM_1 \frac{\lambda^i}{\alpha^\sigma} \sum_{s=-\tau}^{0} \left(\frac{\omega}{4}\right)^{k-\tau-s} \\
&\quad + M \lambda^{i+1} |a| \sum_{s=1}^{k} \left(\frac{\omega}{4}\right)^{k-s} \omega^s + M \lambda^i \frac{p}{\lambda^\sigma} \sum_{s=1}^{k-\tau} \left(\frac{\omega}{4}\right)^{k-\tau-s} \omega^s \\
&\leq M \lambda^i \omega^{k+1} \frac{5}{16} + M \lambda^i \omega^k \frac{\tau + 1}{4\alpha^\sigma} + M \lambda^{i+1} \omega^k \frac{4}{3} |a| + M \lambda^i \omega^k \frac{1}{\lambda^\sigma \omega^\tau} \frac{4}{3} p \\
&\leq M \lambda^i \omega^{k+1} \left[\frac{2}{3} + \frac{1}{\omega} \frac{\tau + 1}{4\alpha^\sigma} + \frac{1}{\lambda^\sigma \omega^\tau \omega} \frac{4p}{3}\right] \leq M \lambda^i \omega^{k+1}.
\end{aligned}$$

The proof is complete. ■

6.4.5 Method of Induction for a Five-Point Delay Equation

In this section, we are concerned with subexponential solutions of the partial difference equations of the form

$$\Delta_1 \Delta_2 \left(a^i b^j u_{ij}\right) + p_{ij} u_{i-\sigma,j-\tau} = 0, \; i, j \in N. \tag{6.90}$$

where a, b and c are real numbers, σ and τ are nonnegative integers, and $\{p_{ij}\}$ is a real sequence defined for $(i, j) \in N \times N$.

A solution of (6.90) is a real bivariate sequence $u = \{u_{ij}\}$ which is defined for

$$\Omega = \{(i,j)|\ i \geq -\sigma, j \geq -\tau\},$$

and satisfies (6.90) respectively.

Next, note that if $a \neq 0$ and $b \neq 0$, then (6.90) can be written as

$$u_{i+1,j+1} = \frac{1}{b}u_{i+1,j} + \frac{1}{a}u_{i,j+1} - \frac{1}{ab}u_{ij} - p_{ij}u_{i-\sigma,j-\tau},\ ab \neq 0;\ i,j \in N. \tag{6.91}$$

Therefore, it is clear that if we are given initial conditions of the form

$$u_{ij} = \phi_{ij},\ (i,j) \in \{(i,j)|\ i \geq -\sigma, -\tau \leq j \leq 0\}, \tag{6.92}$$

$$u_{ij} = \psi_{ij},\ (i,j) \in \{(i,j)|\ j \geq -\tau, -\sigma \leq i \leq 0\}, \tag{6.93}$$

and the compatibility condition

$$\phi_{ij} = \psi_{ij},\ -\sigma \leq i \leq 0,\ \ -\tau \leq j \leq 0, \tag{6.94}$$

then we may successively calculate $u_{11}; u_{21}, u_{12}; u_{31}, u_{22}, u_{13}; \ldots$ in a unique manner.

We first note the following identity for a bivariate sequence $\{x_{ij}\}$:

$$\sum_{t=0}^{i-1}\sum_{s=0}^{j-1} \Delta_{21}^2 x_{ts} = x_{ij} - x_{0j} - x_{i0} + x_{00}. \tag{6.95}$$

Now let $u = \{u_{ij}\}_{(i,j)\in\Omega}$ be a solution of (6.90). Then from (6.90), we see that

$$a^{-i}b^{-j}\sum_{t=0}^{i-1}\sum_{s=0}^{j-1}\left\{\Delta_{21}^2\left(a^t b^s u_{ts}\right) + p_{ts}u_{t-\sigma,s-\tau}\right\} = 0,\ i,j \in Z^+.$$

Therefore, after substituting (6.95) and rewriting, we obtain

$$\begin{aligned} u_{ij} &= b^{-j}u_{i0} + a^{-i}u_{0j} - a^{-i}b^{-j}u_{00} - a^{-i}b^{-j}\sum_{t=0}^{i-1}\sum_{s=0}^{j-1} p_{ts}u_{t-\sigma,s-\tau} \\ &= b^{-j}u_{i0} + a^{-i}u_{0j} - a^{-i}b^{-j}u_{00} - a^{-i}b^{-j}\sum_{s=-\tau}^{j-1-\tau}\sum_{t=-\sigma}^{i-1-\sigma} p_{t+\sigma,s+\tau}u_{ts}. \end{aligned}$$

In view of the conditions (6.92)–(6.94), we see that when $i \geq 1$ and $1 \leq j \leq 1+\tau$,

$$u_{ij} = b^{-j}\phi_{i0} + a^{-i}\psi_{0j} - a^{-i}b^{-j}\phi_{00} - a^{-i}b^{-j}\sum_{s=-\tau}^{j-1-\tau}\sum_{t=-\sigma}^{i-1-\sigma} p_{t+\sigma,s+\tau}\phi_{ts}, \tag{6.96}$$

when $1 \leq i \leq 1+\sigma$ and $j \geq 1$,

$$u_{ij} = b^{-j}\phi_{i0} + a^{-i}\psi_{0j} - a^{-i}b^{-j}\phi_{00} - a^{-i}b^{-j}\sum_{s=-\tau}^{j-1-\tau}\sum_{t=-\sigma}^{j-1-\sigma} p_{t+\sigma,s+\tau}\psi_{ts}, \tag{6.97}$$

and when $i \geq 2+\sigma$ as well as $j \geq 2+\tau$,

$$u_{ij} = b^{-j}\phi_{i0} + a^{-i}\psi_{0j} - a^{-i}b^{-j}\phi_{00} - a^{-i}b^{-j}\sum_{s=-\tau}^{0}\sum_{t=-\sigma}^{i-1-\sigma} p_{t+\sigma,s+\tau}\phi_{ts}$$

$$-a^{-i}b^{-j}\sum_{s=1}^{j-1-\tau}\sum_{t=-\sigma}^{0} p_{t+\sigma,s+\tau}\psi_{ts} - a^{-i}b^{-j}\sum_{s=1}^{j-1-\tau}\sum_{t=1}^{i-1-\sigma} p_{t+\sigma,s+\tau}u_{ts}.$$

Theorem 104 *Suppose $\sigma \neq 0$ or $\tau \neq 0$. Suppose the bivariate sequences $\phi = \{\phi_{ij}\}$ and $\psi = \{\psi_{ij}\}$ are subexponential such that*

$$|\phi_{ij}| \leq M_1\alpha^i,\ i \geq -\sigma, -\tau \leq j \leq 0,$$

and

$$|\psi_{ij}| \leq M_2\beta^j,\ -\sigma \leq i \leq 0, j \geq -\tau.$$

Suppose further that either one of the following holds: (i) $|a| \geq 1, |b| \geq 1$ and $|p_{ij}| \leq p$, (ii) $|a| < 1, |b| < 1$ and $|p_{ij}| \leq p\alpha_1^i\beta_1^j$ where $\alpha_1 \leq |a|, \beta_1 \leq |b|$, (iii) $|a| < 1, |b| \geq 1$ and $|p_{ij}| \leq p\alpha_1^i$ where $\alpha_1 \leq |a|$, and (iv) $|a| \geq 1, |b| < 1$ and $|p_{ij}| \leq p\beta_1^j$ where $\beta_1 \leq |b|$. Then the solution of (6.91) subject to the conditions (6.92)–(6.94) is also subexponential.

Proof. Let

$$M_3 = p\sum_{s=-\tau}^{0}\sum_{t=-\sigma}^{0}|\phi_{ts}|.$$

Let λ_0 and ω_0 be sufficiently large numbers such that for $\lambda \geq \lambda_0$ and $\omega \geq \omega_0$,

$$\left|\frac{p}{\lambda^\sigma\omega^\tau}\right| \leq \frac{1}{2}.$$

Let $\alpha_2 = \max\{\alpha, 2\}$, $\beta_2 = \max\{\beta, 2\}$,

$$\lambda = \max\left\{\alpha, \frac{2\alpha_2}{|a|}, \frac{2}{\alpha_1}+1, \frac{8}{a}, \lambda_0\right\},$$

$$\omega = \max\left\{\beta, \frac{2\beta_2}{|b|}, \frac{2}{\beta_1}+1, \frac{8}{b}, \omega_0\right\},$$

and

$$M = \max\left\{M_1, M_2, M_3, \frac{pM_1(\tau+1)}{\alpha_2^\sigma(\alpha_2-1)}, \frac{pM_2(\sigma+1)}{\beta_2^\tau(\beta_2-1)}\right\}.$$

We will show that

$$|u_{ij}| \leq M\lambda^i\omega^j \tag{6.98}$$

holds for $i \geq 0$ and $j \geq 1$. There are four cases to consider: (i) $1 \leq i \leq 1+\sigma$, $1 \leq j \leq 1+\tau$; (ii) $i \geq 2+\sigma$, $1 \leq j \leq 1+\tau$; (iii) $1 \leq i \leq 1+\sigma$, $j \geq 2+\tau$; and (iv)

$i \ge 1+\sigma, j \ge 1+\tau$. CASE (i): Consider first the case where $1 \le 1+\sigma, 1 \le j \le 1+\tau$. In view of (6.96), we have

$$\begin{aligned}
&|u_{ij}| \\
\le\ & |b|^{-j} M_1 \alpha^i + |a|^{-i} M_2 \beta^j + M_1 |a|^{-i} |b|^{-j} + |a|^{-i} |b|^{-j} p \sum_{s=-\tau}^{0} \sum_{t=-\sigma}^{0} |\phi_{ts}| \\
\le\ & M \left[\lambda^i \left(\frac{\omega}{8}\right)^j + \left(\frac{\lambda}{8}\right)^i \omega^j + 2\left(\frac{\lambda}{8}\right)^i \left(\frac{\omega}{8}\right)^j \right] \\
\le\ & M \lambda^i \omega^j.
\end{aligned}$$

CASE (ii): In view of (6.96), we have

$$\begin{aligned}
|u_{ij}| \le\ & |b|^{-j} M_1 \alpha^i + |a|^{-i} M_2 \beta^j + M_1 |a|^{-i} |b|^{-j} \\
& + |a|^{-i} |b|^{-j} \sum_{s=-\tau}^{j-1-\tau} \sum_{t=-\sigma}^{0} |p_{t+\sigma,s+\tau}| |\phi_{ts}| \\
& + |a|^{-i} |b|^{-j} \sum_{s=-\tau}^{j-1-\tau} \sum_{t=1}^{i-1-\sigma} |p_{t+\sigma,s+\tau}| |\phi_{ts}| \\
\le\ & M\lambda^i \left(\frac{\omega}{8}\right)^j + M\left(\frac{\lambda}{8}\right)^i \omega^j + M\left(\frac{\lambda}{8}\right)^i \left(\frac{\omega}{8}\right)^j \\
& + M\left(\frac{\lambda}{8}\right)^i \left(\frac{\omega}{8}\right)^j + p|a|^{-i} |b|^{-j} M_1(\tau+1) \sum_{t=1}^{i-1-\sigma} \alpha^t \\
\le\ & M\lambda^i \omega^j \left(\frac{1}{8} + \frac{1}{8} + 2 \cdot \frac{1}{8} \cdot \frac{1}{8}\right) + pM_1(\tau+1)|a|^{-i} |b|^{-j} \sum_{t=1}^{i-1-\sigma} \alpha_2^t \\
\le\ & M\lambda^i \omega^j \frac{9}{32} + \frac{pM(\tau+1)}{\alpha_2^\sigma(\alpha_2 - 1)} \left(\frac{\alpha_2}{|a|}\right)^i |b|^{-j} \\
\le\ & M\lambda^i \omega^j \left(\frac{9}{32} + \frac{1}{2} \cdot \frac{1}{8}\right) \\
\le\ & M\lambda^i \omega^j.
\end{aligned}$$

CASE (iii): In view of (6.97), we have

$$\begin{aligned}
|u_{ij}| \le\ & |b|^{-j} M_1 \alpha^i + |a|^{-i} M_2 \beta^j + |a|^{-i} |b|^{-j} M_1 \\
& + |a|^{-i} |b|^{-j} \sum_{s=-\tau}^{0} \sum_{t=-\sigma}^{i-1-\sigma} |p_{t+\sigma,s+\tau}| |\psi_{ts}| \\
& + |a|^{-i} |b|^{-j} \sum_{s=1}^{j-1-\tau} \sum_{t=-\sigma}^{i-1-\tau} |p_{t+\sigma,s+\tau}| |\psi_{ts}| \\
\le\ & M \left[\lambda^i \left(\frac{\omega}{8}\right)^j + \omega^j \left(\frac{\lambda}{8}\right)^i + \left(\frac{\lambda}{8}\right)^i \left(\frac{\omega}{8}\right)^j \right]
\end{aligned}$$

$$+\left(\frac{\lambda}{8}\right)^i\left(\frac{\omega}{8}\right)^j p\sum_{s=-\tau}^{0}\sum_{t=-\sigma}^{0}|\psi_{ts}|+|a|^{-i}|b|^{-j}(\sigma+1)pM_2\sum_{s=1}^{j-1-\tau}\beta_2^s$$
$$\leq M\lambda^i\omega^j\left(\frac{1}{4}+\frac{1}{32}\right)+\frac{(\sigma+1)pM_2}{\beta_2^\tau(\beta_2-1)}\left(\frac{\beta_2}{|b|}\right)^j|a|^{-i}$$
$$\leq M\lambda^i\omega^j\left(\frac{1}{4}+\frac{1}{32}+\frac{1}{2}\cdot\frac{1}{8}\right)$$
$$\leq M\lambda^i\omega^j.$$

CASE (iv): We have already shown that (6.89) holds for $1\leq i\leq 1+\sigma$ and $1\leq j\leq 1+\tau$. Assume by induction that (6.89) holds for $1\leq i\leq I$ and $1\leq j\leq J$, where $I>1+\sigma$ and $J>1+\tau$. Let (m,n) be a point which satisfies either $m=I$ and $1+\tau\leq n\leq J$, or, $n=J$ and $1+\sigma\leq m\leq I$. We need to show that (6.89) holds for $i=m+1$ and $j=n+1$. To see this, note first that

$$|u_{m+1,n+1}|$$
$$\leq |b|^{-(n+1)}M_1\alpha^{m+1}+|a|^{-(m+1)}M_2\beta^{n+1}+M_1|a|^{-(m+1)}|b|^{-(n+1)}$$
$$+|a|^{-(m+1)}|b|^{-(n+1)}\sum_{s=-\tau}^{0}\sum_{t=-\sigma}^{0}|p_{t+\sigma,s+\tau}||\phi_{ts}|$$
$$+|a|^{-(m+1)}|b|^{-(n+1)}\sum_{s=-\tau}^{0}\sum_{t=1}^{m-\sigma}|p_{t+\sigma,s+\tau}||\phi_{ts}|$$
$$+|a|^{-(m+1)}|b|^{-(n+1)}\sum_{s=1}^{n-\tau}\sum_{t=-\sigma}^{0}|p_{t+\sigma,s+\tau}||\psi_{ts}|$$
$$+|a|^{-(m+1)}|b|^{-(n+1)}\sum_{s=1}^{n-\tau}\sum_{t=1}^{m-\sigma}|p_{t-\sigma,s+\tau}||u_{ts}|.$$

Therefore, when $|a|\geq 1$, $|b|\geq 1$ and $|p_{ij}|\leq p$, we have

$$|u_{m+1,n+1}| \leq M\lambda^{m+1}\left(\frac{\omega}{8}\right)^{n+1}+M\left(\frac{\lambda}{8}\right)^{m+1}\omega^{n+1}+M\left(\frac{\lambda}{8}\right)^{m+1}\left(\frac{\omega}{8}\right)^{n+1}$$
$$+\left(\frac{\lambda}{8}\right)^{m+1}\left(\frac{\omega}{8}\right)^{n+1}p\sum_{s=-\tau}^{0}\sum_{t=-\sigma}^{0}|\phi_{ts}|$$
$$+|a|^{-(m+1)}\left(\frac{\omega}{8}\right)^{n+1}p\sum_{s=-\tau}^{0}\sum_{t=1}^{m-\sigma}M_1\alpha_2^t$$
$$+\left(\frac{\lambda}{8}\right)^{m+1}|b|^{-(n+1)}p\sum_{s=-\tau}^{n-\tau}\sum_{t=-\sigma}^{0}M_2\beta_2^s+p\sum_{s=1}^{n-\tau}\sum_{t=1}^{m-\sigma}M\lambda^t\omega^s$$
$$\leq \frac{1}{8}M\lambda^{m+1}\omega^{n+1}+\frac{1}{8}M\lambda^{m+1}\omega^{n+1}+\frac{1}{8}\cdot\frac{1}{8}\cdot M\lambda^{m+1}\omega^{n+1}$$
$$+\frac{1}{8}\frac{1}{8}M\lambda^{m+1}\omega^{n+1}+\frac{1}{8}(\tau+1)pM_1|a|^{-(m+1)}\omega^{n+1}\frac{\alpha_2^{m+1-\sigma}}{\alpha_2-1}$$
$$+\frac{1}{8}(1+\sigma)pM_2\lambda^{m+1}|b|^{-(n+1)}\frac{\beta_2^{n+1-\tau}}{\beta_2-1}+pM\frac{\lambda^{m+1}\omega^{n+1}}{\lambda^\sigma(\lambda-1)\omega^\tau(\omega-1)}$$

$$\begin{aligned}
&\le M\lambda^{m+1}\omega^{n+1}\left(\frac{1}{8}+\frac{1}{8}+2\cdot\frac{1}{8}\cdot\frac{1}{8}+\frac{1}{8}\cdot\frac{1}{2}+\frac{1}{8}\cdot\frac{1}{2}+\frac{1}{2}\right)\\
&\le M\lambda^{m+1}\omega^{n+1}\left(\frac{13}{32}+\frac{1}{2}\right)\le M\lambda^{m+1}\omega^{n+1}.
\end{aligned}$$

When $|a|<1, |b|<1$ and $|p_{ij}|\le p\alpha_1^i\beta_1^j$ where $\alpha_1\le|a|$ and $\beta_1\le|b|$, we have

$$\begin{aligned}
|u_{m+1,n+1}| &\le M\lambda^{m+1}\omega^{n+1}\frac{13}{32}+|a|^{-(m+1)}|b|^{-(n+1)}\sum_{s=1}^{n-\tau}\sum_{t=1}^{m-\sigma}p\alpha_1^{t+\sigma}\beta_1^{s+\tau}M\lambda^t\omega^s\\
&\le M\lambda^{m+1}\omega^{n+1}\frac{13}{32}\\
&\quad+|a|^{-(m+1)}|b|^{-(n+1)}pM\alpha_1^\sigma\beta_1^\tau\frac{(\alpha_1\lambda)^{m+1}}{(\alpha_1\lambda)^\sigma(\alpha_1\lambda-1)}\frac{(\beta\omega)^{n+1}}{(\beta_1\omega)^\tau(\beta_1\omega-1)}\\
&\le M\lambda^{m+1}\omega^{n+1}\frac{13}{32}+\left(\frac{\alpha_1}{|a|}\right)^{m+1}\left(\frac{\beta_1}{|b|}\right)^{n+1}pM\frac{\lambda^{m+1}\omega^{n+1}}{\lambda^\sigma\omega^\tau}\\
&\le M\lambda^{m+1}\omega^{n+1}\left(\frac{13}{32}+\frac{1}{2}\right)\le M\lambda^{m+1}\omega^{n+1}.
\end{aligned}$$

When $|a|<1, |b|\ge 1$ and $|p_{ij}|\le p\alpha_1^i$ where $\alpha_1\le|a|$, we have

$$\begin{aligned}
|u_{m+1,n+1}| &\le M\lambda^{m+1}\omega^{n+1}\frac{13}{32}+|a|^{-(m+1)}\sum_{s=1}^{n-\tau}\sum_{t=1}^{m-\sigma}p\alpha_1^{t+\sigma}M\lambda^t\omega^s\\
&\le M\lambda^{m+1}\omega^{n+1}\frac{13}{32}+M\left(\frac{\alpha_1}{|a|}\right)^{n+1}p\frac{\lambda^{m+1}\omega^{n+1}}{\lambda^\sigma\omega^\tau(\alpha_1\lambda-1)(\omega-1)}\\
&\le M\lambda^{m+1}\omega^{n+1}\left(\frac{13}{32}+\frac{1}{2}\right)\le M\lambda^{m+1}\omega^{n+1}.
\end{aligned}$$

When $|a|\ge 1, |b|<1$ and $|p_{ij}|\le p\beta_1^j$ where $\beta_1\le|b|$, the proof is similar to the previous case. ■

We remark that when $\sigma=\tau=0$, equation (6.91) becomes

$$u_{i+1,j+1}-\frac{1}{b}u_{i+1,j}-\frac{1}{a}u_{i,j+1}+\left(\frac{1}{ab}+\frac{p_{ij}}{a^{i+1}b^{j+1}}\right)u_{ij}=0,\ ab\ne 0;\ i,j\in N.$$

Such an equation has been discussed in a previous section, and bounds for its solutions have been obtained. By means of these bounds, exponentially boundedness criteria are easily obtained.

6.5 Equations Over Finite Domains

Solutions of partial difference equations over finite domains can have a variety of qualitative properties. In this section, we will find explicit maxima of the Green's function associated with the boundary value problem (5.82)–(5.83). Recall that the Green's function has been defined to be the unique solution of the boundary value problem

$$\Xi x_{ij}\equiv\Delta_1^2x_{i-1,j}+\Delta_2^2x_{i,j-1}=-\delta_{ij}^{(u,v)},\ (i,j)\in\Omega, \tag{6.99}$$

$$x_{ij} = 0,\ (i,j) \in \partial\Omega, \tag{6.100}$$

where Ω is a nonempty finite domain of the lattice plane, and $\delta_{ij}^{(u,v)}$ is the Dirac delta function defined by $\delta_{ij}^{(u,v)} = 1$ if $(i,j) = (u,v)$ and $\delta_{ij}^{(u,v)} = 0$ if $(i,j) \neq (u,v)$. This solution has been denoted by $G^{(u,v)} = \{G_{ij}^{(u,v)}\}_{(i,j)\in\Omega+\partial\Omega}$, and it was shown that $G_{st}^{(u,v)} = G_{uv}^{(s,t)}$ for $(u,v),(s,t) \in \Omega$, and

$$\max_{(u,v),(i,j)\in\Omega} G_{ij}^{(u,v)} = \max_{(u,v)\in\Omega} G_{u,v}^{(u,v)}.$$

The explicit forms of the Green's function when Ω is a chain, a cycle or a rectangle have also been given. In case Ω is a chain of the general form $\{(i_1,j_1),...,(i_n,j_n)\}$ such that (i_s,j_s) and (i_t,j_t) are neighbors if and only if $|s-t| = 1$, the corresponding Green's function can be expressed as

$$G^{(i_s,j_s)} = \left\{G_{i_t,j_t}^{(i_s,j_s)}\right\}_{t=1}^{n},$$

and we can write

$$G(t|s) = G_{i_t,j_t}^{(i_s,j_s)},\ s,t = 1,...,n.$$

It was shown that

$$G(t|s) = \begin{cases} X_{s-1}X_{n-t}X_n^{-1} & 1 \le s \le t \le n \\ X_{n-s}X_{t-1}X_n^{-1} & 1 \le t \le s \le n \end{cases}.$$

where $X_{-1} = 0, X_1 = 1$ and

$$X_k = \frac{1}{2\sqrt{3}}\left(\gamma^{k+1} - \frac{1}{\gamma^{k+1}}\right),\ \gamma = 2+\sqrt{3}, k \in N.$$

In view of these facts, we see that

$$\max_{1\le s,t\le n} G(t|s) \le \max_{1\le i\le n} G(i|i) = \begin{cases} X_{m-1}X_mX_n^{-1} & n = 2m \\ X_m^2X_n^{-1} & n = 2m+1 \end{cases}. \tag{6.101}$$

Next, when Ω be a cycle of the general form $\{(i_1,j_1),...,(i_n,j_n)\}$ such that (i_1,j_1) has exactly two neighbors (i_n,j_n) and (i_2,j_2), (i_2,j_2) has exactly two neighbors (i_1,j_1) and (i_3,j_3), etc., the corresponding Green's function is given by

$$G(t|s) = G_{i_t,j_t}^{(i_s,j_s)} = \frac{\left(Y_{|s-t|-1} + Y_{n-1-|t-s|}\right)}{Y_n},\ s,t = 1,2,...,n,$$

where

$$Y_k = \begin{cases} X_k & -1 \le k \le n-1 \\ 4X_{n-1} - 2X_{n-2} - 2 & k = n \end{cases}.$$

Note that

$$G(t|t) = \frac{Y_{n-1}}{Y_n},\ t = 1,2,...,n,$$

so that

$$\max_{1\le s,t\le n} G(t|s) \le \max_{1\le i\le n} G(i|i) = Y_{n-1}Y_n^{-1}. \tag{6.102}$$

Finally, when Ω is a rectangular domain $\{1,2,...,m\} \times \{1,2,...,n\}$, an explicit formula for the associated Green's function has also been given in Section 6 of Chapter 5. In view of this formula, we see that

$$G_{ab}^{(a,b)} = \frac{2}{m+1}\sum_{r=1}^{m}\left(\sin\frac{ar\pi}{m+1}\right)^2 \frac{\sinh(b\beta_r)}{\sinh(\beta_r)}\frac{\sinh((n+1-b)\beta_r)}{\sinh((n+1)\beta_r)}.$$

where $\beta_1,...,\beta_m$ are the roots of the equation

$$\cos\frac{r\pi}{m+1} + \cosh(\beta_r) = 2.$$

When n is odd, note that

$$\begin{aligned} &\sinh(b\beta_r)\sinh\left((n+1-b)\beta_r\right) \\ = \ &\frac{1}{2}\left\{\cosh\left((b+n+1-b)\beta_r\right) - \cosh\left((b-n-1+b)\beta_r\right)\right\} \\ = \ &\frac{1}{2}\left\{\cosh\left((n+1)\beta_r\right) - \cosh\left((n+1-2b)\beta_r\right)\right\}. \end{aligned}$$

Since $\cosh x$ is an even function and increasing for $x \geq 0$, the minimum of

$$\sinh(b\beta_r)\sinh\left((n+1-b)\beta_r\right)$$

occurs when $2b = n+1$. In view of the explicit formula for $G_{ab}^{(a,b)}$ and the fact that $\sinh(\beta_r)\sinh((n+1)\beta_r) > 0$, we see that

$$\max_{1\leq b\leq n} G_{ab}^{(a,b)} = G_{a,(n+1)/2}^{(a,(n+1)/2)}.$$

Similarly, when n is even, we may show that for any fixed a,

$$\max_{1\leq b\leq n} G_{ab}^{(a,b)} = G_{a,n/2}^{(a,n/2)}.$$

By symmetry considerations, when m is odd, then for any b,

$$\max_{1\leq a\leq m} G_{ab}^{(a,b)} = G_{(m+1)/2,b}^{((m+1)/2,b)},$$

and for m even,

$$\max_{1\leq a\leq m} G_{ab}^{(a,b)} = G_{m/2,b}^{(m/2,b)}.$$

Consequently, the maximum of $G_{ab}^{(a,b)}$ for $(a,b) \in \{1,...,m\} \times \{1,...,n\}$ occurs at

$$(a,b) = \begin{cases} (m/2,(n+1)/2) & m \text{ is even, } n \text{ is odd} \\ ((m+1)/2, n/2) & m \text{ is odd, } n \text{ is even} \\ (m/2, n/2) & m \text{ is even, } n \text{ is even} \\ ((m+1)/2,(n+1/)2) & m \text{ is odd, } n \text{ is odd} \end{cases}.$$

Once the maximum of the Green's function associated to the boundary value problem (6.99)–(6.100) is found, then in view of (5.85), we see that a solution $\{x_{ij}\}$ of (6.99)–(6.100) will satisfy the inequality

$$|x_{ij}| \leq \max_{(i,j)\in\Omega} G_{ij}^{(i,j)} \sum_{(u,v)\in\Omega} |p_{uv}|, \ (i,j) \in \Omega.$$

6.6 Notes and Remarks

The method of energies is a powerful one. Unfortunately, the construction of an appropriate energy function is usually difficult. In case of a discrete time cellular neural network with symmetric feedback coefficients, an energy function has been constructed (see e.g. pages 20–21 in [83]) and it is employed to show a dichotomy of either ultimate convergence or ultimate period two oscillation.

Section 6.2.5 is based on Cheng and Lin [32]. There is an error in Lemma 3 of [32], however. The condition $ac \neq 0$ must be replaced by $a, c > 0$. The material in Section 6.2.6 is new. However, the reference [102] cited in this section contains incomplete proofs to correct statements.

Section 6.3.1 is based on Cheng and Lin [31], Section 6.3.2 is based on Cheng and Lu [39], and Section 6.3.3 on Cheng and Medina [25]. The DuFort Frankel and the Richardson schemes mentioned in Section 6.3.2 can be found in Forsythe and Wasow [67].

The results in Section 6.4.1 are not much different from those in Section 6.3.1, but are included for illustrations. In Sections 4.2, 4.3, 4.4 and 4.5 of this chapter, induction methods are used extensively for obtaining stability criteria. The success of these methods is due to the various possible orderings of the lattice points. For original sources of these materials, the reader may consult [109], [110], [111], and [35].

The last section is based on Cheng *et al.* [20].

We have only barely touched upon the stability theory of partial difference equations in this chapter. There are some additional references in the literature such as Cheng and Lin [33], Gregor [76], Thomee [152], Ortega [133], Kuo and Trudinger [101], Stetter [149], Veit [157], Zhang and Tian [174], etc. However, systematic investigations have yet to be made.

Sensitive dependence of solutions on nonstructural variables has not been discussed in this book. The interested reader may consult Bunimovich and Sinai [7], Cahn *et al.* [8], Chow and Shen [46], Wachholz and Bruch [160], as well as Mitchell and Bruch [124].

Chapter 7

Existence

7.1 Introduction

We have seen solutions of partial difference equations in the previous chapters. Under appropriate conditions, some or all of these solutions possess additional qualitative properties which can reflect the physical phenomena observed in real models. As an example, we have found the general solutions of the discrete heat equation

$$u_m^{(n+1)} = au_{m-1}^{(n)} + bu_m^{(n)} + cu_{m+1}^{(n)},\ m \in Z, n \in N,$$

and shown that under the condition $|a| + |b| + |c| < 1$, all its solutions $\{u_m^{(n)}\}$ which satisfy $\left|u_m^{(0)}\right| \leq \Gamma$ for $m \in Z$ will also satisfy

$$\lim_{n\to\infty} u_m^{(n)} = 0,\ m \in Z.$$

In some situations, we will also be interested in the existence of one or perhaps a few solutions which have the desired properties. In this chapter, we will obtain several such existence criteria mainly for solutions which are positive, bounded, or traveling waves. There are other types of solutions which are of interest (e.g. solutions that are symmetric, monotone, stable, periodic, chaotic, etc.) but we have restricted our attention for the sake of simplicity. Indeed, the methods for finding our solutions are quite standard and include contraction methods, monotone methods, matrix methods, root locations of polynomials, etc.

7.2 Traveling Waves

Consider the discrete heat equation which is rewritten here for the sake of convenience

$$u_n^{(t+1)} = au_{n-1}^{(t)} + bu_n^{(t)} + cu_{n+1}^{(t)},\ n \in Z, t \in N, \tag{7.1}$$

where the coefficients a, b and c are real numbers. One type of solution that stands out in applications is the type of solutions of the form

$$u_n^{(t)} = \psi(n - \rho t),\ (n, t) \in \Omega, \rho \in Z, \tag{7.2}$$

where the integer ρ is *a priori* unknown. Such a solution is called a *traveling wave* solution of (7.1) and is so named since the sequence $\{\psi(k-t)\}_{k\in Z}$ is obtained from $\{\psi(k)\}_{k\in Z}$ by translation. The integer ρ is called the velocity of the traveling wave. When $\rho > 0$ (or $\rho < 0$), the wave is said to be traveling in the positive direction (respectively negative direction). When $\rho = 0$, we have the so-called stationary or standing solution.

In this section, we will be interested in finding *positive* traveling wave solutions of the form (7.2) for our equation (7.1). A complete set of necessary and sufficient conditions for the existence of these solutions will be obtained.

Substituting (7.2) into (7.1), we see that

$$\phi(n-\rho t-\rho) = a\phi(n-\rho t-1) + b\phi(n-\rho t) + c\phi(n-\rho t+1).$$

Letting $k = n - \rho t$, we obtain the difference equation

$$\phi(k-\rho) = a\phi(k-1) + b\phi(k) + c\phi(k+1),\ k \in Z.$$

According to Theorem 22, the above difference equation has a positive solution if, and only if, its characteristic equation

$$c\lambda^{\rho+1} + b\lambda^{\rho} + a\lambda^{\rho-1} - 1 = 0$$

has a positive root λ. Thus, equation (7.1) has a positive traveling wave solution if, and only if, it has a positive traveling wave of the form

$$u_n^{(t)} = \lambda^{n-\rho t},\ \rho \in Z; \lambda > 0; (n,t) \in \Omega. \tag{7.3}$$

We will therefore look for positive roots of the above characteristic equation. For the cases where $\rho = -1, 0$ or 1, the corresponding characteristic equations are quadratic equations and therefore the existence of positive roots are easily characterized in terms of the coefficients a, b and c. When $\rho \le -2$ or $\rho \ge 2$, the corresponding equations are more complicated. Fortunately, a complete set of sufficient and necessary conditions can still be derived by elementary means.

When $\rho \ge 2$, the corresponding characteristic equation can be written in the form

$$H(\lambda) \equiv c\lambda^{\rho+1} + b\lambda^{\rho} + a\lambda^{\rho-1} - 1 = 0. \tag{7.4}$$

Theorem 105 *Suppose $\rho \ge 2$. Then the characteristic equation (7.4) has a positive root if, and only if, the coefficients a, b, c satisfy one of the following conditions: (i) $c > 0$; (ii) $c = 0, b > 0$; (iii) $c = 0, b = 0, a > 0$; (iv) $c = 0, b < 0$,*

$$a \ge \left[\frac{\rho^{\rho}}{(\rho-1)^{\rho-1}}\right]^{1/\rho} (-b)^{(\rho-1)/\rho}; \tag{7.5}$$

(v) $c < 0, a > 0, \psi(a,b,c,\rho) \ge 1$; or (vi) $c < 0, a \le 0, b^2\rho^2 \ge 4ac(\rho^2-1), \psi(a,b,c,\rho) \ge 1$, where

$$\psi(a,b,c,\rho) = \frac{1}{\rho+1}\left(-\frac{1}{2c(\rho+1)}\left[b\rho + \sqrt{(b^2\rho^2 - 4ac\rho^2 + 4ac)}\right]\right)^{\rho-1}$$
$$\times\left(-\frac{1}{2}\frac{b}{c(\rho+1)}\left[b\rho + \sqrt{(b^2\rho^2 - 4ac\rho^2 + 4ac)}\right] + 2a\right).$$

Proof. First note that $H(0) = -1$. Consider three cases: $c > 0, c = 0$ and $c < 0$. If $c > 0$, since $\lim_{\lambda\to\infty} H(\lambda) = \infty$, we see that $H(\lambda)$ will have a positive root. If $c = 0$,

$$H(\lambda) = b\lambda^\rho + a\lambda^{\rho-1} - 1,$$

we consider three subcases: $b > 0, b = 0, b < 0$. Suppose $b > 0$, then $\lim_{\lambda\to\infty} H(\lambda) = \infty$. Thus $H(\lambda)$ has a positive root. If $b = 0$, then $H(\lambda) = a\lambda^{\rho-1} - 1$. Hence $H(\lambda)$ has a positive root if, and only if, $a > 0$. Suppose $b < 0$, since $H(0) = -1, H(\infty) = -\infty$, and

$$\begin{aligned} H'(\lambda) &= b\rho\lambda^{\rho-1} + a(\rho-1)\lambda^{\rho-2} \\ &= b\rho\lambda^{\rho-2}\left(\lambda + \frac{\rho-1}{\rho}\frac{a}{b}\right), \end{aligned}$$

we see that $\lambda_1 = 0$ and

$$\lambda_2 = -\frac{\rho-1}{\rho}\frac{a}{b}$$

are the (only real) roots of $H'(\lambda)$. If $a \le 0$, then $\lambda_1, \lambda_2 \le 0$, thus $H'(\lambda) \le 0$ for $\lambda > 0$ and $H(\lambda)$ cannot have any positive roots. If $a > 0$, then $\lambda_2 > 0$ so that $H'(\lambda) > 0$ for $\lambda \in (0, \lambda_2)$, $H'(\lambda) < 0$ for $\lambda \in (\lambda_2, \infty)$, and $H(\lambda)$ has a local maximum

$$H(\lambda_2) = a^\rho\left(-\frac{\rho-1}{b}\right)^{\rho-1}\frac{1}{\rho^\rho} - 1.$$

Thus when $c = 0$ and $b < 0$, $H(\lambda)$ will have a positive root if, and only if, $H(\lambda_2) \ge 0$ which is equivalent to (7.5).

If $c < 0$, then $\lim_{\lambda\to\infty} H(\lambda) = -\infty$, and

$$H'(\lambda) = c(\rho+1)\lambda^{\rho-2}\left(\lambda^2 + \frac{b}{c}\frac{\rho}{\rho+1}\lambda + \frac{a}{c}\frac{\rho-1}{\rho+1}\right).$$

The polynomial $H'(\lambda)$ has the roots

$$\mu_1 = \frac{-1}{2c(\rho+1)}\left[b\rho - \sqrt{b^2\rho^2 - 4ac(\rho^2-1)}\right],$$

$$\mu_2 = \frac{-1}{2c(\rho+1)}\left[b\rho + \sqrt{b^2\rho^2 - 4ac(\rho^2-1)}\right],$$

and when $\rho \ne 2$, the additional root

$$\mu_3 = 0.$$

We now consider two subcases: $a > 0, a \le 0$. In the first subcase, we have $\mu_1 \le \mu_3 = 0 < \mu_2$, so that $H'(\lambda) > 0$ for $\lambda \in (0, \mu_2)$, $H'(\lambda) < 0$ for $\lambda \in (\mu_2, \infty)$ and $H(\lambda)$ over the interval $(0, \infty)$ has the global maximum

$$H(\mu_2) = \frac{1}{\rho+1}\mu_2^{\rho-1}(b\mu_2 + 2a) - 1. \tag{7.6}$$

Hence $H(\lambda)$ has a positive root if, and only if, $H(\mu_2) \ge 0$, that is, $\psi(a, b, c, \rho) \ge 1$. In the second subcase, that is, $a \le 0$, we first note that when $b^2\rho^2 < 4(\rho^2-1)ac$,

μ_1 and μ_2 are imaginary, and hence $H(\lambda)$ cannot have any positive roots. When $b^2\rho^2 \geq 4(\rho^2-1)ac$, we see that $0 < \mu_1 < \mu_2$ and hence $H'(\lambda) < 0$ for $\lambda \in (0, \mu_1)$, $H'(\lambda) > 0$ for $\lambda \in (\mu_1, \mu_2)$, $H'(\lambda) < 0$ for $\lambda \in (\mu_2, \infty)$. Hence $H(\lambda)$ over $(0, \infty)$ has a local maximum $H(\mu_2)$ as given by (7.6) and has a positive root if, and only if $H(\mu_2) \geq 0$. The proof is complete. ■

We remark that in the very last case, i.e., $c < 0, a \leq 0$ and $|b| \geq 2\rho^{-1}\sqrt{(\rho^2-1)}$, since

$$\begin{aligned} & b\mu_2 + 2a \\ = & \frac{-2ab(\rho-1)}{b\rho - \sqrt{b^2\rho^2 - 4ac(\rho^2-1)}} + 2a \\ = & -2a\frac{\sqrt{b^2\rho^2 - 4ac(\rho^2-1)} - b}{b\rho - \sqrt{b^2\rho^2 - 4ac(\rho^2-1)}} \\ = & \frac{-2a(\rho^2-1)(b^2-4ac)}{\left[b\rho - \sqrt{b^2\rho^2 - 4ac(\rho^2-1)}\right]\left[\sqrt{b^2\rho^2 - 4ac(\rho^2-1)} + b\right]}, \end{aligned}$$

which is nonnegative if, and only if, $|b| \geq 2\sqrt{ac}$. Thus when $|b| < 2\sqrt{ac}$, $H(\mu_2) < 0$ and $H(\lambda)$ cannot have any positive roots. In other words, in this case, we need to assume that

$$|b| \geq \max\left\{2\sqrt{ac}, 2\rho^{-1}\sqrt{ac(\rho^2-1)}\right\} = 2\sqrt{ac}$$

in order to find a positive root of $H(\lambda)$.

When $\rho \leq -2$, we can rewrite the corresponding characteristic equation as

$$\lambda^{-\rho+1} - c\lambda^2 - b\lambda - a = 0. \tag{7.7}$$

By symmetric arguments similar to those of the proof of Theorem 105, we can derive the following necessary and sufficient condition.

Theorem 106 *Suppose $\rho \leq -2$. Then the characteristic equation has a positive root if, and only if, the coefficients a, b, c satisfy one of the following conditions: (i) $a > 0$; (ii) $a = 0, b > 0$; (iii) $a = 0, b = 0, c > 0$; (iv) $a = 0, b < 0$,*

$$c \geq -\rho\left(\frac{b}{\rho+1}\right)^{(\rho+1)/\rho};$$

(v) $a < 0, c \geq 0, \phi(a,b,c,\rho) \geq 1$; or (vi) $a < 0, c < 0, b^2\rho^2 \geq 4ac(\rho^2-1), \phi(a,b,c,\rho) \geq 1$, where

$$\phi(a,b,c,\rho) = \left[\frac{2a(\rho-1)}{\sqrt{b^2\rho^2 - 4ac(\rho^2-1)} - b\rho}\right]^{\rho} \frac{\sqrt{b^2\rho^2 - 4ac(\rho^2-1)} - b}{\rho^2-1}.$$

Again, we remark that in case $a < 0$ and $c < 0$, we need to assume that $|b| \geq 2\sqrt{ac}$ in order to find a root of the corresponding characteristic equation.

When $\rho = 1$, the characteristic equation can be written as

$$c\lambda^2 + b\lambda + a - 1 = 0; \tag{7.8}$$

when $\rho = 0$, the characteristic equation can be written as

$$c\lambda^2 + (b-1)\lambda + a = 0; \tag{7.9}$$

and when $\rho = -1$, the corresponding equation can be written as

$$(c-1)\lambda^2 + b\lambda + a = 0. \tag{7.10}$$

These are standard quadratic equations and hence can be dealt with easily. We summarize our investigations as follows.

Theorem 107 *Suppose $\rho = 1$. Then the characteristic equation (7.8) has a positive root if, and only if, the coefficients a, b, c satisfy either one of the following conditions: (1) $c = 0$, and $b(a-1) < 0$; (2) $c > 0, b > 0$ and $a < 1$; (3) $c > 0, b \le 0$ and $b^2 \ge 4c(a-1)$; (4) $c < 0, b < 0$ and $a > 1$; (5) $c < 0, b \ge 0$ and $b^2 \ge 4c(a-1)$.*

Theorem 108 *Suppose $\rho = 0$. Then the characteristic equation (7.9) has a positive root if, and only if, the coefficients a, b, c satisfy either one of the following conditions: (1) $c = 0$ and $a(b-1) < 0$; (2) $c > 0, b \ge 1$ and $a < 0$; (3) $c > 0, b < 1$ and $a \le (b-1)^2/(4c)$; (4) $c < 0, b \le 1$ and $a > 0$; (5) $c < 0, b > 1$ and $a \ge (b-1)^2/(4c)$.*

Theorem 109 *Suppose $\rho = -1$. Then the characteristic equation (7.10) has a positive root if, and only if, the coefficients a, b, c satisfy either one of the following conditions: (1) $c = 1$ and $ab < 0$; (2) $c > 1, b \ge 0$ and $a < 0$; (3) $c > 1, b < 0$ and $a \le b^2/(4c-4)$; (4) $c < 1, b \le 0$ and $a > 0$; (5) $c < 1, b > 0$ and $a \ge b^2/(4c-4)$.*

7.3 Positive and Bounded Solutions

If the discrete heat equation

$$u_m^{(n+1)} - u_m^{(n)} = r\left(u_{m-1}^{(n)} - u_m^{(n)}\right) + r\left(u_{m+1}^{(n)} - u_m^{(n)}\right), \quad m \in Z, n \in N, r > 0,$$

represents a real model, it will be reasonable to expect that it has a positive bounded solution for appropriate initial temperature distributions. Indeed, the bivariate sequence $\left\{u_m^{(n)}\right\} \equiv \{1\}$ is such a solution. The question then arises as to whether a more general nonhomogeneous rod subject to delayed feedback has a positive and bounded solution. In this section, we will consider a more general equation of the form

$$u_m^{(n+1)} - u_m^{(n)} = \alpha u_{m-1}^{(n)} + \beta u_m^{(n)} + \gamma u_{m+1}^{(n)} + q u_m^{(n-\sigma)}, \tag{7.11}$$

where $n \in N$ and $m \in Z$ and σ is a nonnegative integer.

Given an arbitrary set of initial values $u_m^{(n)}$, $-\sigma \le n \le 0, m \in Z$, we can successively calculate $u_0^{(1)}; u_{-1}^{(1)}, u_0^{(2)}, u_1^{(1)}; u_{-2}^{(1)}, u_{-1}^{(2)}, u_0^{(3)}, u_1^{(2)}, u_2^{(1)}; ...$ in a unique manner. Such a bivariate sequence $u = \{u_m^{(n)} | \, m \in Z, n = -\sigma, -\sigma+1, ...\}$ is called a solution of (7.11). Suppose there is some nonnegative integer T such that $u_m^{(n)} > 0$ for $m \in Z$ and $n \ge T$, then u is said to be eventually positive.

By designating the doubly infinite sequence $\left\{..., u_{-1}^{(n)}, u_0^{(n)}, u_1^{(n)}, ...\right\}$ as the column vector $u^{(n)}$ in l^Z, we see that a solution of (7.11) can also be regarded as a vector sequence $\{u^{(n)}\}_{n=-\sigma}^{\infty}$. Furthermore, such a sequence satisfies the delay vector recurrence relation

$$u^{(n+1)} - u^{(n)} = Au^{(n)} + qu^{(n-\sigma)}, \ n \in N, \tag{7.12}$$

where $A = (a_{ij})$ is an infinite matrix with diagonal elements all equal to β, subdiagonal elements all equal to α, and superdiagonal elements to γ, that is, $a_{ii} = \beta, a_{i,i-1} = \alpha$ and $a_{i,i+1} = \gamma$ for $i \in Z$, and zero elsewhere.

If we call a vector v in l^Z positive (denoted by $v > 0$) when all its components are positive, then clearly a solution u of (7.11) is eventually positive if, and only if the vector sequence $\{u^{(n)}\}$ is eventually positive. Therefore, (7.11) has an eventually positive solution if, and only if, the relation (7.12) has an eventually positive solution. Next, we observe that if there is a number λ and a corresponding vector v such that

$$Av = \lambda v, \tag{7.13}$$

then for any solution $\{x_n\}_{n=-\sigma}^{\infty}$ of the scalar difference equation

$$x_{n+1} - x_n = \lambda x_n + qx_{n-\sigma}, \ n \in N, \tag{7.14}$$

we have

$$x_{n+1}v - x_n v = \lambda x_n v + qx_{n-\sigma}v = x_n Av + qx_{n-\sigma}v, n \in N,$$

that is, $\{x_n v\}$ is a solution of (7.12).

In view of these, in order to find an eventually positive and bounded solution of (7.12), it suffices to find a number λ and a corresponding positive and bounded vector v such that (7.13) is satisfied, as well as an eventually positive and bounded solution of (7.14).

Clearly, when $\lambda = \alpha + \beta + \gamma$, (7.13) is satisfied by the positive and bounded constant sequence $v = \{1\}$. Now, the corresponding scalar difference equation is

$$x_{n+1} - x_n = (\alpha + \beta + \gamma)x_n + qx_{n-\sigma}, \ n \in Z. \tag{7.15}$$

To find an eventually positive and bounded solution of (7.15), we may look for one that is of the form $\{t^n\}$ and $0 < t \le 1$. Substituting the unknown solution into (7.15), we obtain the following characteristic equation

$$t^{\sigma+1} - (1 + \alpha + \beta + \gamma)t^{\sigma} - q = 0.$$

Therefore, it suffices to find a root of this equation in $(0, 1]$. This is a relatively easy problem. Indeed, consider the polynomial

$$h(t) = t^{\sigma+1} + at^{\sigma} + b, \ t \in R. \tag{7.16}$$

When $\sigma = 0$, the unique root is $t = -a - b$. Hence it has a root in $(0, 1]$ if, and only if, $-1 \le a + b < 0$. Suppose now $\sigma > 0$. Note that $h(0) = b$ and $\lim_{t\to\infty} h(t) = \infty$, we see that (7.16) has a positive root when $b < 0$. In such a case, it has a root in $(0, 1]$ if, and only if, $h(1) = 1 + a + b \ge 0$. If $b = 0$, then (7.16) has the unique roots 0 and $-a$, hence it has a root in $(0, 1]$ if, and only if, $-1 \le a < 0$. If $b > 0$ and $a \ge 0$, then $h(t) > 0$ for $t > 0$, so that $h(t)$ does not have any positive roots. Finally, when $b > 0$ and $a < 0$, the function $h(t)$ has a minimum at $t^* = -a\sigma/(\sigma + 1)$, and

$$h(t^*) = b - \frac{(-a)^{\sigma+1}\sigma^{\sigma}}{(\sigma + 1)^{\sigma+1}}.$$

Hence $h(t)$ has a root in $(0,1]$ if, and only if,

$$b \le \frac{(-a)^{\sigma+1}\sigma^\sigma}{(\sigma+1)^{\sigma+1}}$$

and

$$0 < t^* = \frac{-a\sigma}{\sigma+1} \le 1.$$

In particular, when $\sigma = 0$, (7.15) has a solution of the form $\{t^n\}$ where $t \in (0,1]$ if, and only if,

$$0 \ge \alpha+\beta+\gamma+q > -1;$$

and when $\sigma \ge 1$, (7.15) has a solution of the form $\{t^n\}$ where $t \in (0,1]$ if, and only if,

$$0 < \alpha+\beta+\gamma \le \frac{1}{\sigma} \text{ and } -\frac{(1+\alpha+\beta+\gamma)^{\sigma+1}\sigma}{(\sigma+1)^{\sigma+1}} \le q < 0, \tag{7.17}$$

or

$$-1 < \alpha+\beta+\gamma \le 0 \text{ and } -\frac{(1+\alpha+\beta+\gamma)^{\sigma+1}\sigma^\sigma}{(\sigma+1)^{\sigma+1}} \le q \le -(\alpha+\beta+\gamma), \tag{7.18}$$

or

$$\alpha+\beta+\gamma \le -1 \text{ and } 0 < q \le -(\alpha+\beta+\gamma). \tag{7.19}$$

Theorem 110 *When $\sigma = 0$, if $-1 < \alpha+\beta+\gamma+q \le 0$, then (7.11) has an eventually positive and bounded solution. When $\sigma > 1$, if (7.17) or (7.18) or (7.19) holds, then (7.11) has an eventually positive and bounded solution.*

We remark that if an eventually positive and zero convergent solution $\{u^{(n)}\}$ of (7.12) is desired, we need only to modify the conditions in the above theorem so that (7.15) has a solution of the form $\{t^n\}$ where $t \in (0,1)$ instead of $(0,1]$.

7.4 Monotone Method for a Finite Laplace Equation

Recall the steady state discrete Laplace equation in Section 2.5 of Chapter 1:

$$v_{m-1,n} + v_{m+1,n} + v_{m,n-1} + v_{m,n+1} - 4v_{mn} = 0, \ (m,n) \in Z^2.$$

It is interesting to note that this equation has bounded and also positive solutions, namely, $\{c\}$. The question then arises whether the corresponding steady state equation for a more general heat conducting material such as

$$\Xi v_{mn} + f(m,n,v_{mn}) = 0, \ m,n \in Z. \tag{7.20}$$

has a bounded and positive solution.

By a *supersolution* of (7.20), we mean a multiple sequence $\{w_{mn}\}_{m,n\in Z}$ which satisfies the functional inequality

$$\Xi w_{mn} + f(m,n,w_{mn}) \le 0, \ m,n \in Z, \tag{7.21}$$

and a *subsolution* is a multiple sequence $\{u_{mn}\}_{m,n\in Z}$ which satisfies

$$\Xi u_{mn} + f(m,n,u_{mn}) \ge 0, \ m,n \in Z. \tag{7.22}$$

Theorem 111 *Let $f : Z^2 \times R \to R$ be a function such that $f(m,n,y)$ is continuous with respect to y for each $(m,n) \in Z^2$. If there exist a supersolution w and a subsolution of (7.20) such that $u_{mn} \le w_{mn}$ for $(m,n) \in Z^2$, and that there is a positive constant Γ such that*

$$f(m,n,s) - f(m,n,t) + (s-t)\Gamma > 0$$

for $(m,n) \in Z^2$ and $\inf_{(m,n)\in Z^2} u_{mn} \le s,t \le \sup_{(m,n)\in Z^2} w_{mn}$, then (7.20) has a solution v such that $u_{mn} \le v_{mn} \le w_{mn}$ for $(m,n) \in Z^2$.

Proof. Let B_k be the sphere containing lattice points within k units from the origin, i.e.

$$B_k = \{(i.,j) \in Z^2 |\ |i| + |j| \le k\},\ k \in N,$$

then

$$\partial B_k = \{(i,j) \in Z^2 |\ |(i,j)| = k+1\}.$$

Pick an arbitrary multiple sequence $\varphi = \{\varphi_{mn}\}_{m,n\in Z}$ such that $u_{mn} \le \varphi_{mn} \le w_{mn}$ for $(m,n) \in Z^2$. For each $k \ge 1$, consider the boundary value problem

$$\Xi v_{mn} + f(m,n,v_{mn}) = 0,\ (m,n) \in B_k, \tag{7.23}$$

$$v_{mn} = \varphi_{mn},\ (m,n) \in \partial B_k. \tag{7.24}$$

Let Γ_k be a positive constant such that

$$f(m,n,s) - f(m,n,t) + \Gamma_k(s-t) > 0$$

for $(m,n) \in B_k$ and $\min_{(m,n)\in B_k} u_{mn} \le s,t \le \max_{(m,n)\in B_k} w_{mn}$. Consider the iteration scheme:

$$\Xi w_{mn} - \Gamma_k w_{mn}^{(t)} = -f(m,n,w_{mn}^{(t-1)}) - \Gamma_k w_{mn}^{(t-1)},\ (m,n) \in B_k,$$

$$w_{mn}^{(t)} = \varphi_{mn},\ (m,n) \in \partial B_k$$

for $t \in Z^+$. If we put $w_{mn}^{(0)} = w_{mn}$ for $(m,n) \in Z^2$, then in view of Theorem 3.53, $w^{(1)}$ exists and is unique. Furthermore,

$$\begin{aligned}&(\Xi - \Gamma_k)\left(w_{mn}^{(1)} - w_{mn}^{(0)}\right)\\ \ge\ & -(f(m,n,w_{mn}) + \Gamma_k w_{mn}) + (f(m,n,w_{mn}) + \Gamma_k w_{mn}) = 0,\end{aligned}$$

for $(m,n) \in B_k$, and

$$w_{mn}^{(1)} - w_{mn}^{(0)} = \varphi_{mn} - w_{mn} \le 0, \quad (m,n) \in \partial B_k.$$

Thus by Theorem 3.52, we have $w_{mn}^{(1)} \le w_{mn}^{(0)} = w_{mn}$ for $x \in B_k$. Note also that

$$\begin{aligned}&(\Xi - \Gamma_k)\left(u_{mn} - w_{mn}^{(1)}\right)\\ \ge\ & f(m,n,w_{mn}^{(1)}) - f(m,n,u_{mn}) + \Gamma_k(w_{mn}^{(1)} - u_{mn}) \ge 0,\end{aligned}$$

for $(m,n) \in B_k$, and

$$u_{mn} - w_{mn}^{(1)} \le 0,\ (m,n) \in \partial B_k.$$

Thus by Theorem 3.52 again, we have $u_{mn} \le w_{mn}^{(1)}$ for $(m,n) \in B_k$. By induction, it is not difficult to see that

$$u_{mn} \le w_{m,n}^{(t+1)} \le w_{mn}^{(t)} \le w_{mn},\ (m,n) \in B_k$$

for all $t \ge 0$.

Similarly, the iteration scheme

$$\Xi u_{mn}^{(t)} = -f(m,n,u_{mn}^{(t-1)}),\ (m,n) \in B_k,$$

$$u_{mn}^{(t)} = \varphi_{mn},\ (m,n) \in \partial B_k$$

for $t > 0$, and

$$u_{mn}^{(0)} = u_{mn},\ (m,n) \in Z^2,$$

will give rise to a sequence $\{u^{(t)}\}$ which satisfies

$$u_{mn} \le u_{mn}^{(t)} \le u_{mn}^{(m+1)} \le w_{mn},\ (m,n) \in B_k$$

for all $t \ge 0$.

By similar reasoning, it is not difficult to see that

$$u_{mn}^{(t)} \le w_{mn}^{(t)},\ (m,n) \in B_k$$

for all $t > 0$. We have thus shown that

$$u_{mn} = u_{mn}^{(0)} \le u_{mn}^{(1)} \le u_{mn}^{(2)} \le \dots \le w_{mn}^{(1)} \le w_{mn}^{(0)} = w_{mn}, \tag{7.25}$$

for $(m,n) \in B_k + \partial B_k$. Therefore, $\{u^{(t)}\}$ and $\{w^{(t)}\}$ converge pointwise to some $u^{(k)}$ and $w^{(k)}$ respectively in B_k, and $u_{mn}^{(k)} \le w_{mn}^{(k)}$ for $(m,n) \in B_k$. By taking limits on both sides of the above iteration schemes, we see further that $u^{(k)}$ and $w^{(k)}$ are solutions of the boundary value problems (7.23)-(7.24).

We now assert that the sequence $\{u^{(i)}\}$ has a subsequence $\{u^{(i_j)}\}$ which converges pointwise to an entire solution of (7.23)-(7.24). Indeed, the sequence $\{u_{00}^{(i)}\}$ is bounded between the numbers u_{00} and w_{00}, hence $\{u_{00}^{(i)}\}$ has a convergent subsequence $\{u_{00}^{(i_k)}\}$, where we may assume that $i_1 > 1$. Next, since the subsequence $\{u_{10}^{(i_k)}\}_{k=2}^{\infty}$, is bounded between u_{10} and w_{10}, it has a converging subsequence also. Noting that Z^2 is countable, we may proceed inductively to conclude that $\{u^{(i)}\}$ has a subsequence $\{u^{(i_j)}\}$ where $i_j > j$ and which converges pointwise to a function v. Clearly, $u_{mn} \le v_{mn} \le w_{mn}$ for $(m,n) \in Z^2$. Furthermore, by taking limits as $j \to \infty$ on both sides of

$$\Xi u_{mn}^{(i_j)} + f(m,n,u_{mn}^{(i_j)}) = 0,\ (m,n) \in Z^2,$$

we see that v is a solution of (7.23)-(7.24). Our proof is complete. ■

We remark that if $f : Z^2 \times R \to R$ is a function such that $f(m,n,y) \ge 0$ for $y \ge 0$ and any $(m,n) \in Z^2$, then the functional inequality (7.21) has a nontrivial and nonnegative solution if, and only if, it has a positive solution. To see this, suppose w is a nontrivial and nonnegative solution such that, say, $w_{00} > 0$. If $w_{ab} = 0$ for some $(a,b) \ne (0,0)$, then $\Xi w_{ab} \ge 0$. On the other hand, $\Xi w_{ab} \le -f(a,b,w_{ab}) \le 0$. Thus $\Xi w_{ab} = 0$. But then the values of w at the neighbors of y are all equal to zero.

If one of the neighbors is not the origin, we may repeat the same arguments a finite number of times to conclude that $w(0) = 0$, which is a contradiction.

We illustrate our previous results by an example. Suppose $f : Z^2 \times R \to R$ is continuous and satisfies $f(m,n,b) \geq 0$ and $f(m,n,c) \leq 0$ for some constants b and c such that $b < c$. Then $u = \{b\}$ is a subsolution and $w = \{c\}$ is a supersolution, and hence there is a solution h of (7.20) such that $b \leq h_{mn} \leq c$ for $(m,n) \in Z^2$. In case $b = 0$, and f satisfies the additional condition that $f(m,n,y) \geq 0$ for $(m,n) \in Z^2$ and $y \geq 0$, then in view of the above remark, h is also positive. An example of such a function is

$$f(m,n,y) = y(y-c)^{2k},$$

where $c > 0$ and k is a positive integer. Since $|\partial f/\partial y|$ is bounded below for $0 \leq y \leq c$, therefore, for such functions, equation (7.20) has at least one positive and bounded solution.

In the rest of this section, we consider another equation which is more general than the discrete Laplace equation:

$$\alpha_{ij}u_{i-1,j} + \beta_{ij}u_{i+1,j} + \gamma_{ij}u_{i,j-1} + \delta_{ij}u_{i,j+1} - \sigma_{ij}u_{ij} + f(i,j,u_{ij}) = 0, \quad (7.26)$$

where $\alpha_{ij}, \beta_{ij}, \gamma_{ij}, \delta_{ij}, \sigma_{ij} > 0$ for $(i,j) \in Z^2$.

We will establish an existence criterion for bounded solutions of (7.26). Let $l^{Z\times Z}$ linear space of all bivariate sequences $\{u_{ij}\}$ under the usual addition and real multiplication, and let $S : l^{Z\times Z} \to l^{Z\times Z}$ be defined as follows: for $u = \{u_{ij}\} \in l^{Z\times Z}$,

$$(Su)_{ij} = \frac{\alpha_{ij}u_{i-1,j}}{\sigma_{ij}} + \frac{\beta_{ij}u_{i+1,j}}{\sigma_{ij}} + \frac{\gamma_{ij}u_{i,j-1}}{\sigma_{ij}} + \frac{\delta_{ij}u_{i,j+1}}{\sigma_{ij}} + \frac{f(i,j,u_{ij})}{\sigma_{ij}}.$$

Consider the following sequence of successive approximations: $u^{(0)} = \{B^*\}$ and

$$u^{(n+1)} = Su^{(n)}, \; n \in N. \quad (7.27)$$

Note that

$$u_{ij}^{(1)} = B^*\left\{\frac{\alpha_{ij}}{\sigma_{ij}} + \frac{\beta_{ij}}{\sigma_{ij}} + \frac{\gamma_{ij}}{\sigma_{ij}} + \frac{\delta_{ij}}{\sigma_{ij}}\right\} + \frac{f(i,j,B^*)}{\sigma_{ij}}, \; (i,j) \in Z^2.$$

Thus, if we impose the condition

$$B^*\left\{\frac{\alpha_{ij}}{\sigma_{ij}} + \frac{\beta_{ij}}{\sigma_{ij}} + \frac{\gamma_{ij}}{\sigma_{ij}} + \frac{\delta_{ij}}{\sigma_{ij}}\right\} + \frac{f(i,j,B^*)}{\sigma_{ij}} \leq B^* \quad (7.28)$$

then $u^{(1)} \leq u^{(0)} = 1$. Similarly, if we define $v^{(0)} = \{B_*\}$, and $v^{(n+1)} = Sv^{(n)}$ for $n \in N$, then under the condition

$$B_*\left\{\frac{\alpha_{ij}}{\sigma_{ij}} + \frac{\beta_{ij}}{\sigma_{ij}} + \frac{\gamma_{ij}}{\sigma_{ij}} + \frac{\delta_{ij}}{\sigma_{ij}}\right\} + \frac{f(i,j,B_*)}{\sigma_{ij}} \geq B_*, \quad (7.29)$$

we have $v^{(0)} \leq v^{(1)}$. Next, note that if we assume f is nondecreasing in the third variable, then $x \leq y$ implies $Sx \leq Sy$. Thus when $B_* \leq B^*$, we have

$$B_* = v^{(0)} \leq v^{(1)} \leq \ldots \leq v^{(n)} \leq \ldots \leq u^{(n)} \leq \ldots \leq u^{(0)} \leq B^*.$$

It follows that for each $(i,j) \in Z^2$, $u_{ij}^{(n)}$ converges to a nonnegative limit u_{ij} and $v_{ij}^{(n)}$ to v_{ij} as $n \to \infty$. By taking limits as $n \to \infty$ on both sides of (7.27), we see that the bivariate sequence $u = \{u_{ij}\}$ and $v = \{v_{ij}\}$ are solutions of $w = Sw$, and hence bounded solutions of (7.26). The following is now clear.

Theorem 112 *Suppose $\alpha_{ij}, \beta_{ij}, \gamma_{ij}, \delta_{ij}, \sigma_{ij} > 0$ holds for $(i,j) \in Z^2$. Suppose further that $f : Z^2 \times R \to R$ satisfies (7.28) and (7.29) where $B_* \leq B^*$, and is continuous and nondecreasing in the third variable for any $(i,j) \in Z^2$. Then (7.26) has a solution bounded between B_* and B^*.*

In particular, for the case where $f \equiv 0$, $\alpha_{ij} = \beta_{ij} = \gamma_{ij} = \delta_{ij} = 1$ and $\sigma_{ij} = 4$, a positive and bounded solution exists.

7.5 Contraction Method for a Finite Laplace Equation

Let us consider the equation (7.26) again,

$$\alpha_{ij}u_{i-1,j} + \beta_{ij}u_{i+1,j} + \gamma_{ij}u_{i,j-1} + \delta_{ij}u_{i,j+1} - \sigma_{ij}u_{ij} + f(i,j,u_{ij}) = 0, \quad (7.30)$$

but $\alpha_{ij}, \beta_{ij}, \gamma_{ij}, \delta_{ij}, \sigma_{ij}$ are only required to be defined for $(i,j) \in Z^2$ such that $\sigma_{ij} \neq 0$ for $(i,j) \in Z^2$. Let us further assume that $f : Z^2 \times R \to R$ satisfies

$$|f(i,j,u)| \leq |\sigma_{ij}|\,\omega(|u|),\ (i,j) \in Z^2, u \in R,$$

for some function $\omega : R \to R$ which is bounded on $[0,\infty)$, and

$$|f(i,j,u) - f(i,j,v)| \leq \lambda_{ij}\,|u - v|,\ (i,j) \in Z^2; u, v \in R,$$

for some nonnegative bivariate sequence $\{\lambda_{ij}\}$.

Let $l^{Z\times Z}$ be the Banach space of all bounded bivariate sequences $\{x_{ij}\}$ defined on Z^2 under the norm

$$\|\{x_{ij}\}\| = \sup\left\{|x_{ij}| \mid (i,j) \in Z^2\right\}.$$

Let $T : l^{Z\times Z} \to l^{Z\times Z}$ be defined as follows: for $u = \{u_{ij}\} \in l^{Z\times Z}$,

$$(Tu)_{ij} = \frac{\alpha_{ij}u_{i-1,j}}{\sigma_{ij}} + \frac{\beta_{ij}u_{i+1,j}}{\sigma_{ij}} + \frac{\gamma_{ij}u_{i,j-1}}{\sigma_{ij}} + \frac{\delta_{ij}u_{i,j+1}}{\sigma_{ij}} + \frac{f(i,j,u_{ij})}{\sigma_{ij}}$$

for $(i,j) \in Z^2$. Then

$$|(Tu)_{ij}| \leq \left\{\left|\frac{\alpha_{ij}}{\sigma_{ij}}\right| + \left|\frac{\beta_{ij}}{\sigma_{ij}}\right| + \left|\frac{\gamma_{ij}}{\sigma_{ij}}\right| + \left|\frac{\delta_{ij}}{\sigma_{ij}}\right|\right\}\|u\| + \omega\left(|u_{ij}|\right)$$

for $(i,j) \in Z^2$, and

$$\|Tu - Tv\| \leq \sup_{i,j\in Z}\left\{\left|\frac{\alpha_{ij}}{\sigma_{ij}}\right| + \left|\frac{\beta_{ij}}{\sigma_{ij}}\right| + \left|\frac{\gamma_{ij}}{\sigma_{ij}}\right| + \left|\frac{\delta_{ij}}{\sigma_{ij}}\right| + \left|\frac{\lambda_{ij}}{\sigma_{ij}}\right|\right\}\|u - v\|$$

for $u = \{u_{ij}\}, v = \{v_{ij}\} \in l^{Z\times Z}$.

By Banach's contraction principle, we see that if the additional condition

$$\sup_{i,j\in Z}\left\{\left|\frac{\alpha_{ij}}{\sigma_{ij}}\right| + \left|\frac{\beta_{ij}}{\sigma_{ij}}\right| + \left|\frac{\gamma_{ij}}{\sigma_{ij}}\right| + \left|\frac{\delta_{ij}}{\sigma_{ij}}\right| + \left|\frac{\lambda_{ij}}{\sigma_{ij}}\right|\right\} < 1$$

is imposed, then equation (7.30) will have a unique bounded solution which is nontrivial if we assume the additional condition $f(i,j,0) \neq 0$ for some $(i,j) \in Z^2$.

7.6 Monotone Method for Evolutionary Equations

The prototype linear difference equation

$$x_{n+1} - x_n + p(n)x_{n-v} = 0, \; n \in N,$$

is much studied in the literature due to its relations with many mathematical models of reality. A natural multivariate analog is the following partial difference equation

$$x_{m+1,n} + x_{m,n+1} - x_{mn} + p(m,n)x_{m-\sigma,n-\tau} = 0, \; m,n \in N. \tag{7.31}$$

In this section, we will establish an existence criterion for eventually positive solutions of these equations. Equation (7.31) can be regarded as a special case of the following functional inequality

$$x_{m+1,n} + x_{m,n+1} - x_{mn} + p(m,n)x_{m-\sigma,n-\tau} \leq 0, \; m,n \in N. \tag{7.32}$$

Furthermore, it is clear that finding an eventually positive solution of (7.32) is easier than finding one for equation (7.31). The idea, therefore, is to show that once an eventually positive solution of (7.32) is found, an eventually positive solution of (7.31) can then be constructed.

In (7.31) and (7.32), we assume that σ and τ are nonnegative integers, and $\{p(m,n)\}$ is a nonnegative bivariate sequence. By a solution of (7.32), we mean a real bivariate sequence

$$x = \{x_{mn} |\; m \geq -\sigma, n \geq -\tau\}, \; \sigma = \max(\sigma_1, ..., \sigma_K), \tau = \max(\tau_1, ..., \tau_K),$$

which satisfies (7.32). It is not difficult to formulate and prove an existence theorem for the solutions of (7.32) when appropriate initial conditions are given. As is customary, we say that a solution $x = \{x_{mn}\}$ of (7.32) is eventually positive if $x_{mn} > 0$ for all large m and large n.

Let $x = \{x_{mn}\}$ be an eventually positive solution of (7.32) such that $x_{mn} > 0$ for $m \geq M - \sigma$ and $n \geq J - \tau$, where M and J are nonnegative integers. Since

$$x_{m+1,n} + x_{m,n+1} - x_{mn} \leq 0$$

for $m \geq M$ and $n \geq J$, we see that $\Delta_1 x_{mn}$ and $\Delta_2 x_{mn}$ are nonpositive for all large m and n. Next note that the functional inequality (7.32) is equivalent to

$$\Delta_2\Delta_1 x_{mn} \geq x_{m+1,n+1} + p(m,n)x_{m-\sigma,n-\tau}, \; m,n \in N. \tag{7.33}$$

Thus summing (7.33) from (m, n) to ∞, we see that

$$\sum_{(i,j)=(m,n)}^{\infty} \Delta_2\Delta_1 x_{ij} \geq \sum_{(i,j)=(m,n)}^{\infty} x_{i+1,j+1} + \sum_{(i,j)=(m,n)}^{\infty} p(i,j)x_{i-\sigma,j-\tau}.$$

Since

$$\sum_{(i,j)=(m,n)}^{(\bar{m},\bar{n})} \Delta_2\Delta_1 x_{ij} = \sum_{i=m}^{\bar{m}} \Delta_1 x_{i,\bar{n}+1} - \sum_{i=m}^{\bar{m}} \Delta_1 x_{in}$$

$$\leq -\sum_{i=m}^{\bar{m}} \Delta_1 x_{in} = -x_{\bar{m}+1,n} + x_{mn} \leq x_{mn},$$

thus we have

$$x_{mn} \geq \sum_{(i,j)=(m,n)}^{\infty} x_{i+1,j+1} + \sum_{(i,j)=(m,n)}^{\infty} p(i,j)x_{i-\sigma,j-\tau} \tag{7.34}$$

for $m \geq M$ and $n \geq J$. Let Ω be the set of all real bivariate sequences of the form $y = \{y_{mn} | \; m \geq M - \sigma, n \geq J - \tau\}$. Define an operator $T : \Omega \to \Omega$ by

$$(Ty)_{mn} = \sum_{(i,j)=(m,n)}^{\infty} y_{i+1,j+1} + \sum_{(i,j)=(m,n)}^{\infty} p(i,j)y_{i-\sigma,j-\tau}$$

for $m \geq M$ and $n \geq J$, and

$$(Ty)_{mn} = x_{MJ}$$

elsewhere. Consider the following sequence of successive approximations $\{y^{(t)}\}$ contained in Ω :

$$y^{(0)}_{mn} = \begin{cases} x_{mn} & m \geq M, n \geq J \\ x_{MJ} & elsewhere \end{cases},$$

and

$$y^{(t+1)} = Ty^{(t)}, \; t \in N.$$

Clearly, in view of (7.34),

$$0 \leq y^{(t+1)}_{mn} \leq y^{(t)}_{mn} \leq y^{(0)}_{mn}, \; m \geq M, n \geq J, t \geq 1.$$

Thus as $t \to \infty$, $y^{(t)}$ converges pointwise to some nonnegative sequence $w = \{w_{mn}\}$ which satisfies

$$w_{mn} = \sum_{(i,j)=(m,n)}^{\infty} w_{i+1,j+1} + \sum_{(i,j)=(m,n)}^{\infty} p(i,j)w_{i-\sigma,j-\tau} \tag{7.35}$$

for $m \geq M$ and $n \geq J$, and $w_{mn} = x_{MJ}$ elsewhere. By taking partial differences on both sides of the above equation, we see that w is an eventually nonnegative solution of (7.31). Finally, we show that w is eventually positive, provided that $\min(\sigma, \tau) > 0$ and $p(m, n) > 0$ for $m, n \in N$. Indeed, suppose to the contrary that there exists a pair of integers $m^* \geq M$ and $n^* \geq J$ such that $w_{mn} > 0$ for

$$(m, n) \in \{M - \sigma, ..., m^*\} \times \{J - \tau, ..., n^*\} \backslash \{(m^*, n^*)\}$$

but $w_{m^*n^*} = 0$. Then in view of (7.35),

$$0 = \sum_{(i,j)=(m^*,n^*)}^{\infty} w_{i+1,j+1} + \sum_{(i,j)=(m^*,n^*)}^{\infty} p(i,j)w_{i-\sigma,j-\tau},$$

which implies $w_{ij} = 0$ for $i > m^*$ and $j > n^*$, and

$$p(i,j)w_{i-\sigma,j-\tau} = 0, \quad i \geq m^*, j \geq n^*,$$

which is the desired contradiction. The following is now clear.

Theorem 113 *Suppose $\{p(m,n)\}$ is a positive bivariate sequences, and σ and τ are nonnegative integers such that $in(\sigma,\tau) > 0$. If the functional inequality (7.32) has an eventually positive solution, then so does equation (7.31).*

We now seek an eventually positive solution of (7.32). To this end, let us first seek an eventually positive solution of the following functional inequality

$$x_{m+1,n} + x_{m,n+1} - x_{mn} + Px_{m-\sigma,n-\tau} \leq 0, \; m,n \geq 0, \tag{7.36}$$

where $P \geq 0$, and σ, τ are nonnegative integers. A possible candidate is the bivariate sequence $y = \{y_{mn}\}$ defined by

$$y_{mn} = \left(\frac{1-\lambda}{2}\right)^{m+n}, \; m \geq -\sigma, n \geq -\tau,$$

where λ is a number to be determined. For y to be eventually positive, it is necessary that $\lambda < 1$. Furthermore, since

$$\begin{aligned} y_{m+1,n} + y_{m,n+1} - y_{mn} &= -\lambda\left(\frac{1-\lambda}{2}\right)^{m+n} \\ &= -\lambda\left(\frac{1-\lambda}{2}\right)^{\sigma+\tau}\left(\frac{1-\lambda}{2}\right)^{m+n-\sigma-\tau} \\ &= -\lambda\left(\frac{1-\lambda}{2}\right)^{\sigma+\tau} y_{m-\sigma,n-\tau}, \end{aligned}$$

if we impose the additional condition that

$$\lambda\left(\frac{1-\lambda}{2}\right)^{\sigma+\tau} \geq P, \tag{7.37}$$

then y will be an eventually positive solution of (7.36). We now need to decide when the above two conditions can be satisfied. Note that

$$\max_{0\leq\lambda<1} \lambda(1-\lambda)^{\sigma+\tau} = \frac{(\sigma+\tau)^{\sigma+\tau}}{(\sigma+\tau+1)^{\sigma+\tau+1}}.$$

Thus when

$$P \leq \frac{1}{2^{\sigma+\tau}} \frac{(\sigma+\tau)^{\sigma+\tau}}{(\sigma+\tau+1)^{\sigma+\tau+1}},$$

we can find $\lambda \in [0,1)$ such that the conditions $\lambda < 1$ and (7.37) can be satisfied simultaneously. We summarize these in the following theorem.

Theorem 114 *Suppose $\{p(m,n)\}$ is a nonnegative bivariate sequence, and σ, τ are nonnegative integers. Suppose further that*

$$p(m,n) \le P \le \frac{1}{2^{\sigma+\tau}} \frac{(\sigma+\tau)^{\sigma+\tau}}{(\sigma+\tau+1)^{\sigma+\tau+1}}, \quad m,n \in N.$$

Then the functional inequality (7.32) has an eventually positive solution.

Indeed, this follows from the fact that under the conditions of Theorem 114, an eventually positive solution of (7.36) is also an eventually positive solution of (7.32).

7.7 Eigenvalue Method for a Boundary Problem

In this section, we consider the problem

$$\Xi u_{ij} + \lambda q_{ij} u_{ij} = 0, \ (i,j) \in \Omega, \tag{7.38}$$

subject to the Dirichlet boundary condition

$$u_{ij} = 0, \ (i,j) \in \partial\Omega, \tag{7.39}$$

where $\lambda \in C$, Ω is a nonempty finite domain, and the bivariate sequence $q = \{q_{ij}\}_{(i,j)\in\Omega}$ is nonnegative and nontrivial. Our boundary value problem may be regarded as a linear eigenvalue problem. One way to see this is as follows. First of all, since Ω is finite, we may assume without loss of generality that

$$\Omega = \{(i_1, j_1), ..., (i_n, j_n)\}. \tag{7.40}$$

Let $A = (\alpha_{st})$ be the "adjacency matrix" of Ω defined by $\alpha_{st} = 1$ if (i_s, j_s) and (i_t, j_t) are neighbors, and zero otherwise. Then our boundary value problem can be written as the following matrix eigenvalue problem

$$(A - 4I)u + \lambda \operatorname{diag}(q_{i_1 j_1}, ..., q_{i_n j_n})\, u = 0, \tag{7.41}$$

where I is the identity matrix and $u = \operatorname{col}(u_{i_1 j_1}, ..., u_{i_n j_n})$.

We first note that $\lambda = 0$ cannot be an eigenvalue, for otherwise, its corresponding eigensolution $u = \{u_{ij}\}$ satisfies

$$\begin{aligned} \Xi u_{ij} &= 0, \ (i,j) \in \Omega \\ u_{ij} &= 0, \ (i,j) \in \partial\Omega, \end{aligned}$$

and hence is trivial in view of Theorem 53. Next, our eigenvalue problem has real eigenvalues only. To see this, let λ be an eigenvalue and u a corresponding eigensolution. Note that we may write our eigenvalue problem, in view of Theorem 76, as the following equivalent eigenvalue problem

$$u_{ij} = \lambda \sum_{(s,t)\in\Omega} G_{ij}^{(s,t)} q_{st} u_{st}, \ (i,j) \in \Omega. \tag{7.42}$$

By means of the ordering introduced in (7.40), the above eigenvalue problem can also be written as

$$u = \lambda G \operatorname{diag}(q_{i_1 j_1}, ..., q_{i_n j_n})\, u, \tag{7.43}$$

where G is a symmetric and positive matrix (see Section 7 of Chapter 5). Multiplying both sides of (7.43) by the transpose of the complex conjugate u' of u, we see that

$$u'u = \lambda \{u'G \operatorname{diag}(q_{i_1j_1}, ..., q_{i_nj_n}) u\}.$$

The left hand side is real and positive, while the product in the brace bracket is real. Thus λ must be a real number.

Next, we show that (7.41) has an eigenvalue which is smaller than or equal to any other eigenvalue and there corresponds to it an eigensolution $u = \{u_{ij}\}$ such that $u_{ij} > 0$ for $(i,j) \in \Omega$. Indeed, since the matrix $G \operatorname{diag}(q_{i_1j_1}, ..., q_{i_nj_n})$ is nontrivial and nonnegative, by Perron's Theorem 18 for nonnegative matrices, we know it has a nonnegative (and hence positive) eigenvalue which is equal to the spectral radius ρ of $G \operatorname{diag}(q_{i_1j_1}, ..., q_{i_nj_n})$ and there is a nonnegative eigenvector corresponding to it.

Next, we assert that the spectral radius ρ obtained above is simple. Indeed, if $u = \{u_{ij}\}$ and $v = \{v_{ij}\}$ are two solutions of (7.38)–(7.39) corresponding to the same positive eigenvalue and $u_{ij} > 0$ for $(i,j) \in \Omega$, then by the discrete Wirthinger's inequality condition stated in Theorem 62, v must be a constant multiple of u.

We have thus obtained the following result.

Theorem 115 *The eigenvalue problem (7.38)–(7.39) has real eigenvalues only. Furthermore, a simple positive eigenvalue $\lambda[q]$ exists (which is smaller than any other eigenvalues), and there corresponds to it an eigenvector $u = \{u_{ij}\}_{(i,j)\in\Omega+\partial\Omega}$ which satisfies $u_{ij} > 0$ for $(i,j) \in \Omega$.*

7.8 Contraction Method for a Boundary Problem

We have already seen solutions of steady state equations over the lattice plane. We can also obtain existence criteria for equations such as the following

$$\Xi v_{ij} + f(i,j,v_{ij}) = 0,\ (i,j) \in \Omega, \tag{7.44}$$

subject to the Dirichlet condition

$$v_{ij} = 0,\ (i,j) \in \partial\Omega, \tag{7.45}$$

where Ξ is the discrete Laplacian and Ω is now a nonempty finite domain.

We first note, in view of Theorem 76, that our boundary problem is equivalent to the fixed point problem

$$u_{ij} = \sum_{(s,t)\in\Omega} G_{ij}^{(s,t)} f(s,t,u_{st}),\ (i,j) \in \Omega. \tag{7.46}$$

Theorem 116 *Suppose $f : Z^2 \times R \to R$ satisfies the Lipschitz condition*

$$|f(i,j,u) - f(i,j,v)| \le q_{ij} |u - v|,\ (i,j) \in \Omega,$$

where $q = \{q_{ij}\}$ is a nontrivial and nonnegative bivariate sequence. Suppose further that the least positive eigenvalue $\lambda[q]$ of (7.38)–(7.39) satisfies $\lambda[q] > 1$, then the boundary value problem (7.44)–(7.45) has a unique solution, and the condition $\lambda[q] > 1$ is sharp.

Proof. Let $v = \{v_{ij}\}$ be the corresponding eigensolution of $\lambda[q]$ such that $v_{ij} > 0$ for $(i,j) \in \Omega$. In view of (7.42), we have

$$v_{ij} = \lambda[q] \sum_{(s,t)\in\Omega} G_{ij}^{(s,t)} q_{st} v_{st}, \ (i,j) \in \Omega.$$

Let B be the Banach space l^Ω which is equipped with the weighted norm

$$\|u\|_v = \max_{(i,j)\in\Omega} \frac{|u_{ij}|}{v_{ij}}, \ u = \{u_{ij}\} \in l^\Omega.$$

Let $T : l^\Omega \to l^\Omega$ be defined by

$$(Tu)_{ij} = \sum_{(s,t)\in\Omega} G_{ij}^{(s,t)} f(s,t,u_{st}), \ (i,j) \in \Omega.$$

Then the fixed point problem

$$u_{ij} = (Tu)_{ij}, \ (i,j) \in \Omega,$$

has a unique solution if, and only if, our problem (7.44)-(7.45) has a unique solution. For any $x = \{x_{ij}\}$ and $y = \{y_{ij}\}$ in l^Ω,

$$\begin{aligned}
\frac{|(Tx - Ty)_{ij}|}{v_{ij}} &\le \frac{1}{v_{ij}} \sum_{(s,t)\in\Omega} G_{ij}^{(s,t)} \left| f(s,t,x_{st}) - f(s,t,y_{st}) \right| \\
&\le \frac{1}{v_{ij}} \sum_{(s,t)\in\Omega} G_{ij}^{(s,t)} q_{st} \left| x_{st} - y_{st} \right| \\
&\le \frac{1}{v_{ij}} \sum_{(s,t)\in\Omega} G_{ij}^{(s,t)} q_{st} v_{st} \frac{|x_{st} - y_{st}|}{v_{st}} \\
&\le \|x - y\|_v \frac{1}{v_{st}} \sum_{(s,t)\in\Omega} G_{ij}^{(s,t)} q_{st} v_{st} \\
&= \frac{\|x - y\|_v}{\lambda[q]}.
\end{aligned}$$

This shows that T is a contraction mapping in B and thus has a unique fixed point.

Finally, the condition $\lambda[q] > 1$ is sharp in the sense that if $\lambda[q] = 1$, then there is a function f such that (7.44)-(7.45) has more than one solution. Indeed, we may take $f(i,j,u_{ij}) = q_{ij} u_{ij}$, then the trivial bivariate sequence and the eigensolution v are solutions of (7.44)-(7.45). The proof is complete. ■

7.9 Monotone Method for Boundary Problems

A real bivariate sequence $w = \{w_{ij}\}_{(i,j)\in\Omega+\partial\Omega}$ is called an upper solution of (7.44)–(7.45) if

$$\Xi w_{ij} + f(i,j,w_{ij}) \le 0, \ (i,j) \in \Omega,$$

and

$$w_{ij} \ge 0, \ (i,j) \in \partial\Omega.$$

Similarly, a real bivariate sequence $u = \{u_{ij}\}_{(i,j)\in\Omega+\partial\Omega}$ is called a lower solution of (7.44)–(7.45) if

$$\Xi u_{ij} + f(i,j,u_{ij}) \geq 0,\ (i,j) \in \Omega,$$

and

$$u_{ij} \leq 0,\ (i,j) \in \partial\Omega.$$

Note that by means of the maximum principle stated in Theorem 52, if $v = \{v_{ij}\}$ is a lower solution of (7.44)–(7.45) where $f(i,j,v) \leq 0$ for $(i,j) \in \Omega$ and $v \geq 0$, then $v_{ij} \leq 0$ for $(i,j) \in \Omega+\partial\Omega$. Next, assume that $f(i,j,v)$ is nonincreasing in v for each $(i,j) \in \Omega$. Then for any lower solution $u = \{u_{ij}\}$ and upper solution $w = \{w_{ij}\}$ of (7.44)–(7.45), we have $u_{ij} \leq w_{ij}$ for $(i,j) \in \Omega$. Indeed, note that

$$\Xi(u_{ij} - w_{ij}) + f(i,j,u_{ij}) - f(i,j,w_{ij}) \geq 0,\ (i,j) \in \Omega, \tag{7.47}$$

and

$$u_{ij} - w_{ij} \leq 0,\ (i,j) \in \partial\Omega.$$

By Theorem 52, we see that $u_{ij} - w_{ij} \leq 0$ for $(i,j) \in \Omega + \partial\Omega$.

We remark that, in general, a lower solution may not be less than or equal to an upper solution. It is, however, useful to know when a lower solution is less than or equal to an upper solution. To this end, let $w = \{w_{ij}\}$ be an upper solution of (7.44)–(7.45) such that there is a positive sequence $z = \{z_{ij}\}$ which satisfies

$$\Xi(\lambda z_{ij}) < f(i,j,w_{ij}) - f(i,j,w_{ij} + \lambda z_{ij}),\ (i,j) \in \Omega,$$

for any $\lambda > 0$. Then $v \leq w$ for any lower solution $v = \{v_{ij}\}$ of (7.44)–(7.45).

Indeed, suppose to the contrary that v is a lower solution of (7.44)–(7.45) such that

$$v_{\alpha\beta} - w_{\alpha\beta} = \max_{(i,j)\in\Omega}\{v_{ij} - w_{ij}\} > 0,$$

then $v_{\alpha\beta} = w_{\alpha\beta} + \lambda^* z_{\alpha\beta}$ for some $\lambda^* > 0$. Furthermore,

$$\begin{aligned} 0 &\geq \Xi(v_{\alpha\beta} - w_{\alpha\beta} - \lambda^* z_{\alpha\beta}) \\ &> -f(i,j,v_{ij}) + f(i,j,w_{ij}) + f(i,j,w_{ij} + \lambda^* z_{ij}) - f(i,j,w_{ij}) = 0, \end{aligned}$$

which is a contradiction.

Another comparison theorem for lower and upper solutions is as follows, the proof of which is elementary.

Theorem 117 *Assume that $f_1(i,j,v) \leq f(i,j,v) \leq f_2(i,j,v)$ for $(i,j) \in \Omega$. Then an upper solution of*

$$\begin{aligned} \Xi w_{ij} + f_2(i,j,w_{ij}) &= 0,\ (i,j) \in \Omega, \\ w_{ij} &= 0,\ (i,j) \in \partial\Omega, \end{aligned}$$

is also an upper solution of (7.44)–(7.45), a lower solution of

$$\begin{aligned} \Xi u_{ij} + f_1(i,j,u_{ij}) &= 0,\ (i,j) \in \Omega, \\ u_{ij} &= 0,\ (i,j) \in \partial\Omega, \end{aligned}$$

is also a lower solution of (7.44)–(7.45).

Next, we derive an existence theorem for upper and lower solutions. Before doing so, let us recall, in view of Theorem 76, that the boundary problem (7.44)-(7.45) is equivalent to the matrix problem (7.46).

Theorem 118 *If $f(i,j,v) \ge L$ (or $f(i,j,v) \le L$) for $(i,j) \in \Omega$, then (7.44)-(7.45) has a lower solution (respectively, an upper solution).*

Proof. Consider the system of linear equations

$$\begin{aligned} \Xi v_{ij} + L &= 0, \ (i,j) \in \Omega, \\ v_{ij} &= 0, \ (i,j) \in \partial\Omega. \end{aligned}$$

This system has a unique solution by Theorem 3.53, which is also the desired lower solution of (7.44)-(7.45). ■

Next, suppose there is a nonnegative constant c such that $f(i,j,c) \le 0$ for $(i,j) \in \Omega$. Then the bivariate sequence $v = \{c\}$ satisfies

$$\Xi v_{ij} + f(i,j,v_{ij}) = f(i,j,c) \le 0$$

for $(i,j) \in \Omega$. That is, v is an upper solution of (7.44)-(7.45). The following is now clear.

Theorem 119 *Suppose there is a nonnegative (or nonpositive) constant c such that $f(i,j,c) \le 0$ (respectively, $f(i,j,c) \ge 0$) for $(i,j) \in \Omega$, then $v = \{c\}$ is an upper solution (respectively, a lower solution) of (7.44)-(7.45).*

Other existence theorems for lower and upper solutions of (7.44)-(7.45) can be obtained by means of the comparison Theorem 117. For instance, we may look for eigen-solutions of the linear eigenvalue problem

$$\Xi x_{ij} + \lambda p_{ij} x_{ij} = 0, \ (i,j) \in \Omega, \tag{7.48}$$

$$x_{ij} = 0, \ (i,j) \in \partial\Omega. \tag{7.49}$$

Theorem 120 *Let λ be an eigenvalue of the eigenvalue problem (7.48)-(7.49) and let $u = \{u_{ij}\}$ be its corresponding eigenvector. Suppose further that $f(i,j,x) \le \lambda q_{ij} x$ ($\lambda q_{ij} x \le f(i,j,x)$) for $(i,j) \in \Omega$. Then u is an upper solution (respectively, a lower solution) of (7.44)-(7.45).*

We now derive several existence theorems for the solutions of the boundary problem (7.44)-(7.45).

Theorem 121 *Suppose $|f(i,j,v)| \le M$ for $(i,j) \in \Omega$ and $v \in R$. Suppose further that $f(i,j,u)$ is continuous in u for each $(i,j) \in \Omega$. Then the boundary problem (7.44)-(7.45) has a solution.*

Proof. Let $T : R^n \to R^n$ is defined by

$$(Tx)_{ij} = \sum_{(u,v)\in\Omega} G_{ij}^{(u,v)} f(u,v,x_{uv}), \ (i,j) \in \Omega.$$

Let

$$K = \max_{(i,j)\in\Omega} \sum_{(u,v)\in\Omega} \left| G_{ij}^{(u,v)} \right|,$$

and let

$$S = \{x = \{x_{ij}\} | \ \|x\|_\infty \le KM\},$$

where $\|x\| = \max_{(i,j)\in\Omega} |x_{ij}|$. It is easy to see that S is a bounded, convex and closed subset of l^Ω. Furthermore, T transforms S into S in a continuous manner in view of the assumptions imposed on f. By the Brouwer fixed point Theorem 2.20, there exists a vector $x^* \in S$ such that $x^* = Tx^*$. ■

Theorem 122 *Suppose that $f(i,j,u)$ is continuous with respect to u for each $(i,j) \in \Omega$. Then for any lower solution $u = \{u_{ij}\}$ and upper solution $w = \{w_{ij}\}$ of (7.44)-(7.45) satisfying $u \le w$, there is a solution $v = \{v_{ij}\}$ of (7.44)–(7.45) which satisfies $u \le v \le w$.*

Proof. Consider the boundary value problem

$$\Xi x_{ij} + \Phi(i,j,x_{ij}) = 0, \ (i,j) \in \Omega, \tag{7.50}$$

$$x_{ij} = 0, \ (i,j) \in \Omega + \partial\Omega, \tag{7.51}$$

where

$$\Phi(i,j,x) = \begin{cases} f(i,j,w_{ij}) + (w_{ij} - x)/(1+x^2) & x > w_{ij} \\ f(i,j,x) & u_{ij} \le x \le w_{ij} \\ f(i,j,u_{ij}) + (u_{ij} - x)/(1+x^2) & x < u_{ij} \end{cases}, \tag{7.52}$$

for $(i,j) \in \Omega$. Clearly, the function Φ is bounded for $(i,j) \in \Omega$ and $x \in R$, and is continuous in x. Thus by Theorem 7.121, there exists a solution $v = \{v_{ij}\}$ of the boundary problem (7.50)-(7.51). We assert that the solution v satisfies $u \le v \le w$ so that it is also a solution of (7.44)-(7.45) in view of the definition of Φ. Indeed, suppose to the contrary that $v_{\alpha\beta} - w_{\alpha\beta} = \max_{(i,j)\in\Omega}\{v_{ij} - w_{ij}\} > 0$. Then

$$0 \ge \Xi(v_{\alpha\beta} - w_{\alpha\beta}) \ge f(\alpha,\beta,w_{\alpha\beta}) - \Phi(\alpha,\beta,v_{\alpha\beta}) = \frac{v_{\alpha\beta} - w_{\alpha\beta}}{1 + v_{\alpha\beta}^2} > 0,$$

which is a contradiction. Similarly, we may show that $u \le v$. ■

As an example, suppose

$$0 \le f(i,j,v) \le 4\left\{\sin^2\frac{\pi}{4m+2} + \sin^2\frac{\pi}{4n+2}\right\}v, \ 1 \le i \le m, 1 \le j \le n.$$

Then the zero sequence is a lower solution of (7.44)–(7.45), and the sequence $w = \{w_{ij}\}$ defined by

$$w_{ij} = \sin\frac{\pi i}{2m+1}\sin\frac{\pi j}{2n+1}, \ 0 \le i \le m+1, 0 \le j \le n+1,$$

is an upper solution of (7.44)–(7.45) since it satisfies

$$\Xi w_{ij} + 4\left\{\sin^2\frac{\pi}{4m+2} + \sin^2\frac{\pi}{4n+2}\right\}w_{ij} = 0, \ 1 \le i \le m, 1 \le j \le n.$$

and

$$\begin{aligned} w_{i0} &= w_{i,n+1} = 0, \ 0 \le i \le m+1, \\ w_{0j} &= w_{m+1,j} = 0, \ 0 \le j \le n+1. \end{aligned}$$

In view of Theorem 122, (7.44)–(7.45) has a solution $v = \{v_{ij}\}$ which satisfies

$$0 \le v_{ij} \le \sin\frac{\pi i}{2m+1}\sin\frac{\pi j}{2n+1}, \ 1 \le i \le m, 1 \le j \le n.$$

Next, we derive an existence theorem when a Lipschitz condition is satisfied.

Theorem 123 *Suppose $f(i,j,u)$ is continuous with respect to u for each $(i,j) \in \Omega$. Suppose further that there exist a lower solution $u = \{u_{ij}\}$, an upper solution $w = \{w_{ij}\}$ and a positive sequence $\{p_{ij}\}_{(i,j)\in\Omega}$ such that $u \le w$ and for each $(i,j) \in \Omega$, the following one-sided Lipschitz condition*

$$f(i,j,x) - f(i,j,y) \ge -p_{ij}(x-y), \tag{7.53}$$

holds whenever $u_{ij} \le y \le x \le w_{ij}$. Then the boundary problem (7.44)–(7.45) has a solution u^* and a solution w^* such that $u^* \le w^*$.

Proof. For any sequence η which satisfies $u \le \eta \le w$, consider the following boundary problem

$$\Xi v_{ij} + f(i,j,\eta_{ij}) - p_{ij}(v_{ij} - \eta_{ij}) = 0, \ (i,j) \in \Omega, \tag{7.54}$$

$$v_{ij} = 0, \ (i,j) \in \partial\Omega. \tag{7.55}$$

This problem has a unique solution in view of Theorem 3.53. Let us define the sector

$$S = \{\eta = \{\eta_{ij}\} | \ u \le \eta \le w\}$$

and let $\Gamma : S \to l^{\Omega+\partial\Omega}$ be defined by

$$\Gamma\eta = \mu$$

where $\mu = \{\mu_{ij}\}$ is the unique solution of the boundary problem (7.54)-(7.55). We assert that $u \le \Gamma u$ and $\Gamma w \le w$. Indeed, let $\Gamma u = \xi = \{\xi_{ij}\}$ and suppose to the contrary that

$$v_{\alpha\beta} - \xi_{\alpha\beta} = \max_{(i,j)\in\Omega}\{v_{ij} - \xi_{ij}\} > 0,$$

then

$$\begin{aligned} 0 &\ge \Xi(v_{\alpha\beta} - \xi_{\alpha\beta}) \ge -f(\alpha,\beta,v_{\alpha\beta}) + \left(f(\alpha,\beta,v_{\alpha\beta}) - p_{\alpha\beta}(\xi_{\alpha\beta} - v_{\alpha\beta})\right) \\ &= p_{\alpha\beta}(v_{\alpha\beta} - \xi_{\alpha\beta}) > 0, \end{aligned}$$

which is a contradiction. Similarly, we may show that $\Gamma w \le w$. Next, we assert that for any ξ, $\psi \in \Omega$ and $\xi \le \psi$, we have $\Gamma\xi \le \Gamma\psi$. Indeed, let $\Gamma\xi = \tau = \{\tau_{ij}\}$ and $\Gamma\psi = \rho = \{\rho_{ij}\}$, and suppose to the contrary that

$$\tau_{\alpha\beta} - \rho_{\alpha\beta} = \max_{(i,j)\in\Omega}\{\tau_{ij} - \rho_{ij}\} > 0,$$

then we have

$$\begin{aligned} 0 &\geq \Xi(\tau_{\alpha\beta}-\rho_{\alpha\beta}) \\ &= \{-f(\alpha,\beta,\xi_{\alpha\beta})+p_{\alpha\beta}(\tau_{\alpha\beta}-\xi_{\alpha\beta})\}+\{f(\alpha,\beta,\psi_{\alpha\beta})-p_{\alpha\beta}(\rho_{\alpha\beta}-\psi_{\alpha\beta})\} \\ &= \{f(\alpha,\beta,\psi_{\alpha\beta})-f(\alpha,\beta,\xi_{\alpha\beta})\}+p_{\alpha\beta}\{\tau_{\alpha\beta}-\xi_{\alpha\beta}-\rho_{\alpha\beta}+\psi_{\alpha\beta}\} \\ &\geq -p_{\alpha\beta}(\psi_{\alpha\beta}-\xi_{\alpha\beta})+p_{\alpha\beta}(\tau_{\alpha\beta}-\rho_{\alpha\beta})+p_{\alpha\beta}(\psi_{\alpha\beta}-\xi_{\alpha\beta}) \\ &= p_{\alpha\beta}(\tau_{\alpha\beta}-\rho_{\alpha\beta})>0, \end{aligned}$$

which is a contradiction. Therefore, if we define two sequences as follows:

$$u^{(0)}=u,\ u^{(j+1)}=\Gamma u^{(j)},\ j\in N,$$

and

$$w^{(0)}=w,\ w^{(j+1)}=\Gamma w^{(j)},\ j\in N,$$

then we have

$$u=u^{(0)}\leq u^{(1)}\leq\ldots\leq w^{(1)}\leq w^{(0)}=w.$$

It follows that the limits $\lim_{j\to\infty}u^{(j)}=u^*$ and $\lim_{j\to\infty}w^{(j)}=w^*$ exist. Furthermore, by means of the continuity of the function $f(i,j,u)$, we easily see that they are solutions of (7.44)-(7.45). ■

We remark that the solutions u^* and w^* of (7.44)–(7.45) found in the above Theorem are minimal and maximal in the sense that if v is any solution of (7.44)–(7.45) which satisfies $u\leq v\leq w$, then $u^*\leq v\leq w^*$. Indeed, note that $\Gamma v=v$. Thus $u^{(1)}=\Gamma u^{(0)}\leq\Gamma v=v$, and by induction, $u^{(j)}\leq v$ for $j\geq 1$. This shows that $u^*\leq v$. Similarly, $v\leq w^*$.

7.10 Notes and Remarks

Traveling waves for partial differential equations have been studied to some extent, but studies of those for partial difference equations are relatively unknown except for those by Afraimovich and Pesin [2], Jennings [85] and a few others. In Section 7.2, we are only concerned with positive traveling waves, while there are many other types which are important. The material in this section is based on Cheng *et al.* [36].

Theorem 110 is due to Cheng and Medina [27]. Theorem 111 is new, while Theorem 112 is taken from Cheng and Medina [26].

Section 7.6 is based on Liu and Cheng [113].

Theorems 115 and 116 are contained in Cheng and Lin [21]. Similar results can be obtained for boundary value problems involving ordinary difference equations, see e.g. Cheng and Zhang [169].

The last section is based on Zhuang *et al.* [177]. Monotone methods for obtaining positive solutions to partial difference equations have been considered by many authors. See for example, Pao [134], [135, 136, 137, 138], Sheng and Agarwal [148], and others.

Chapter 8

Nonexistence

8.1 Introduction

Besides the sufficient conditions for the existence of solutions with desired qualitative properties, it is also of interest to know when these solutions cease to exist. In this chapter, we will derive a number of conditions mainly for the nonexistence of solutions that are "eventually positive". In case our partial difference equations are linear, such nonexistence criteria become oscillation theorems.

We have already seen several definitions of eventually positive sequences. For the sake of completeness, we will briefly go through these definitions again and add a few more. First of all, given a real sequence $\{u_k\}_{k\in N}$ in l^N, it is natural to say that it is eventually positive when $u_k > 0$ for k larger than or equal to some integer in N. Similarly, given a sequence $\left\{u^{(j)}\right\}_{j\in N}$ of finite vectors $u^{(j)}$ in $l^{\{0,1,\ldots,n\}}$ or a bivariate sequence of the form

$$\left\{u_i^{(j)} \mid i = 0, 1, ..., n+1; j \in N\right\},$$

it is natural to say that it is eventually positive if $u^{(j)} > 0$ for all large j. For bivariate sequences in $l^{N\times N}$, the corresponding concepts are not unique. Indeed, one possible concept can be defined as follows: $\{u_{ij}\}_{i,j\in N}$ is eventually positive if $u_{ij} > 0$ for all large i and large j. Yet another definition is as follows: $\{u_{ij}\}_{i,j\in N}$ is eventually positive if $u_{ij} > 0$ for $i + j \geq T$, where $T \in N$. In general, if $\{\psi_k\}$ is an enumeration (with respect to a linear ordering $\preccurlyeq$ of the lattice subset Ω) of the multiple sequence $u \in l^\Omega$, we may say that u is eventually positive (with respect to the ordering $\preccurlyeq$) if $\psi_k > 0$ for all large k.

To avoid confusion, we will repeat some of these definitions in later sections.

8.2 Equations Over The Plane

Let S_k be the k-sphere in Z^2 defined by

$$S_k = \{(i,j) \in Z^2 \mid \ |i| + |j| = k\}, \ k \in N.$$

Note that the size $|S_k|$ of a k-sphere is $4k$ when $k \geq 1$. Note further that the four corners of a k-sphere are $(k,0)$, $(-k,0)$, $(0,k)$ and $(0,-k)$ when $k \geq 1$. For the sake of convenience, we collect these corners into a set and denote it by C_k, where $k \geq 1$.

Consider the partial difference equation

$$\Xi u_{ij} + q(i,j,u_{ij}) = f_{ij},\ (i,j) \in Z^2, \tag{8.1}$$

where Ξ is the discrete Laplacian, $q(i,j,u)$ is a real function defined for $(i,j) \in Z^2$ and $u \in R$, and f_{ij} is a real function defined for $(i,j) \in Z^2$.

A solution of (8.1) is a bivariate sequence $\{u_{ij}\}_{i,j\in Z}$ which satisfies (8.1) for $(i,j) \in Z^2$. It is not difficult to construct solutions of (8.1). Indeed, since (8.1) can be written in the form

$$u_{i,j+1} = -u_{i-1,j} + 4u_{ij} - q(i,j,u_{ij}) - u_{i+1,j} - u_{i,j-1},$$

if we impose conditions such as

$$u_{ij} = \phi_{ij},\ i \in Z;\ j = -1, 0,$$

then we can calculate successively

$$u_{01};\ u_{0,-2}; u_{-1,1}, u_{02}, u_{11};\ u_{-1,-2}, u_{0,-3}, u_{1,-2}; u_{-2,1}, u_{-1,2}, u_{03}, u_{12}, u_{21};\ \ldots,$$

from (8.1) in a unique manner. It is not clear, however, whether a positive solution, i.e. $u_{ij} > 0$ for all $(i,j) \in Z^2$, can exist. In this section, we will show that, if a positive solution exists, certain conditions must be imposed on the functions $q(i,j,u)$ and f_{ij}. For now, we will assume that $q(i,j,u)$ satisfies a commonly seen requirement, namely,

(H1) $q(i,j,u) \geq p(|i| + |j|)\psi(u)$ for all $(i,j) \in Z^2$ and $u \geq 0$, where p is a nonnegative function defined on Z^+ and ψ is a nonnegative and convex function defined on $(0,\infty)$.

We will employ a technique which is called the technique of averaging.

Lemma 1 *Let $\{u_{ij}\}_{i,j\in Z}$ be a real bivariate sequence. Then for $k \geq 1$, we have*

$$\begin{aligned}\sum_{(i,j)\in S_{k+1}} \Xi u_{ij} &= 3\sum_{(i,j)\in C_k} u_{ij} + \sum_{(i,j)\in S_k\backslash C_k} 2u_{ij} \\ &\quad - \sum_{(i,j)\in S_{k+1}} 4u_{ij} + \sum_{(i,j)\in C_{k+2}} u_{ij} + \sum_{(i,j)\in S_{k+2}\backslash C_{k+2}} 2u_{ij}.\end{aligned}$$

Proof. Note that the sum

$$\sum_{(i,j)\in S_{k+1}} \Xi u_{ij}$$

is a sum of u_{ij} where (i,j) belongs to the spheres S_k or S_{k+1} or S_{k+2} only. Furthermore, it is not difficult to see that each u_{ij} in the sum appears twice for $(i,j) \in S_k\backslash C_k$, thrice for $(i,j) \in C_k$, -4 times for $(i,j) \in S_{k+1}$, twice for $(i,j) \in S_{k+2}\backslash C_{k+2}$, and once in C_{k+2}. ■

Let $u = \{u_{ij}\}_{i,j\in Z}$ be a real bivariate sequence. We will denote the mean of u over the spheres by

$$U_k = \frac{1}{|S_k|}\sum_{(i,j)\in S_k} u_{ij},\ k \geq 0. \tag{8.2}$$

Lemma 2 *Let $u = \{u_{ij}\}_{i,j \in Z}$ be a solution of (8.1). Then its mean $U = \{U_k\}_{k \in N}$ defined by (8.2) over the spheres satisfies the following recurrence relations:*

$$\begin{aligned} & 4\{2kU_k - 4(k+1)U_{k+1} + (k+2)U_{k+2}\} + \sum_{(i,j)\in S_{k+1}} q(i,j,u_{ij}) \\ = & \sum_{(i,j)\in S_{k+1}} f_{ij} - \sum_{(i,j)\in C_k} u_{ij} - \sum_{(i,j)\in S_{k+2}\setminus C_{k+2}} u_{ij}, \ k \geq 1, \end{aligned} \tag{8.3}$$

and

$$\begin{aligned} & 4\{2kU_k - 4(k+1)U_{k+1} + 2(k+2)U_{k+2}\} + \sum_{(i,j)\in S_{k+1}} q(i,j,u_{ij}) \\ = & \sum_{(i,j)\in S_{k+1}} f_{ij} + \sum_{(i,j)\in C_{k+2}} u_{ij} - \sum_{(i,j)\in C_k} u_{ij}, \ k \geq 1. \end{aligned} \tag{8.4}$$

Proof. Summing (8.1) over S_{k+1}, we have, in view of Lemma 1, that

$$\begin{aligned} & -\sum_{(i,j)\in S_{k+1}} q(i,j,u_{ij}) + \sum_{(i,j)\in S_{k+1}} f_{ij} \\ = & \sum_{C_k} u + 2\sum_{C_k} u + \sum_{S_k\setminus C_k} 2u - \sum_{S_{k+1}} 4u + 2\sum_{C_{k+2}} u + \sum_{S_{k+2}\setminus C_{k+2}} 2u - \sum_{C_{k+2}} u \\ = & \sum_{(i,j)\in S_k} 2u_{ij} - \sum_{(i,j)\in S_{k+1}} 4u_{ij} + \sum_{(i,j)\in S_{k+2}} 2u_{ij} + \sum_{(i,j)\in C_k} u_{ij} - \sum_{(i,j)\in C_{k+2}} u_{ij} \\ = & 2|S_k|U_k - 4|S_{k+1}|U_{k+1} + 2|S_{k+2}|U_{k+2} + \sum_{(i,j)\in C_k} u_{ij} - \sum_{(i,j)\in C_{k+2}} u_{ij} \\ = & 8\{kU_k - 2(k+1)U_{k+1} + (k+2)U_{k+2}\} + \sum_{(i,j)\in C_k} u_{ij} - \sum_{(i,j)\in C_{k+2}} u_{ij}, \end{aligned}$$

which is equivalent to (8.4). Similarly,

$$\begin{aligned} & -\sum_{(i,j)\in S_{k+1}} q(i,j,u_{ij}) + \sum_{(i,j)\in S_{k+1}} f_{ij} \\ = & 2\sum_{S_k\setminus C_k} u + 2\sum_{C_k} u + \sum_{C_k} u - 4\sum_{S_{k+1}} u + \sum_{C_{k+2}} u + \sum_{S_{k+2}\setminus C_{k+2}} u + \sum_{S_{k+2}\setminus C_{k+2}} u \\ = & 2\sum_{S_k} u - 4\sum_{S_{k+1}} u + \sum_{S_{k+2}} u + \sum_{C_k} u + \sum_{S_{k+2}\setminus C_{k+2}} u \\ = & 4\{2kU_k - 4(k+1)U_{k+1} + (k+2)U_{k+2}\} + \sum_{C_k} u + \sum_{S_{k+2}\setminus C_{k+2}} u, \end{aligned}$$

which is equivalent to (8.3). ∎

Let us assume that $u = \{u_{ij}\}_{i,j\in Z}$ is a positive solution of (8.1), and that (H1) holds, then by means of the Jensen's inequality,

$$\begin{aligned} \sum_{(i,j)\in S_k} q(i,j,u_{ij}) & \geq \sum_{(i,j)\in S_k} p(|i|+|j|)\,\psi(u_{ij}) \\ & = p_k \sum_{(i,j)\in S_k} \psi(u_{ij}) \geq p_k\,|S_k|\,\psi(U_k) = 4kp_k\psi(U_k). \end{aligned} \tag{8.5}$$

Let us also denote the mean of the function f_{ij} in (8.1) by

$$F_k = \frac{1}{|S_k|} \sum_{(i,j)\in S_k} f_{ij},\ k \in N. \tag{8.6}$$

Then, in view of (8.3) and (8.5), we have

$$4\{2(k-1)U_{k-1} - 4kU_k + (k+1)U_{k+1}\} + 4kp_k\psi(U_k)$$

$$\leq \sum_{(i,j)\in S_k} f_{ij} - \sum_{(i,j)\in C_{k-1}} u_{ij} - \sum_{(i,j)\in S_{k+1}\setminus C_{k+1}} u_{ij} \leq 4kF(k),\ k \geq 2.$$

The following result is now clear.

Theorem 124 *Suppose (H1) holds. If $u = \{u_{ij}\}_{i,j\in Z}$ is a positive solution of (8.1), then the mean U_k defined by (8.2) satisfies the recurrence relation*

$$2(k-1)U_{k-1} - 4kU_k + (k+1)U_{k+1} + kp_k\psi(U_k) \leq kF_k \tag{8.7}$$

for $k \geq 2$.

Note that if $\Delta_1 u_{i0} \leq 0$ for $i \geq 0$, $\Delta_1 u_{i0} \geq 0$ for $i \leq -1$, $\Delta_2 u_{0j} \leq 0$ for $j \geq 0$ and $\Delta_2 u_{0j} \geq 0$ for $j \leq -1$, then

$$\sum_{(i,j)\in C_{k+1}} u_{ij} \leq \sum_{(i,j)\in C_{k-1}} u_{ij}.$$

The following is now clear from (8.3) and (8.5).

Theorem 125 *Suppose (H1) holds. If $u = \{u_{ij}\}_{i,j\in Z}$ is a positive solution of (8.1) such that $\Delta_1 u_{i0} \leq 0$ for $i \geq 0$, $\Delta_1 u_{i0} \geq 0$ for $i \leq -1$, $\Delta_2 u_{0j} \leq 0$ for $j \geq 0$ and $\Delta_2 u_{0j} \geq 0$ for $j \leq -1$, then the mean U_k defined by (8.2) satisfies the recurrence relations*

$$2(k-1)U_{k-1} - 4kU_k + 2(k+1)U_{k+1} + kp_k\psi(U_k) \leq kF_k \tag{8.8}$$

for $k \geq 2$.

We remark that the recurrence relation (8.7) can also be written as

$$\Delta^2\left((k-1)U_{k-1}\right) - 2kU_k + (k-1)U_{k-1} + kp_k\psi(U_k) \leq kF_k,$$

while the recurrence relation (8.8) can also be written as

$$2\Delta^2\left((k-1)U_{k-1}\right) + kp_k\psi(U_k) \leq kF_k,$$

or as

$$2\Delta\left(k(k-1)\Delta U_{k-1}\right) + k^2 p_k\psi(U_k) \leq k^2 F_k.$$

We have now reduced our original problem to that involving an ordinary difference equation. In the literature, there are quite a few results which provide nonexistence criteria for positive solutions of ordinary difference equations. We will not repeat these criteria here since they do not belong to the main theme of this book.

8.3 Equations Over Quadrants

In this section, we consider several partial difference equations defined on the non-negative quadrant. We will also explain several techniques which are useful for obtaining nonexistence criteria of positive solutions of these equations.

8.3.1 Three-Point Equations with Two Constant Coefficients

In this section, we will be concerned with the following partial difference equation

$$v_{i+1,j} + bv_{i,j+1} + cv_{ij} = 0, \ i, j \in N, \tag{8.9}$$

where b, c are real numbers. If the condition $b \neq 0$ in (8.9) is not imposed, then (8.9) may degenerate into an ordinary difference equation. For this reason, we will assume that $b \neq 0$ in (8.9). Since (8.9) can be written in the form

$$v_{i+1,j} = -bv_{i,j+1} - cv_{ij}, \ i, j \in N,$$

it is clear that if v_{0j} is given for each $j \in N$,

$$v_{0j} = \phi_j, \ j \in N,$$

then we can calculate

$$v_{10}; \ v_{11}, v_{20}; \ v_{12}, v_{21}, v_{30}; \ \ldots$$

successively in a unique manner. An existence and uniqueness theorem for solutions of (8.9) is thus easily formulated and proved. Note that since $b \neq 0$, we can also impose conditions of the form

$$v_{i0} = \psi_i, \ i \in N,$$

and calculate a corresponding solution for (8.9).

We will show that every solution of (8.9) is oscillatory if, and only if, $b > 0$ and $c \geq 0$.

Lemma 3 *Suppose $b \neq 0$. If $b > 0$ and $c \geq 0$, then there cannot be any pair of positive numbers z and w such that*

$$z + bw + c = 0. \tag{8.10}$$

The converse also holds.

Proof. Suppose $b > 0$ and $c \geq 0$. Then for any pair of positive numbers z and w, we have $z + bw + c > 0$. This shows that (8.10) cannot hold. Conversely, we need only to consider three cases: (i) if $b < 0$ and $c \geq 0$, then (8.10) has a solution pair $z = 1$ and $w = -(c+1)/b$; (ii) if $b < 0$ and $c < 0$, then (8.10) has a solution pair $z = -(b+c)$ and $w = 1$; and (iii) if $b > 0$ and $c < 0$, then (8.10) has a solution pair $z = -c/2$ and $w = -c/(2b)$. In all three cases, the solution pairs are positive. ∎

We remark that the "characteristic equation" (8.10) will occur in a natural manner if we seek solutions of the form $v = \{v_{ij}\}$ defined by $v_{ij} = z^i w^j$. Indeed, if we substitute $v_{ij} = z^i w^j$ into (8.9), we obtain

$$z^i w^j \{z + bw + c\} = 0.$$

Theorem 126 *Suppose $b \neq 0$. Then (8.9) cannot have any eventually positive solution if, and only if $b > 0$ and $c \geq 0$.*

Proof. If $b < 0$ or $c < 0$, then by means of Lemma 3, we can find a pair of positive numbers z and w such that $z + bw + c = 0$. Then, as can easily be verified, the bivariate sequence $\{v_{ij}\}$ defined by

$$v_{ij} = z^i w^j, \; i, j \in N,$$

is a positive solution of (8.9). Conversely, suppose $b > 0$ and $c \geq 0$ and that (8.9) has an eventually positive solution $v = \{v_{ij}\}_{i,j \in Z}$. We may assume without loss of any generality that $v_{ij} > 0$ for $i, j \geq 0$. But then the left hand side of the equality (8.9) is strictly greater than 0. This contradiction establishes our proof. ■

Example 65 *Every solution of*

$$v_{i+1,j} + v_{i,j+1} + cv_{ij} = 0, \; i, j \in N. \tag{8.11}$$

is oscillatory if, and only if, $c \geq 0$.

This follows from Theorem 126. A direct proof may also be shown. Assuming a solution of the form $v = \{v_{ij}\}$ defined by $v_{ij} = \lambda^{i+j}$, we are led to the following equation in λ:

$$\lambda^{i+j}(2\lambda + c) = 0,$$

which has a positive root if $c < 0$. Thus if $c < 0$, the sequence $v = \{(-c/2)^{i+j}\}$ is a positive solution of (8.11).

Conversely, if $v = \{v_{ij}\}$ is an eventually positive solution of (8.11), then

$$c = \frac{-v_{i+1,j} - v_{i,j+1}}{v_{ij}} < 0,$$

as required.

8.3.2 Four-Point Equations with Three Constant Coefficients

In this section, we consider the following partial difference equation

$$v_{i+1,j+1} + av_{i+1,j} + bv_{i,j+1} + cv_{ij} = 0, \; i, j \in N. \tag{8.12}$$

where a, b and c are real numbers. We remark that when $a = c = 0$ or $b = c = 0$, equation (8.12) degenerates into an ordinary difference equation. For this reason, we will assume that $|a| + |c| > 0$ and $|b| + |c| > 0$ in (8.12).

Since (8.12) can be written as

$$v_{i+1,j+1} = -av_{i+1,j} - bv_{i,j+1} - cv_{ij}, \; i, j \in N,$$

it is clear that if the conditions

$$v_{0j} = \phi_j, \; j \in N,$$

$$v_{i0} = \psi_i, \; i \in N,$$

and the compatibility condition

$$\psi_0 = \phi_0,$$

are imposed, we can calculate

$$v_{11};\ v_{12}, v_{21};\ v_{13}, v_{22}, v_{31};\ \cdots$$

successively in a unique manner.

Lemma 4 *Suppose $|a| + |c| > 0$ and $|b| + |c| > 0$. If $a \geq 0$, $b \geq 0$ and $c \geq 0$, then there cannot be any pair of positive numbers z and w such that*

$$zw + az + bw + c = 0. \tag{8.13}$$

The converse also holds.

Indeed, if $a \geq 0$, $b \geq 0$ and $c \geq 0$ as well as $z > 0$ and $w > 0$, then clearly (8.13) cannot hold. Conversely, we need to consider twelve cases:

(a) if $a > 0, b \geq 0$ and $c < 0$, then (8.13) has a solution pair $z = -c/(2a), w = -ac/(2ab - c)$;

(b) if $a > 0, b < 0$ and $c < 0$, then (8.13) has a solution pair $z = -b/2 - c/(2a), w = a$;

(c) if $a > 0, b < 0$ and $c \geq 0$, then (8.13) has a solution pair $z = -b/2, w = a - 2c/b$;

(d) if $a = 0, b \geq 0$ and $c < 0$, then (8.13) has a solution pair $z = b + 1, w = -c/(2b + 1)$;

(e) if $a = 0, b < 0$ and $c > 0$, then (8.13) has a solution pair $z = -b/2, w = -2c/b$;

(f) if $a = 0, b < 0$ and $c < 0$, then (8.13) has a solution pair $z = -2b, w = c/b$;

(g) if $a < 0, b > 0$ and $c < 0$, then (8.13) has a solution pair $z = b, w = -a/2 - c/2b$;

(h) if $a < 0, b > 0$ and $c \geq 0$, then (8.13) has a solution pair $z = -c/a + (ab - c)/a, w = -a/2$;

(i) if $a < 0, b = 0$ and $c < 0$, then (8.13) has a solution pair $z = -c, w = -a + 1$;

(j) if $a < 0, b = 0$ and $c > 0$, then (8.13) has a solution pair $z = -2c/a, w = -a/2$;

(k) if $a < 0, b < 0$ and $c > ab$, then (8.13) has a solution pair $z = -b/2, w = -c/b + (ab - c)/b$;

(l) if $a < 0, b < 0$ and $c \leq ab$, then (8.13) has a solution pair $z = -2b, w = -a - (ab - c)/b$.

In all these cases, the solution pairs are positive.

Theorem 127 *Suppose $|a| + |c| > 0$ and $|b| + |c| > 0$. Equation (8.12) cannot have any eventually positive solution if, and only if, $a \geq 0, b \geq 0$ and $c \geq 0$.*

Proof. If one of the numbers a, b and c is negative, then by Lemma 4, we can find a pair of positive numbers z and w such that $zw + az + bw + c = 0$. Then, as can easily be verified, the bivariate sequence $v = \{v_{ij}\}$ defined by $v_{ij} = z^i w^j$ is a positive solution of (8.12). Conversely, suppose $a \geq 0, b \geq 0$ and $c \geq 0$ and that (8.12) has an eventually positive solution $v = \{v_{ij}\}_{i,j \in Z}$. We may assume without loss of generality that $v_{ij} > 0$ for $i, j \geq 0$. But then the left hand side of (8.12) is strictly positive, which is a contradiction. ■

We remark that the "characteristic equation" (8.13) will occur in a natural manner if we seek solutions of the form $v = \{v_{ij}\}$ defined by $v_{ij} = z^i w^j$. Indeed, if we substitute $v_{ij} = z^i w^j$ into (8.12), we obtain

$$z^i w^j \{zw + az + bw + c\} = 0.$$

8.3.3 Characteristic Initial Value Problems

In this section, we consider the following partial difference equation

$$\Delta_2 \Delta_1 u_{ij} + q(i, j, u_{ij}) = f_{ij},\ i, j \in Z^+, \tag{8.14}$$

subject to the conditions

$$\begin{aligned} \Delta_2 u_{0j} &= g_j,\ j \in N & (8.15)\\ \Delta_1 u_{i0} &= h_i,\ i \in N, & (8.16) \end{aligned}$$

where $f = \{f_{ij}\}_{i,j \in Z^+}$ is real, $\{g_j\}_{j \in N}$ and $\{h_j\}_{j \in N}$ are real, and q is a real function defined on $Z^+ \times Z^+ \times R$. Given an arbitrary number γ, it is easily seen that a unique solution $u = \{u_{ij}\}_{i,j \in N}$ of (8.14)–(8.16) exists and satisfies $u_{00} = \gamma$.

Let $w = \{w_{ij}\}_{i,j \in N}$ be a bivariate sequence. Suppose there is some nonnegative integer T such that $w_{ij} > 0$ for $i+j \geq T$ and $i, j \geq 1$, then w is said to be eventually positive. An eventually negative w is defined in a similar manner.

Associated with each solution $\{u_{ij}\}$ of (8.14), we define a sequence $U = \{U_n\}$ by

$$U_n = \frac{1}{n+1} \sum_{k=0}^{n} u_{n-k,k},\ n \in Z^+. \tag{8.17}$$

Then it is easily verified that

$$\Delta((n+1)U_n) = u_{0,n+1} + \sum_{k=0}^{n} \Delta_1 u_{n-k,k},$$

and

$$\Delta^2((n+1)U_n) = \Delta_2 u_{0,n+1} + \Delta_1 u_{n+1,0} + \sum_{k=0}^{n} \Delta_2 \Delta_1 u_{n-k,k}.$$

Suppose $q(i, j, u) \geq p_{i+j}\phi(u)$ for $i, j \geq 1$ and $u \in R$, where $p_n \geq 0$ for $n \in N$, and ϕ is nonnegative and convex on $(0, \infty)$. Let $u = \{u_{ij}\}$ be an eventually positive solution of (8.14)–(8.16), then the function $U = \{U_n\}$ defined by (8.17) satisfies the recurrence relation

$$\Delta^2((n+1)U_n) + (n+1)p_n\phi(U_n) \leq G_n$$

for all large n, where

$$G_n = g_{n+1} + h_{n+1} + \sum_{k=0}^{n} f_{n-k,k},\ n \in Z^+.$$

In addition, suppose $q(i,j,u) = -q(i,j,-u)$ for $i,j \geq 1$ and $u \in R$, then the following recurrence relation

$$\Delta^2((n+1)w_n) + (n+1)p_n\phi(w_n) \leq -G_n,\ n \in N, \tag{8.18}$$

has an eventually positive solution.

From (8.14), we obtain

$$\begin{aligned} \Delta_2\Delta_1 u_{n-k,k} &= -q(n-k,k,u_{n-k,k}) + f_{n-k,k} \\ &\leq -p_n\phi(u_{n-k,k}) + f_{n-k,k} \\ &= -(n+1)p_n\frac{\phi(u_{n-k,k})}{n+1} + f_{n-k,k}. \end{aligned}$$

Substituting the above inequality into (8.18), we obtain

$$\Delta^2((n+1)U_n) \leq g_{n+1} + h_{n+1} - (n+1)p_n\phi(U_n) + \sum_{k=0}^{n} f_{n-k,k},$$

as desired.

Under the additional condition on q, if $v = \{v_{ij}\}$ is an eventually negative solution, then $-v$ is an eventually positive solution of

$$\begin{aligned} \Delta_2\Delta_1 v_{ij} + q(i,j,v_{ij}) &= -f_{ij},\ i,j \in Z^+, \\ \Delta_2 v_{0j} &= -g_j,\ j \in N, \\ \Delta_1 v_{i0} &= -h_i,\ i \in N. \end{aligned}$$

The proof now follows by applying the first part of our theorem.

In view of the above result, in order to obtain oscillation theorems for (8.14)–(8.16), we only need to find nonexistence criteria for eventually positive solutions of recurrence relations of the form

$$\Delta^2((n+1)U_n) + (n+1)p_n\phi(U_n) \leq \psi_n,\ n \in Z^+.$$

Such criteria can be found in many articles dealing with oscillation of ordinary difference equations.

8.3.4 Delay Partial Difference Equations

In this section, we consider a class of partial difference equations of the form

$$u_{i+1,j} + u_{i,j+1} - u_{ij} + p_{ij}u_{i-\sigma,j-\tau} = 0,\ i,j \in N, \tag{8.19}$$

where $\{p_{ij}\}_{i,j\in N}$ is nonnegative, and σ, τ are positive integers. Given a bivariate sequence $\psi = \{\psi_{ij}\}$ defined on

$$\{(i,j) \in Z^2 |\ -\sigma \leq i < \infty, -\tau \leq j < \infty\} \backslash \{(i,j) \in Z^2 |\ i \geq 0, j \geq 1\},$$

we can calculate successively $u_{01}; u_{11}, u_{02}; u_{21}, u_{12}, u_{03}; \ldots$ in a unique manner. Such a bivariate sequence is a solution of (8.19). A solution $u = \{u_{ij}\}$ is said to be eventually positive if $u_{ij} > 0$ for all large i and j.

Note that if $u = \{u_{ij}\}$ is an eventually positive solution of (8.19), then u is nonincreasing in both of its independent variables. Indeed, from equation (8.19), we have

$$u_{i+1,j} + u_{i,j+1} - u_{ij} \le 0,$$

for all large i and j, which implies $u_{i+1,j} \le u_{ij}$ and $u_{i,j+1} \le u_{ij}$ for all large i and j.

Theorem 128 *Suppose*

$$\sum_{i=0}^{\sigma-1} \sum_{j=0}^{\tau-1} p_{m+i,n+j} \ge M > 0$$

for all large m and all large n. Then for any eventually positive solution $\{u_{mn}\}_{m,n=0}^{\infty}$ of (8.19), we have $u_{mn} > M^{\sigma+\tau} u_{m-\sigma,n-\tau}$ for all large m and all large n.

Indeed, if $\{u_{mn}\}$ is eventually positive, then it is nonincreasing in either one of its independent variables. Summing (8.19), we then obtain

$$\sum_{i=m}^{m+\sigma} \sum_{j=n+1}^{n+\tau} u_{i+1,j} + \sum_{i=m}^{m+\sigma} u_{i,n+\tau+1} + u_{m+\sigma+1,n} - u_{mn} + \sum_{i=m}^{m+\sigma} \sum_{j=n}^{n+\tau} p_{ij} u_{i-\sigma,j-\tau} = 0,$$

so that

$$\begin{aligned} u_{mn} &> \sum_{i=m}^{m+\sigma} \sum_{j=n}^{n+\tau} p_{ij} u_{i-\sigma,j-\tau} \\ &\ge \max\left\{ \sum_{i=m}^{m+\sigma} \sum_{j=n}^{n+\tau} p_{ij} u_{i-1,j}, \sum_{i=m}^{m+\sigma} \sum_{j=n}^{n+\tau} p_{ij} u_{i,j-1} \right\} \\ &\ge M \max\{u_{m-1,n}, u_{m,n-1}\} \end{aligned}$$

for $m, n \ge T_1$, where T_1 is some large integer. If we repeat this argument a number of times, we see that

$$u_{mn} > M^{\sigma+\tau} u_{m-\sigma,n-\tau}$$

for all large m and large n.

Theorem 129 *Suppose*

$$\liminf_{m,n\to\infty} \frac{1}{\sigma\tau} \sum_{i=m-\sigma}^{m-1} \sum_{j=n-\tau}^{n-1} p_{ij} > \frac{1}{2^\lambda} \frac{\lambda^\lambda}{(\lambda+1)^{\lambda+1}}, \tag{8.20}$$

where $\lambda = 2\sigma\tau/(\sigma+\tau)$. Then (8.19) cannot have any eventually positive solutions.

Proof. Suppose to the contrary that $u = \{u_{ij}\}$ is an eventually positive solution of (8.19). Then u is nonincreasing in its independent variables, so that

$$u_{i+1} + u_{i,j+1} - u_{ij} = -p_{ij} u_{i-\sigma,j-\tau} \le -p_{ij} u_{ij}.$$

Thus

$$p_{ij} \leq 1 - \frac{u_{i+1,j} + u_{i,j+1}}{u_{ij}} \leq 1 - \frac{2(u_{i+1,j}u_{i,j+1})^{1/2}}{u_{ij}}.$$

In view of (8.20), we can choose a constant Ψ such that for sufficiently large m and n,

$$\begin{aligned}
\frac{1}{2^\lambda}\frac{\lambda^\lambda}{(\lambda+1)^{\lambda+1}} &< \Psi \\
&\leq \frac{1}{\sigma\tau}\sum_{i=m-\sigma}^{m-1}\sum_{j=n-\tau}^{n-1} p_{ij} \\
&\leq \frac{1}{\sigma\tau}\sum_{i=m-\sigma}^{m-1}\sum_{i=n-\tau}^{n-1}\left\{1 - \frac{2(u_{i+1,j}u_{i,j+1})^{1/2}}{u_{ij}}\right\}.
\end{aligned}$$

By the well known inequality between the arithmetic and geometric means, we have

$$\begin{aligned}
&\sum_{i=m-\sigma}^{m-1}\sum_{i=n-\tau}^{n-1}\frac{(u_{i+1,j}u_{i,j+1})^{1/2}}{u_{ij}} \\
\geq\ & \sum_{i=m-\sigma}^{m-1}\tau\left\{\prod_{j=n-\tau}^{n-1}\frac{(u_{i+1,j}u_{i,j+1})^{1/2}}{u_{ij}}\right\}^{1/\tau} \\
=\ & \tau\sum_{i=m-\sigma}^{m-1}\left\{\frac{u_{in}}{u_{i,n-\tau}}\prod_{j=n-\tau}^{n-1}\frac{u_{i+1,j}}{u_{ij}}\right\}^{1/2\tau} \\
=\ & \sigma\tau\prod_{i=m-\sigma}^{m-1}\left\{\frac{u_{in}}{u_{i,n-\tau}}\right\}^{1/2\sigma\tau}\prod_{i=m-\sigma}^{m-1}\prod_{j=n-\tau}^{n-1}\left\{\frac{u_{i+1,j}}{u_{ij}}\right\}^{1/2\sigma\tau} \\
=\ & \sigma\tau\prod_{i=m-\sigma}^{m-1}\left\{\frac{u_{in}}{u_{i,n-\tau}}\right\}^{1/2\sigma\tau}\prod_{j=n-\tau}^{n-1}\left\{\frac{u_{mj}}{u_{m-\sigma,j}}\right\}^{1/2\sigma\tau} \\
\geq\ & \sigma\tau\left\{\frac{u_{mn}}{u_{m-\sigma,n-\tau}}\right\}^{(\sigma+\tau)/2\sigma\tau},
\end{aligned}$$

where the last inequality follows from the monotonicity of u. Thus,

$$1 - \Psi \geq 2\left\{\frac{u_{mn}}{u_{m-\sigma,n-\tau}}\right\}^{(\sigma+\tau)/2\sigma\tau} = 2\left\{\frac{u_{mn}}{u_{m-\sigma,n-\tau}}\right\}^{1/\lambda}.$$

In particular, this implies $0 < \Psi < 1$. In view of the inequality

$$1 - \Psi \leq \left\{\frac{\lambda^\lambda}{(\lambda+1)^{\lambda+1}}\right\}^{1/\lambda}\Psi^{-1/\lambda}, 0 < \Psi \leq 1,$$

we see further that

$$\frac{u_{ij}}{u_{i-\sigma,j-\tau}} \leq \frac{1}{2^\lambda}\frac{\Gamma}{\Psi}$$

for $i \geq M_1, j \geq N_1$, where M_1, N_1 are large positive integers and where we have used Γ to denote $\lambda^\lambda/((\lambda+1)^{\lambda+1}2^\lambda)$. If we apply the above inequality to (8.19), we obtain

$$u_{i+1,j} + u_{i,j+1} - u_{ij} \leq -p_{ij}\frac{\Psi}{\Gamma}u_{ij}, i \geq M_1, j \geq N_1.$$

A similar procedure then leads to

$$\frac{u_{ij}}{u_{i-\sigma,j-\tau}} \leq \left(\frac{\Gamma}{\Psi}\right)^2,$$

say, for $i \geq M_2 \geq M_1$ and $j \geq N_2 \geq N_1$. Inductively, we see that for any positive integer k, there are integers M_k and N_k such that

$$\frac{u_{ij}}{u_{i-\sigma,j-\tau}} \leq \left(\frac{1}{2\lambda}\frac{\Gamma}{\Psi}\right)^k, i \geq M_k, j \geq N_k.$$

On the other hand, in view of Theorem 128, we see that

$$\frac{u_{ij}}{u_{i-\sigma,j-\tau}} > M^{\sigma+\tau}$$

for all large i and j. Since $\Gamma < \Psi$, we may choose k so large that $(\Gamma/(2\lambda\Psi))^k < M^{\sigma+\tau}$. This is then the desired contradiction. The proof is complete. ■

8.3.5 Frequently Positive Solutions

By means of the frequency measures, we have seen in Section 2.8 how the concept of eventually positive univariate sequence can be generalized. For bivariate sequences, similar concepts have been defined and are repeated here for the sake of convenience.

Let $x = \{x_{ij}\}_{i,j\in N}$ be a real bivariate sequence. If $\mu^*(x \leq 0) = 0$, then the sequence x is said to be frequently positive. If $\mu^*(x \geq 0) = 0$, then x is said to be frequently negative. The sequence x is said to be frequently oscillatory if it is neither frequently positive nor frequently negative.

We remark that if a bivariate sequence is eventually positive, then it is frequently positive; and if it is eventually negative, then it is also frequently negative.

Let $x = \{x_{ij}\}_{i,j\in N}$ be a real bivariate sequence. If the upper frequency measure μ^* of $(x \leq 0)$ is less than or equal to ω, then x is said to be frequently positive of upper degree ω. Similarly, if $\mu^*(x \geq 0) \leq \omega$, then x is said to be frequently negative of upper degree ω. The sequence x is said to be frequently oscillatory of upper degree ω if it is neither frequently positive nor frequently negative of the same upper degree ω. The concepts of frequently positive of lower degree, etc. are similarly defined by means of μ_*.

Clearly, if a bivariate sequence is frequently oscillatory of lower degree ω, then it is also frequently oscillatory of upper degree ω; and if it is frequently oscillatory of any upper degree, then it is frequently oscillatory.

We will derive an oscillation criterion for partial difference equation of the form

$$x_{i+1,j} + x_{i,j+1} - x_{ij} + p_{ij}x_{ij} = 0,\ i,j \in N, \tag{8.21}$$

where $p = \{p_{ij}\}_{i,j\in N}$ is a real bivariate sequence.

Let ω be a nonnegative number such that $\mu_*(p \geq 1) > 4\omega$. Then every solution of (8.21) is frequently oscillatory of lower degree ω.

Indeed, let $x = \{x_{ij}\}_{i,j \in N}$ be a solution such that $\mu_*(x \leq 0) \leq \omega$. In view of Theorem 15,

$$\mu^*\left(\sum_{i=-1}^{0}\sum_{j=-1}^{0} E_x^i E_y^j (x \leq 0)\right) \leq 4\mu_*(x \leq 0) \leq 4\omega.$$

Thus,

$$\begin{aligned}
&\mu^*\left(N^2 \backslash \sum_{i=-1}^{0}\sum_{j=-1}^{0} E_x^i E_y^j (x \leq 0)\right) \\
&= 1 - \mu_*\left(\sum_{i=-1}^{0}\sum_{j=-1}^{0} E_x^i E_y^j (x \leq 0)\right) \\
&\geq 1 - 4\omega,
\end{aligned}$$

which implies

$$\mu^*\left(N^2 \backslash \sum_{i=-1}^{0}\sum_{j=-1}^{0} E_x^i E_y^j (x \leq 0)\right) + \mu_*(p \geq 1) > 1.$$

By means of Theorem 13, the intersection

$$\left(N^2 \backslash \sum_{i=-1}^{0}\sum_{j=-1}^{0} E_x^i E_y^j (x \leq 0)\right) \cdot (p \geq 1)$$

is an infinite subset of N^2, which, together with (2.19), implies that there is a lattice point (m, n) such that $p_{mn} \geq 1$, and $x_{ij} > 0$ for $m \leq i \leq m+1$ and $n \leq j \leq n+1$. But then,

$$0 \geq 1 - p_{mn} = \frac{x_{m+1,n} + x_{m,n+1}}{x_{mn}} > 0,$$

which is a contradiction.

8.4 Equations Over Cylinders

In this section, we will show that the techniques of averaging, Wirtinger's inequalities and separation of variables can be used to obtain nonexistence criteria for a number of partial difference equations.

8.4.1 Linear Discrete Heat Equation With Constant Coefficients

To begin with, let us consider a simple discrete reaction-diffusion equation of the form

$$u_i^{(t+1)} - u_i^{(t)} = p\left(u_{i-1}^{(t)} - 2u_i^{(t)} + u_{i+1}^{(t)}\right) + qu_i^{(t-\sigma)},\ i = 1, ..., n; t \in N, \tag{8.22}$$

where σ is a nonnegative integer and p, q are real numbers. Dirichlet boundary conditions of the form

$$u_0^{(t)} = 0 = u_{n+1}^{(t)}, \ t \in N, \tag{8.23}$$

will be imposed. Given an arbitrary set of initial values $u_i^{(t)}$, $-\sigma \le t \le 0, 1 \le i \le n$, we can successively calculate $u_1^{(1)}, ..., u_n^{(1)}; u_1^{(2)}, ..., u_n^{(2)}; ...$ in a unique manner. Such a bivariate sequence $u = \{u_i^{(t)} \mid i = 1, ..., n; t = -\sigma, -\sigma+1, ...\}$ is called a solution of (8.22)–(8.23). Suppose there is some nonnegative integer T such that $u_i^{(t)} > 0$ for $1 \le i \le n$ and $t \ge T$, then u is said to be eventually positive. An eventually negative solution is similarly defined. The bivariate sequence u is said to be oscillatory if it is neither eventually positive nor eventually negative.

By designating col $\left(u_1^{(t)}, u_2^{(t)}, ..., u_n^{(t)}\right)$ as the vector $u^{(t)}$, we see that a solution of (8.22)–(8.23) can also be regarded as a vector sequence $\{u^{(t)}\}_{t=-\sigma}^{\infty}$. Furthermore, such a sequence satisfies the delay vector recurrence relation

$$u^{(t+1)} - u^{(t)} = pAu^{(t)} + qu^{(t-\sigma)}, \ t \in N, \tag{8.24}$$

where

$$A = \begin{bmatrix} -2 & 1 & 0 & ... & 0 \\ 1 & -2 & 1 & ... & 0 \\ & \cdot & \cdot & \cdot & \\ 0 & ... & 1 & -2 & 1 \\ 0 & ... & 0 & 1 & -2 \end{bmatrix}_{n \times n}.$$

Clearly a solution u of (8.22)–(8.23) is eventually positive if, and only if the vector sequence $\{u^{(t)}\}$ is eventually positive. Therefore, (8.22)–(8.23) has an eventually positive solution if, and only if, the relation (8.24) has an eventually positive solution.

We first obtain a necessary and sufficient condition for the existence of an eventually positive solution of (8.24). To this end, let us first recall that the eigenvalue problem

$$Av = \lambda v$$

has eigenvalues

$$\lambda_j = -2 + 2\cos\frac{j\pi}{n+1}, \ j = 1, ..., n,$$

with corresponding eigenvectors

$$\psi_j = \text{col}\left(\sin\frac{j\pi}{n+1}, \sin\frac{2j\pi}{n+1}, ..., \sin\frac{nj\pi}{n+1}\right), \ j = 1, ..., n.$$

Note that $\lambda_1 < 0$ and that its corresponding eigenvector ψ_1 has positive components. Next, let us denote the transpose of a vector v by v', then the inner product of two vectors u and v is given by $v'u$.

We first show that if $\{u^{(t)}\}$ is a solution of (8.24), then $\{\psi_j' u^{(t)}\}$ is a solution of

$$x_{t+1} - x_t = p\lambda_j x_t + qx_{t-\sigma}, \ t \in N, \tag{8.25}$$

and if $\{x_t\}$ is a solution of (8.25), then $\{x_t \psi_j\}$ is a solution of (8.24).

Indeed, taking the inner product of (8.24) with the vector ψ_j, we see that

$$\begin{aligned}\psi_j' u^{(t+1)} - \psi_j' u^{(t)} &= p\psi_j' A u^{(t)} + q\psi_j' u^{(t-\sigma)} \\ &= p(A\psi_j)' u^{(t)} + q\psi_j' u^{(t-\sigma)} \\ &= p\lambda_j \psi_j' u^{(t)} + q\psi_j' u^{(t-\sigma)}\end{aligned}$$

for $t \geq -\sigma$. In other words, the scalar sequence $\{x_t\}_{t=-\sigma}^{\infty} \equiv \{\psi_j' u^{(t)}\}_{t=-\sigma}^{\infty}$ is a solution of the scalar difference equation (8.25).

Conversely, if $\{x_t\}_{t=-\sigma}^{\infty}$ is a solution of (8.25), then we assert that the "separable" sequence $\{u^{(t)}\}_{t=-\sigma}^{\infty}$ defined by

$$u^{(t)} = x_t \psi_j, \ t \geq -\sigma,$$

is an eventually positive solution of (8.24). Indeed,

$$\begin{aligned}& u^{(t+1)} - u^{(t)} \\ = \ & x_{t+1}\psi_j - x_t\psi_j = p\lambda_j x_t \psi_j + q x_{t-\sigma}\psi_j \\ = \ & p x_t(\lambda_j \psi_j) + q x_{t-\sigma}\psi_j = pA(x_t\psi_j) + q x_{t-\sigma}\psi_j \\ = \ & pAu^{(t)} + qu^{(t)}\end{aligned}$$

as required.

In particular, note that since $\psi_1 > 0$, (8.22)–(8.23) has an eventually positive solution if, and only if, (8.25) has an eventually positive solution. In other words, we have shown the following.

Theorem 130 *Every solution of (8.22)–(8.23) oscillates if, and only if, every solution of*

$$x_{t+1} - x_t = p\lambda_1 x_t + q x_{t-\sigma}, \ t \in N, \tag{8.26}$$

oscillates, where $\lambda_1 = -2 + 2\cos(\pi/(n+1))$.

The usefulness of the above theorem lies in the fact that we have reduced the problem of oscillation of a partial difference equation into one involving an ordinary difference equation, and that much is known about the ordinary difference equations. Indeed, when $\sigma = 0$, it is known and easily shown that every solution of (8.26) oscillates if, and only if, $p\lambda_1 + q < -1$. When σ is an arbitrary positive integer, in view of Theorem 22, every solution of (8.26) oscillates if, and only if its characteristic equation

$$z^{1+\sigma} - (1 + p\lambda_1)z^{\sigma} - q = 0 \tag{8.27}$$

does not have any positive roots. Such a condition is easy to find. Indeed, consider the polynomial

$$H(z) = z^{\sigma+1} + az^{\sigma} + b, \ z \in R. \tag{8.28}$$

Note that $H(0) = b$ and $\lim_{z\to\infty} H(z) = \infty$, we see that (8.28) has a positive root when $b < 0$. If $b = 0$, then (8.28) has the unique roots 0 and $-a$, hence it has a positive root if, and only if, $a < 0$. If $b > 0$ and $a \geq 0$, then $H(z) > 0$ for $z > 0$, so that (8.28) does not have any positive roots. Finally, when $b > 0$ and $a < 0$, the function $H(z)$ has a minimum at $z^* = -a\sigma/(\sigma+1)$, and

$$H(z^*) = b - \frac{(-a)^{\sigma+1}\sigma^\sigma}{(\sigma+1)^{\sigma+1}}.$$

Hence $H(z)$ has a positive root if, and only if,

$$b \leq \frac{(-a)^{\sigma+1}\sigma^\sigma}{(\sigma+1)^{\sigma+1}}.$$

In particular, (8.27) does not have any positive roots if, and only if, $1+p\lambda_1 \leq 0$ and $q \leq 0$, or, $1+p\lambda_1 > 0$ and

$$q < -\frac{(1+p\lambda_1)^{\sigma+1}\sigma^\sigma}{(\sigma+1)^\sigma}.$$

8.4.2 Parabolic Type Equations with Variable Coefficients

In this section, we consider two classes of parabolic type partial difference equations. First, we consider

$$\Delta_2 u_i^{(j)} = a_j \Delta_1^2 u_{i-1}^{(j)} - q_i^{(j)} f\left(u_i^{(j-\sigma)}\right),\ i = 1,2,...,n;\ j \in N, \tag{8.29}$$

where the delay σ is a nonnegative integer, $a_j > 0$ for $j \in N$ and f is a real function defined on R. We will assume that subsidiary conditions

$$u_0^{(j)} + \alpha_j u_1^{(j)} = 0,\ j \in N, \tag{8.30}$$

$$u_{n+1}^{(j)} + \beta_j u_n^{(j)} = 0,\ j \in N, \tag{8.31}$$

where $\alpha_j + 1 \geq 0$ and $\beta_j + 1 \geq 0$ for $j \in N$, are imposed.

Given an arbitrary bivariate sequence $\rho = \{\rho_i^{(j)}\}$ which is defined for $j = -\sigma, ..., 0$ and $i = 0, ..., n+1$, it is easily seen that a unique solution

$$u = \{u_i^{(j)} \mid 0 \leq i \leq n+1, j \geq -\sigma\}$$

of (8.29)–(8.31) exists which satisfies

$$u_i^{(j)} = \rho_i^{(j)},\ -\sigma \leq j \leq 0, 0 \leq i \leq n+1. \tag{8.32}$$

We now assume that $q_i^{(j)} \geq 0$ for $1 \leq i \leq n$ and $j \in N$. Let

$$Q_j = \min_{1\leq i\leq n} q_i^{(j)},\ j \in N. \tag{8.33}$$

Assume further that $f(x)$ is a nonnegative, nondecreasing and convex function on $(0,\infty)$. If $u = \{u_i^{(j)}\}$ is an eventually positive solution of (8.29)–(8.31) such that $u_i^{(j)} > 0$ for $1 \leq i \leq n$ and $j \geq T$. Then from (8.29), we have

$$\frac{1}{n}\sum_{i=1}^n \Delta_2 u_i^{(j)} = \frac{a_j}{n}\sum_{i=1}^n \Delta_1^2 u_{i-1}^{(j)} - \sum_{i=1}^n \frac{q_{ij}}{n} f\left(u_i^{(j-\sigma)}\right).$$

Since f is convex,

$$\sum_{i=1}^{n} \frac{q_i^{(j)}}{n} f\left(u_i^{(j-\sigma)}\right) \geq Q_j f\left(\frac{1}{n}\sum_{i=1}^{n} u_i^{(j-\sigma)}\right).$$

Thus

$$\Delta w_j + Q_j f(w_{j-\sigma}) \leq 0,\ j \in N, \tag{8.34}$$

has an eventually positive solution $w = \{w_j\}$ defined by

$$w_j = \frac{1}{n}\sum_{i=1}^{n} u_i^{(j)},\ j \geq -\sigma.$$

We have thus shown the following result.

Theorem 131 *Assume that* $a_j \geq 0, \alpha_j \geq -1$ *and* $\beta_j \geq -1$ *for* $j \in N$, *that* $q_i^{(j)} \geq 0$ *for* $1 \leq i \leq n$ *and* $j \in N$, *and that* f *is a nonnegative, nondecreasing and convex function on* $(0, \infty)$. *If equation (8.29)–(8.31) has an eventually positive solution, so does equation (8.34), where* $Q = \{Q_j\}$ *is defined by (8.33).*

Next, we consider

$$\Delta_2 u_i^{(j)} = b_j \Delta_1^2 u_{i-1,j} - p_j u_i^{(j-\sigma)} + d_i^{(j)},\ i = 1, ..., n, j \in N, \tag{8.35}$$

subject to the conditions

$$u_{0j} = g_j,\ j \in Z^+, \tag{8.36}$$

$$u_{n+1,j} = h_j,\ j \in Z^+. \tag{8.37}$$

We assume that σ is a nonnegative integer, $\{b_j\}$ and $\{p_j\}$ are real sequences, and $\{d_i^{(j)}\}$ is a doubly indexed sequence defined for $i = 1, ..., n$ and $j \in N$. The main difference between (8.29) and (8.35) is that the latter contains a nonhomogeneous term. As before, a solution of (8.35)–(8.37) which satisfies (8.32) exists and is unique.

Let $u = \{u_{ij}\}$ be an eventually positive solution of (8.35)–(8.37), then in view of (8.35),

$$\sum_{i=1}^{n} \Delta_2 u_i^{(j)} = b_j \sum_{i=1}^{n} \Delta_1^2 u_{i-1}^{(j)} - p_j \sum_{i=1}^{n} u_i^{(j-\sigma)} + \sum_{i=1}^{n} d_i^{(j)}.$$

Since

$$\begin{aligned}\sum_{i=1}^{n} \Delta_1^2 u_{i-1,j} &= \Delta_1 u_n^{(j)} - \Delta_1 u_0^{(j)} \\ &= \left\{h_j - u_n^{(j)}\right\} - \left\{u_1^{(j)} - g_j\right\} \\ &\leq h_j + g_j,\end{aligned}$$

thus when $b_j \geq 0$ for all large j, we have

$$\Delta v_j + p_j v_{j-\sigma} \leq b_j \{h_j + g_j\} + \sum_{i=1}^{n} d_i^{(j)} \tag{8.38}$$

for all large j, where

$$v_j = \sum_{i=1}^{n} u_i^{(j)}.$$

Theorem 132 *Suppose $b_j \geq 0$ for all large j. If (8.35)–(8.37) has an eventually positive solution, then so does (8.38).*

Similarly, let $G = (g_{ij})$ be the inverse of the Jacobi matrix defined by (2.21). If we multiply (8.35) by g_{ti} and then sum from $i = 1$ to $i = n$, we obtain

$$\Delta_2 \sum_{i=1}^{n} g_{ti} u_i^{(j)} = b_j \sum_{i=1}^{n} g_{ti} \Delta_1^2 u_{i-1}^{(j)} - p_j \sum_{i=1}^{n} g_{ti} u_i^{(j-\sigma)} + \sum_{i=1}^{n} g_{ti} d_i^{(j)}. \tag{8.39}$$

Note that, in view of (2.22),

$$\sum_{i=1}^{n} g_{ti} \Delta_1^2 u_{i-1}^{(j)} = -u_t^{(j)} + \Gamma_t^{(j)} \leq \Gamma_t^{(j)},$$

where

$$\Gamma_1^{(j)} = \frac{ng_j}{n+1}, \ \Gamma_n^{(j)} = \frac{nh_j}{n+1}, \ \Gamma_t^{(j)} = 0, \ 2 \leq t \leq n-1.$$

Thus when $b_j \geq 0$ for all large j, we obtain from (8.39) that

$$\Delta_2 \sum_{i=1}^{n} g_{ti} u_i^{(j)} + p_j \sum_{i=1}^{n} g_{ti} u_i^{(j-\sigma)} \leq b_j \Gamma_t^{(j)} + \sum_{i=1}^{n} g_{ti} f_i^{(j)} \tag{8.40}$$

for $1 \leq i \leq n$ and all large j. Summing (8.40) from $t = 1$ to $t = n$, we then see that

$$\Delta x_j + p_j x_{j-\sigma} \leq \frac{nb_j(g_j + h_j)}{n+1} + \sum_{t=1}^{n} \sum_{i=1}^{n} g_{ti} f_i^{(j)} \tag{8.41}$$

for all large j, where

$$x_j = \sum_{t=1}^{n} \sum_{i=1}^{n} g_{ti} u_i^{(j)}, \ j \in N.$$

Theorem 133 *Suppose $a_j \geq 0$ for all large j. If (8.35)–(8.37) has an eventually positive solution then so does equation (8.41).*

Next, let $\{w_k\}_{k=0}^{n+1}$ be the sequence defined by (2.26). Let $u = \{u_{ij}\}$ be a solution of (8.35)–(8.37). Then the sequence $v = \{v_j\}$ defined by

$$y_j = \sum_{i=1}^{n} u_i^{(j)} w_i$$

satisfies

$$\Delta y_j = b_j \sum_{i=1}^{n} w_i \Delta_1^2 u_{i-1}^{(j)} - p_j y_{j-\sigma} + \sum_{i=1}^{n} f_i^{(j)} w_i \tag{8.42}$$

for all large j. Note that in view of (35),

$$\begin{aligned}
\sum_{i=1}^{n} w_i \Delta_1^2 u_{i-1}^{(j)} &= \sum_{i=1}^{n} \left\{ \mu \sum_{s=1}^{n} g_{si} w_s \Delta_1^2 u_{i-1}^{(j)} \right\} \\
&= 4 \sin^2 \frac{\pi}{2n+2} \sum_{s=1}^{n} \left\{ \sum_{i=1}^{n} g_{si} \Delta_1^2 u_{i-1}^{(j)} \right\} w_s \\
&= 4 \sin^2 \frac{\pi}{2n+2} \sum_{s=1}^{n} \left\{ -u_s^{(j)} + \delta_s^{(j)} \right\} w_s \\
&= 4 \sin^2 \frac{\pi}{2n+2} \left\{ y_j + \sum_{s=1}^{n} \delta_s^{(j)} w_s \right\} \\
&= -4 \sin^2 \frac{\pi}{2n+2} y_j + \delta_1^{(j)} w_1 + \delta_n^{(j)} w_n \\
&= -4 \sin^2 \frac{\pi}{2n+2} y_j + \frac{n}{n+1} \left\{ u_0^{(j)} w_1 + u_{n+1}^{(j)} w_n \right\} \\
&= -4 \sin^2 \frac{\pi}{2n+2} y_j + \frac{n}{n+1} \{ g_j + h_j \} \sin \frac{\pi}{n+1}.
\end{aligned}$$

Thus from (8.42), we have

$$\begin{aligned}
&\Delta y_j + 4 \sin^2 \frac{\pi}{2n+2} b_j y_j + p_j y_{j-\sigma} \qquad (8.43) \\
= &\frac{n}{n+1} b_j \{ g_j + h_j \} \sin \frac{\pi}{n+1} + \sum_{i=1}^{n} f_i^{(j)} w_i.
\end{aligned}$$

Theorem 134 *If (8.35)-(8.37) has an eventually positive solution, then so does equation (8.43).*

8.4.3 Discrete Elliptic Equations

Oscillation criteria for elliptic differential equations have been obtained by many authors since thirty years ago (see Kreith [103]). Surprisingly, little is known for discrete elliptic type equations. This section is concerned with such a discrete equation of the form

$$\Xi u_{ij} + q(i,j) u_{ij} = 0, \; 1 \le i \le n, j \in Z^+ \qquad (8.44)$$

under the boundary condition

$$u_{0j} = 0 = u_{n+1,j}, \; j \in Z^+, \qquad (8.45)$$

where Ξ is the discrete Laplacian, $\{q_{ij}\}$ is a real bivariate sequence defined for $1 \le i \le n$ and $j \in Z^+$. A solution of (8.44) is a bivariate sequence $\{u_{ij}\}$ defined for $0 \le i \le n+1$ and $j \in N$ which satisfies (8.44) for $1 \le i \le n$ and $j \in Z^+$. It is not difficult to construct solutions of (8.44)–(8.45). A solution $\{u_{ij}\}$ of (8.44)–(8.45) is said to be eventually positive (eventually negative) if $u_{ij} > 0$ (respectively $u_{ij} < 0$) for $1 \le i \le n$ and all large j, and it is said to be oscillatory if it is neither eventually positive nor eventually negative. In this note, we shall derive an oscillation criterion which ensures that all solutions of (8.44)–(8.45) are oscillatory.

We first derive a Sturm type comparison theorem for (8.44).

Lemma 5 *Suppose (8.44) has a solution $\{u_{ij}\}$ defined for $0 \le i \le n+1$ and $s-1 \le j \le t+1$ which is positive for $1 \le i \le n$ and $s \le j \le t$ and satisfies the boundary conditions*

$$u_{0j} = 0 = u_{n+1,j},\ s \le j \le t, \tag{8.46}$$

$$\Delta_2 u_{it} = -\beta_i u_{it},\ 1 \le i \le n, \tag{8.47}$$

$$\Delta_2 u_{i,s-1} = \alpha_i u_{is},\ 1 \le i \le n \tag{8.48}$$

where $\alpha_i \ge 1$ and $\beta_i \ge 1$ for $1 \le i \le n$. Suppose further that

$$Q_{ij} \ge q_{ij},\ 1 \le i \le n, s \le j \le t. \tag{8.49}$$

Then no solution $\{v_{ij}\}$ of the boundary problem

$$\Xi v_{ij} + Q_{ij} v_{ij} = 0,\ 1 \le i \le n, j \ge 1 \tag{8.50}$$

$$v_{0j} = 0 = v_{n+1,j},\ s \le j \le t, \tag{8.51}$$

can remain positive for $1 \le i \le n$ and $s-1 \le j \le t+1$.

Proof. Let $\{u_{ij}\}$ be a solution of (8.44) such that $u_{ij} > 0$ for $1 \le i \le n$ and $s \le j \le t$ and satisfies (8.46)–(8.48). Further, assume to the contrary that $\{v_{ij}\}$ is a solution of (8.50)-(8.51) and $v_{ij} > 0$ for $1 \le i \le n$ and $s-1 \le j \le t+1$. Then we can write

$$\Delta_2 v_{i,s-1} = \sigma_i v_{is},\ 1 \le i \le n \tag{8.52}$$

where

$$\sigma_i = 1 - \frac{v_{i,s-1}}{v_{is}} < 1,\ 1 \le i \le n,$$

and similarly

$$\Delta_2 v_{it} = -\tau_i v_{it},\ 1 \le i \le n, \tag{8.53}$$

where $\tau_i < 1$ for $1 \le i \le n$. In view of (8.44) and (8.50), we have

$$\begin{aligned} 0 &= v_{ij}\left\{\Delta_1^2 u_{i-1,j} + \Delta_2^2 u_{i,j-1} + q_{ij} u_{ij}\right\} - u_{ij}\left\{\Delta_1^2 v_{i-1,j} + \Delta_2^2 v_{i,j-1} + Q_{ij} v_{ij}\right\} \\ &= v_{ij}\Delta_1^2 u_{i-1,j} - u_{ij}\Delta_1^2 v_{i-1,j} + v_{ij}\Delta_2^2 u_{i,j-1} - u_{ij}\Delta_2^2 v_{i,j-1} - (Q_{ij} - q_{ij}) u_{ij} v_{ij}. \end{aligned}$$

Note that, in view of the boundary conditions (8.46) and (8.51),

$$\sum_{i=1}^{n}\left[v_{ij}\Delta_1^2 u_{i-1,j} - u_{ij}\Delta_1^2 v_{i-1,j}\right] = \sum_{i=1}^{n}\Delta_1\left[v_{ij}\Delta_1 u_{i-1,j} - u_{ij}\Delta_1 v_{i-1,j}\right] = 0.$$

Note further that, in view of the boundary conditions (8.47), (8.48), (8.52) and (8.53),

$$\begin{aligned} &\sum_{j=s}^{t}\left[v_{ij}\Delta_2^2 u_{i,j-1} - u_{ij}\Delta_2^2 v_{i,j-1}\right] \\ =\ &\sum_{j=s}^{t}\Delta_2\left[v_{ij}\Delta_2 u_{i,j-1} - u_{ij}\Delta_2 v_{i,j-1}\right] \\ =\ & v_{i,t+1}\Delta_2 u_{it} - u_{i,t+1}\Delta_2 v_{it} - v_{is}\Delta_2 u_{i,s-1} + u_{is}\Delta_2 v_{i,s-1} \\ =\ & (1-\tau_i)v_{it}\left(-\beta_i u_{it}\right) - (1-\beta_i)\, u_{it}(-\tau_t)v_{it} - v_{is}\left(\alpha_i u_{is}\right) + u_{is}\left(\sigma_i v_{is}\right) \\ =\ & (\tau_i - \beta_i)\, u_{it} v_{it} + (\sigma_i - \alpha_i)\, u_{is} v_{is}. \end{aligned}$$

Thus

$$0 \le \sum_{i=1}^{n}\sum_{j=s}^{t}(Q_{ij} - q_{ij})\, u_{ij}v_{ij} = \sum_{i=1}^{n}\{(\tau_i - \beta_i)\, u_{it}v_{it} + (\sigma_i - \alpha_i)\, u_{is}v_{is}\} < 0,$$

which is a contradiction. ∎

We now recall that the eigenvalue problem

$$\Delta^2 x_{i-1} + \lambda x_i = 0, \ 1 \le i \le n$$

$$x_0 = 0 = x_{n+1}.$$

has an eigenvalue

$$\lambda_1 = 4\sin^2\left(\frac{\pi}{2n+2}\right) \tag{8.54}$$

and its corresponding eigensolution is given by

$$x_i = \begin{cases} 0 & i = 0, n+1 \\ \sin(i\pi/(n+1)) & 1 \le i \le n \end{cases},$$

which is positive for $1 \le i \le n$. If y_j is a solution of

$$\Delta^2 y_{j-1} + (q_j - \lambda_1)\, y_j = 0, \ j \ge 1, \tag{8.55}$$

then, as can be verified easily, $u_{ij} = x_i y_j$ is a solution of the partial difference equation

$$\Delta_1^2 u_{i-1,j} + \Delta_2^2 u_{i,j-1} + q_j u_{ij} = 0, \ 1 \le i \le n, j \ge 1, \tag{8.56}$$

and satisfies the boundary condition

$$u_{0j} = 0 = u_{n+1,j}, \ j \ge 1.$$

As a consequence, if (8.55) has an oscillatory solution y_j, then for any positive integer M, we can find two integers s and t such that $M < s \le t$, $y_{s-1}y_s \le 0, y_t y_{t+1} \le 0$ and $y_j > 0$ for $s \le j \le t$. Since we can write

$$\Delta y_{s-1} = \alpha y_s,$$

and

$$\Delta y_t = -\beta y_t$$

where $\alpha, \beta \ge 1$, by means of our previous lemma, we see that if $Q_{ij} \ge q_{ij}$ for $1 \le i \le n$ and $s \le j \le t$, then no solution of the boundary problem (8.50)–(8.51) can remain positive for $1 \le i \le n$ and $s - 1 \le j \le t + 1$. We have thus derived the following oscillation criterion.

Theorem 135 *Suppose the difference equation*

$$\Delta^2 y_{j-1} + \left(q_j - 4\sin^2\left(\frac{\pi}{2n+2}\right)\right) y_j = 0, \ j \ge 1$$

has a nontrivial oscillatory solution, and suppose further that

$$Q_{ij} \geq q_j, \ 1 \leq i \leq n, j \geq 1,$$

then every solution of the boundary problem

$$\Delta_1^2 v_{i-1,j} + \Delta_2^2 v_{i,j-1} + Q_{ij} v_{ij} = 0, \ 1 \leq i \leq n, j \geq 1$$

$$v_{0j} = 0 = v_{n+1,j}, \ j \geq 1$$

is oscillatory.

8.4.4 Initial Boundary Value Problems

In this section, we consider the following partial difference equation

$$\Delta_2^2 u_i^{(j-1)} - \Delta_1^2 u_{i-1}^{(j)} + q(i,j,u_i^{(j)}) = f_i^{(j)}, \ 1 \leq i \leq n, j \in Z^+, \tag{8.57}$$

subject to the conditions

$$\Delta_1 u_0^{(j)} = g_j, \ j \in N, \tag{8.58}$$

$$\Delta_1 u_n^{(j)} = h_j, \ j \in N, \tag{8.59}$$

where $\{f_i^{(j)}\}$, $\{g_j\}$, and $\{h_j\}$ are real sequences, and q is a real function defined on $\{1, ..., n\} \times N \times R$.

Given an arbitrary function

$$\psi = \{\psi_i^{(j)} | \ 1 \leq i \leq n; j = 0, 1\},$$

we can show easily that a solution $u = \{u_i^{(j)}\}$ of (8.57)–(8.59), which is a bivariate sequence defined for $0 \leq i \leq n+1$ and $j \in N$, exists and satisfies $u_i^{(j)} = \psi_i^{(j)}$ for $1 \leq i \leq n$ and $j = 0, 1$.

Associated with every solution $\{u_i^{(j)}\}$ of (8.57)–(8.59), we define

$$U_j = \frac{1}{n} \sum_{i=1}^{n} u_i^{(j)}, \ j \in N. \tag{8.60}$$

Theorem 136 *Suppose $q(i,j,u) \geq p_j \phi(u)$ for $1 \leq i \leq n, j \in N$ and $u \in R$, where $p_j \geq 0$ for $j \in N$ and ϕ is nonnegative and convex on $(0, \infty)$. Let $u = \{u_i^{(j)}\}$ be an eventually positive solution of (8.57)–(8.59), then the function $U = \{U_j\}$ defined by (8.60) satisfies the recurrence relation*

$$\Delta^2 U_{j-1} + p_j \phi(U_j) \leq F_j$$

for all large j, where

$$F_j = \frac{1}{n} \left\{ \sum_{i=1}^{n} f_i^{(j)} + h_j - g_j \right\}.$$

In addition, suppose $q(i,j,-u) = -q(i,j,u)$ for $1 \leq i \leq n, j \in N$ and $u \in R$. Then when $v = \{v_{ij}\}$ is an eventually negative solution of (8.57)–(8.59), the recurrence relation

$$\Delta^2 w_{j-1} + p_j \phi(w_j) \leq -F_j$$

has an eventually positive solution.

Proof. After summing equation (8.57) with respect to the first variable from $i = 1$ to $i = n$, and then dividing through by n, we obtain

$$\Delta^2 U_{j-1} = \frac{h_j - g_j}{n} + \sum_{i=1}^{n} \frac{f_i^{(j)}}{n} - \sum_{i=1}^{n} \frac{q(i,j,u_i^{(j)})}{n}.$$

By the convexity of ϕ, we have

$$\sum_{i=1}^{n} \frac{q(i,j,u_i^{(j)})}{n} \geq p_j \sum_{i=1}^{n} \frac{\phi(u_i^{(j)})}{n} \geq p_j \phi(U_j),$$

thus

$$\Delta^2 U_{j-1} + p_j \phi(U_j) \leq F_j$$

for all large j.

Under the additional condition on q, if $v = \{v_{ij}\}$ is an eventually negative solution, then $-v$ is an eventually positive solution of

$$\begin{aligned} \Delta_2^2 w_i^{(j-1)} - \Delta_1^2 w_{i-1}^{(j)} + q(i,j,w_i^{(j)}) &= -f_i^{(j)},\ 1 \leq i \leq n, j \in Z^+, \\ \Delta_1 w_0^{(j)} &= -g_j,\ j \in Z^+, \\ \Delta_1 w_n^{(j)} &= -h_j,\ j \in Z^+. \end{aligned}$$

The proof now follows by applying the first part of our Theorem. ■

In view of the above theorem, in order to obtain nonexistence theorems for (8.57)–(8.59), we only need to consider recurrence relations of the form

$$\Delta^2 U_{j-1} + p_j \phi(U_j) \leq \psi_j,\ j \in Z^+.$$

8.4.5 Linear Hybrid Five-Point Equations

In this section, we will be concerned with Sturmian theorems for a class of partial difference equations of the form

$$\Delta_2 \left(A_{j-1} \Delta_2 u_{i,j-1} \right) + B_j \Delta_1^2 u_{i-1,j} + C_j u_{ij} = 0,\ i = 1, ..., n; j \in Z^+, \tag{8.61}$$

where $A_j > 0$ for $j \in N$. When $B_j > 0$ for $j \in Z^+$, this equation can be regarded as a discrete analog of an elliptic equation, while if $B_j < 0$ for $j \in Z^+$, it can be regarded as a discrete analog of a hyperbolic equation.

Let $\alpha, \beta \geq 0$. We will be interested in the existence of a rectangle of the form

$$[1,n] \times \left[s - \frac{1}{1+\alpha}, t + \frac{1}{1+\beta} \right],$$

such that no solution $\{u_{ij}\}$ of (8.61) "can remain positive" over this rectangle. To be more precise, let us say that a sequence $g = \{g_k\}_{k=a}^{b}$ remains positive over a real interval I if its linear interpolating function f^* is positive over I. Similarly, let $f = \{f_{ij}\}$ be defined on a set of lattice points of the form

$$\{(i,j) \in Z^2 |\ i = a, a+1, ..., b; j = c, c+1, ..., d\}.$$

For each fixed j_0 in $\{c, ..., d\}$, we join the points (i, f_{i,j_0}), $a \le i \le b$, by straight line segments, and for each fixed i_0 in $\{a, ..., b\}$, we join the points $(j, f_{i_0,j})$, $c \le j \le d$, by straight line segments to form a broken net. Again, we denote the function which represents this net by f^*, and say that f is positive over a real rectangle Γ if $f^*(s,t) > 0$ for (s,t) in the intersection of Γ and the domain of f^*.

Consider a functional of the form

$$\sum_{k=s}^{t-1} A_k (\Delta y_k)^2 - \sum_{k=s}^{t} (C_k - \mu B_k) y_k^2 + (1+\alpha) A_{s-1} y_s^2 + (1+\beta) A_t y_t^2,$$

where μ is to be specified later, and a minorant partial difference equation of the form

$$\Delta_2 (a_{j-1} \Delta_2 v_{i,j-1}) + b_j \Delta_1^2 v_{i-1,j} + c_j v_{ij} = 0. \tag{8.62}$$

In place of the interval $[\nu, \lambda]$, we consider a rectangle of the form

$$[0, n+1] \times [s - 1/(\alpha+1), t + 1/(\beta+1)]$$

such that (8.62) has a solution $v = \{v_{ij}\}$ which satisfies the conditions

$$v_{0j} = 0 = v_{n+1,j}, \ s \le j \le t,$$

$$v_{i,s-1} + \alpha v_{is} = 0 = v_{i,t+1} + \beta v_{it}, \ 1 \le i \le n.$$

$$v_{ij} > 0, \ 1 \le i \le n, s \le j \le t.$$

For the sake of convenience, we shall call such a rectangle a rectangular nodal domain of (8.62). We assert that if (8.61) majorizes (8.62) in an appropriate manner, then no solution of (8.61) can remain positive or negative over any rectangular nodal domain of (8.62).

Now let $\{u_{ij}\}$ be a solution of (8.61). Let

$$U_j = \sum_{i=1}^{n} u_{ij} w_i, \ j \in N, \tag{8.63}$$

where $w = \{w_j\}$ is defined by (2.26). Multiplying (8.61) by w_i and then summing from $i = 1$ to $i = n$, we have

$$\Delta (A_{j-1} \Delta U_{j-1}) + B_j \sum_{i=1}^{n} w_i \Delta_1^2 u_{i-1,j} + C_j U_j = 0, \ j \in Z^+.$$

In view of Example 35,

$$\begin{aligned}
&\sum_{i=1}^{n} w_i \Delta_1^2 u_{i-1,j} \\
= \ & \sum_{i=1}^{n} \left\{ \mu_n \sum_{s=1}^{n} g_{si} w_s \Delta_1^2 u_{i-1,j} \right\} = \mu_n \sum_{s=1}^{n} \left\{ \sum_{i=1}^{n} g_{si} \Delta_1^2 u_{i-1,j} \right\} w_s \\
= \ & \mu_n \sum_{s=1}^{n} \left\{ -u_{sj} + \delta_s^{(j)} \right\} w_s = -\mu_n U_j + \mu_n \sum_{s=1}^{n} \delta_s^{(j)} w_s \\
= \ & -\mu_n U_j + \delta_1^{(j)} \mu_n w_1 + \delta_n^{(j)} \mu_n w_n \\
= \ & -\mu_n U_j + \frac{n}{n+1} \mu_n \{ u_{0j} w_1 + u_{n+1,j} w_n \},
\end{aligned}$$

where $\delta_1^{(j)} = nu_{0j}/(n+1), \delta_n^{(j)} = nu_{n+1,j}/(n+1)$ and $\delta_i^{(j)} = 0$ for $2 \le i \le n-1$, thus we obtain

$$\Delta\left(A_{j-1}\Delta U_{j-1}\right) + \left(C_j - \mu_n B_j\right) U_j = -\frac{n}{n+1}\mu_n \left\{u_{0j}w_1 + u_{n+1,j}w_n\right\}, j \in Z^+.$$

The following two results are now clear.

Lemma 6 *Let $\{u_{ij}\}$ be a solution of (8.61) which satisfies the boundary conditions*

$$u_{0j} = 0 = u_{n+1,j}, \quad s \le j \le t,$$

then the function $\{U_j\}$ defined by (8.63) satisfies

$$\Delta\left(A_{j-1}\Delta U_{j-1}\right) + \left(C_j - \mu_n B_j\right) U_j = 0 \tag{8.64}$$

for $s \le j \le t$.

Lemma 7 *If $[1,m] \times [s - 1/(\alpha+1), t + 1/(\beta+1)]$ is a rectangular nodal domain of (8.62), then the following equation*

$$\Delta\left(a_{j-1}\Delta V_{j-1}\right) + \left(c_j - \mu_m b_j\right) V_j = 0, \; s \le j \le t \tag{8.65}$$

has a solution $\{V_j\}$ which satisfies

$$V_j > 0, \; s \le j \le t,$$

and

$$V_{s-1} + \alpha V_s = 0 = V_{t+1} + \beta V_t.$$

We now note that (8.64) has the same form as equation (4.7). We let

$$\begin{aligned} & J[y|\; A, C - \mu_n B, s, t] \\ = \; & \sum_{k=s}^{t-1} A_k \left(\Delta y_k\right)^2 - \sum_{k=s}^{t} \left(C_k - \mu_n B_k\right) y_k^2 + (1+\alpha)A_{s-1}y_s^2 + (1+\beta)A_t y_t^2 \end{aligned} \tag{8.66}$$

be the functional associated with equation (8.64). According to the remarks preceding Theorem 59, if there is a nontrivial vector $y = (y_{s-1}, y_s, \ldots, y_{t+1})$ such that

$$y_{s-1} + \alpha y_s = 0 = y_{t+1} + \beta y_t, \tag{8.67}$$

and $J[y|\; A, C - \mu_n B, s, t] \le 0$, then (8.61) cannot have a solution $\{u_{ij}\}$ which satisfies $u_{0j} = 0 = u_{n+1,j}$ for $s \le j \le t$, and remains positive over the rectangle $[1,n] \times [s - 1/(1+\alpha), t + 1/(1+\beta)]$, for otherwise the function U_j defined by (8.63), in view of Lemma 6, would be positive over the interval $[s - 1/(1+\alpha), t + 1/(1+\beta)]$.

Theorem 137 *If there is a nontrivial vector $y = (y_{s-1}, y_s, \ldots, y_{t+1})$ such that (8.67) and $J[y|\; A, C - \mu_n B, s, t] \le 0$ are satisfied, then (8.61) cannot have a solution $\{u_{ij}\}$ which satisfies $u_{0j} = 0 = u_{n+1,j}$ for $s \le j \le t$ and also remains positive over the rectangle $[1,n] \times [s - 1/(1+\alpha), t + 1/(1+\beta)]$.*

There now arises the question as to when the hypotheses of Theorem 137 are satisfied. For this purpose, we will take (8.62) as a minorant equation with associated functional

$$\begin{aligned} & J[y|\ \{a_k\},\{c_k-\mu_m b_k\},s,t] \\ = & \sum_{k=s}^{t-1} a_k(\Delta y_k)^2-\sum_{k=s}^{t}(c_k-\mu_m b_k)y_k^2+(1+\alpha)a_{s-1}y_s^2+(1+\beta)a_t y_t^2 \end{aligned}$$

and associated rectangular nodal domain $[1,m]\times[s-1/(\alpha+1),t+1/(\beta+1)]$. According to Lemma 7, (8.65) has a solution $\{V_j\}$ which satisfies $V_j>0$ for $s\le j\le t$ and

$$V_{s-1}+\alpha V_s=0=V_{t+1}+\beta V_t.$$

In view of Theorem 57, $J[V|\ \{a_k\},\{c_k-\mu_m b_k\},s,t]=0$. Furthermore, if we impose the conditions

$$A_j\le a_j,\ s-1\le j\le t, \tag{8.68}$$

and

$$C_j-\mu_n B_j\ge c_j-\mu_m b_j,\ s\le j\le t, \tag{8.69}$$

then

$$\begin{aligned} & J[V|\ A,C-\mu_n B,s,t] \\ = & J[V|\ A,C-\mu_n B,s,t]-J[V|\ \{a_k\},\{c_k-\mu_m b_k\},s,t] \\ = & \sum_{k=s}^{t-1}(A_k-a_k)(\Delta V_k)^2-\sum_{k=s}^{t}(C_k-\mu_n B_k-c_k+\mu_m b_k)V_k^2 \\ & +(A_{s-1}-a_{s-1})(1+\alpha)V_s^2+(A_t-a_t)(1+\beta)V_t^2 \\ \le & \ 0. \end{aligned}$$

The following result now follows from Theorem 137.

Theorem 138 *Suppose $[1,m]\times[s-1/(\alpha+1),t+1/(\beta+1)]$ is a rectangular nodal domain of (8.62), then under the additional conditions (8.68) and (8.69), (8.61) cannot have a solution $\{u_{ij}\}$ which satisfies $u_{0j}=0=u_{n+1,j}$ for $s\le j\le t$ and also remains positive over the rectangle $[1,n]\times[s-1/(1+\alpha),t+1/(1+\beta)]$.*

There are at least two variants of Theorem 138 which we can obtain without too much effort. First, instead of the partial difference equation (8.62), we may consider a minorant equation of the form

$$\Delta(p_{k-1}\Delta u_{k-1})+q_k u_k=0,\ k\in Z^+, \tag{8.70}$$

where $p_k>0$ for $k\in N$. By means of the same reasoning used in the derivation of Theorem 138, we obtain the following result.

Theorem 139 *Suppose (8.70) has a solution $\{x_k\}$ which satisfies $x_k>0$ for $s\le k\le t$ and*

$$x_{s-1}+\alpha x_s=0=x_{t+1}+\beta x_t. \tag{8.71}$$

Suppose further that

$$A_k \le p_k, \; s-1 \le k \le t,$$

and

$$C_k - \mu_n B_k \ge q_k, \; s \le k \le t.$$

Then (8.61) cannot have a solution $\{u_{ij}\}$ which satisfies $u_{0j} = 0 = u_{n+1,j}$ for $s \le j \le t$ and also remains positive over the rectangle $[1,n]\times[s-1/(1+\alpha), t+1/(1+\beta)]$.

Next, we suppose (8.70) has a nontrivial solution $\{x_k\}$ which satisfies (8.71). Denoting $C_k - \mu_n B_k$ by Q_k, we assert that if

$$\sum_{k=s}^{t}\left\{\left(Q_k - \frac{A_{k-1}q_k}{p_{k-1}}\right)x_k^2 + p_k x_k \Delta x_k \Delta\left(\frac{A_{k-1}}{p_{k-1}}\right)\right\} \ge 0, \tag{8.72}$$

then (8.61) cannot have a solution $\{u_{ij}\}$ which remains positive (or remains negative) for $1 \le i \le n$ and $s \le j \le t$. Indeed, note that

$$\begin{aligned}
&\Delta(A_{k-1}\Delta x_{k-1}) + Q_k x_k \\
=\;& \Delta\left\{\frac{A_{k-1}}{p_k} p_{k-1}\Delta x_{k-1}\right\} + Q_k x_k \\
=\;& \left(\frac{A_{k-1}}{p_{k-1}}\right)\Delta\left(p_{k-1}\Delta x_{k-1}\right) + p_k \Delta x_k \Delta\left(\frac{A_{k-1}}{p_{k-1}}\right) + Q_k x_k \\
=\;& \left(Q_k - \frac{A_{k-1}q_k}{p_{k-1}}\right)x_k + p_k \Delta u_k \Delta\left(\frac{A_{k-1}}{p_{k-1}}\right).
\end{aligned} \tag{8.73}$$

If we multiply the left hand side of (8.73) by x_k and rearrange to obtain

$$\Delta\left(A_{k-1}x_{k-1}\Delta x_{k-1}\right) - A_{k-1}\left(\Delta x_{k-1}\right)^2 + Q_k x_k^2,$$

then after summing from $k = s$ to $k = t$, we will have

$$A_t x_t \Delta x_t - A_{s-1}x_{s-1}\Delta x_{s-1} - \sum_{k=s-1}^{t-1} A_k\left(\Delta x_k\right)^2 + \sum_{k=s}^{t} Q_k x_k^2.$$

If we now substitute $\Delta x_t = -(1+\beta)x_t$ and $\Delta x_{s-1} = (1+\alpha)x_s$ into the above formula, we obtain

$$-J[x|\; A, Q, s, t].$$

On the other hand, if we multiply the right hand side of (8.73) by x_k and then sum from $k = s$ to $k = t$, we obtain the left hand side of (8.72). Our assertion thus follows from Theorem 137.

Theorem 140 *Suppose (8.70) has a solution $\{x_k\}$ which satisfies (8.71). If (8.72) is satisfied, where $Q_k = C_k - \mu_n B_k$, then (8.61) cannot have a solution $\{u_{ij}\}$ which satisfies $u_{0j} = 0 = u_{n+1,j}$ for $s \le j \le t$ and also remains positive over the rectangle $[1,n]\times[s-1/(1+\alpha), t+1/(1+\beta)]$.*

Example 66 *Consider the case of a one dimensional crystal made of $n+2$ atoms of mass m. Each atom interacts with its two adjacent neighbors by means of elastic forces with constant χ. The discrete wave equation which describes this model is*

$$\Delta_2^2 u_{i,j-1} = \omega^2 \Delta_1^2 u_{i-1,j},\ 1 \le i \le n, j \ge 1 \tag{8.74}$$

where ω^2 is directly proportional to χ and inversely proportional to m. If the first and the last atom are fixed, then the corresponding boundary conditions are given by

$$u_{0j} = 0 = u_{n+1,j},\ j \in N.$$

Recall (see Example 36) that the eigenvalue problem

$$\Delta^2 X_{i-1} + \lambda X_i = 0,\ s \le i \le s+J-1$$

$$X_s = 0 = X_{s+J}$$

possesses an eigenvalue

$$\mu_J = 4\sin^2\left(\frac{\pi}{2J+2}\right)$$

and the corresponding eigensolution is given by

$$X_i = \begin{cases} 0 & k = s \text{ or } s+J \\ \sin\left((i-s+1)\pi/(J+1)\right) & s \le i \le s+J-1 \end{cases}.$$

Since $X_i > 0$ for $s \le i \le s+N-1$, thus by means of Theorem 139, if

$$\omega^2 \ge \frac{\mu_J}{\mu_n},$$

then no solution of (8.74) can remain positive (or negative) over a rectangle of the form $[1,n] \times [s-1, s+J]$. Since J and hence μ_J can be observed experimentally, our result provides a lower estimate for the unknown ω^2 in actual crystals.

8.5 Equations Over Finite Domains

Let Ω be a nonempty finite domain and let $\Upsilon(\Omega)$ denote the set of points (w,z) in $\partial\Omega \times \partial'\Omega$ such that w, z are neighbors. In this section, we consider the problem

$$\Xi y_{ij} + p_{ij} y_{ij} = 0,\ (i,j) \in \Omega, \tag{8.75}$$

subject to the Dirichlet boundary condition

$$y_{ij} + \Psi_{ij}^{(u,v)} y_{uv} = 0,\ ((i,j),(u,v)) \in \Upsilon(\Omega). \tag{8.76}$$

where Ξ is the discrete Laplacian, and $\Psi = \Psi_w^z$ is a real function defined on $\Upsilon(\Omega)$. We say that $y = \{y_{ij}\}_{(i,j)\in\Omega+\partial\Omega}$ is a ground state solution of (8.75)–(8.76) if y is a solution such that $y_{ij} > 0$ for $(i,j) \in \Omega$. We will find a necessary condition for the existence of ground state solution of (8.75)–(8.76).

Suppose y is a ground state solution, then in view of (8.75), we have

$$p_{ij} = 4 - \left\{\frac{u_{i+1,j}}{u_{ij}} + \frac{u_{i-1,j}}{u_{ij}} + \frac{u_{i,j+1}}{u_{ij}} + \frac{u_{i,j-1}}{u_{ij}}\right\}$$

for $(i,j) \in \Omega$. Summing over Ω, we obtain

$$\sum_{(i,j)\in\Omega} p_{ij} = 4|\Omega| - \sum_{(i,j)\in\Omega} \left\{ \frac{u_{i+1,j}}{u_{ij}} + \frac{u_{i-1,j}}{u_{ij}} + \frac{u_{i,j+1}}{u_{ij}} + \frac{u_{i,j-1}}{u_{ij}} \right\}.$$

Note that for any term $u(w)/u(z)$ in the second sum above, either w is a neighbor of z in Ω, in which case the term $u(z)/u(w)$ appears simultaneously, or else w is an exterior boundary point of Ω, in which case z is a neighboring interior boundary point of w so that $(w,z) \in \Upsilon(\Omega)$. Thus,

$$\begin{aligned} \sum_{(i,j)\in\Omega} p_{ij} &= 4|\Omega| - \sum_{w,z\in\Omega,|w|+|z|=1} \left\{ \frac{u(w)}{u(z)} + \frac{u(z)}{u(w)} \right\} - \sum_{(w,z)\in\Upsilon(\Omega)} \frac{u(w)}{u(z)} \\ &\leq 4|\Omega| - 2 \sum_{w,z\in\Omega,|w|+|z|=1} 1 + \sum_{(w,z)\in\Upsilon(\Omega)} \Psi_w^z. \end{aligned}$$

The inequality condition is sharp. Indeed, let $x = \{x_{ij}\}_{(i,j)\in\Omega+\partial\Omega}$ be defined by $x_{ij} = 1$ for $(i,j) \in \Omega$ and $x_{ij} = -\Psi_{ij}^{(u,v)}$ for $(i,j) \in \partial\Omega$ and $((i,j),(u,v)) \in \Upsilon(\Omega)$. Also, let

$$p_{ij} = \frac{\Xi x_{ij}}{x_{ij}}, \ (i,j) \in \Omega.$$

Then clearly, $\Xi x_{ij} + p_{ij}x_{ij} = 0$ for $(i,j) \in \Omega$ and $x_{ij} + \Psi_{ij}^{(u,v)} x_{uv} = 0$ for $((i,j),(u,v)) \in \Upsilon(\Omega)$ as required.

As an application, recall that the linear eigenvalue problem (7.38)–(7.39) has a least positive eigenvalue $\lambda[q]$ and a corresponding ground state eigensolution. Thus our result in this section implies the estimate

$$0 < \lambda[q] \leq \frac{4|\Omega| - 2|S|}{\sum_{(i,j)\in\Omega} p_{ij}},$$

where we have used S to denote the set of edges of Ω. It is interesting to note that the above upper bound increases with the number of points in the domain Ω and decreases with the number of edges when the denominator is held fixed or when $p_{ij} = 1$ for $(i,j) \in \Omega$.

In case $\Psi_{ij}^{(u,v)} \equiv 0$ for $((i,j),(u,v)) \in \Upsilon(\Omega)$, our boundary condition (8.76) changes to

$$y_{ij} = 0, \ (i,j) \in \partial\Omega. \tag{8.77}$$

Recall from Theorem 76 that the boundary problem (8.75),(8.77) is equivalent to

$$y_{ij} = \sum_{(u,v)\in\Omega} G_{ij}^{(u,v)} p_{uv} y_{uv}, \ (i,j) \in \Omega, \tag{8.78}$$

where $G_{ij}^{(u,v)} = \left\{G_{ij}^{(u,v)}\right\}_{(i,j)\in\Omega+\partial\Omega}$ is the Green's function associated with the boundary value problem (5.82),(8.77). Now if $\{x_{ij}\}$ is a nontrivial solution of (8.78), let $(\alpha,\beta) \in \Omega$ such that

$$0 < |x_{\alpha\beta}| = \max_{(i,j)\in\Omega} |x_{ij}|.$$

Then

$$\begin{aligned}|x_{\alpha\beta}| &\le \sum_{(u,v)\in\Omega} G_{\alpha\beta}^{(u,v)} |p_{uv}|\,|x_{uv}| \le |x_{\alpha\beta}| \sum_{(u,v)\in\Omega} G_{\alpha\beta}^{(u,v)} |p_{uv}| \\ &\le |x_{\alpha\beta}| \max_{(u,v),(i,j)\in\Omega} G_{ij}^{(u,v)} \sum_{(u,v)\in\Omega} |p_{uv}| \\ &= |x_{\alpha\beta}| \max_{(u,v)\in\Omega} G_{uv}^{(u,v)} \sum_{(u,v)\in\Omega} |p_{uv}|\end{aligned}$$

where the last equality follows from Theorem 77, so that

$$\sum_{(u,v)\in\Omega} |p_{uv}| \ge \frac{1}{\max_{(u,v)\in\Omega} G_{uv}^{(u,v)}}. \tag{8.79}$$

This condition is sharp in the sense that we can find a nonnegative bivariate sequence $\{p_{ij}\}_{(i,j)\in\Omega}$ and a nontrivial solution of (8.75),(8.77) such that the equality in (8.79) is satisfied. Indeed, let

$$\max_{(u,v)\in\Omega} G_{uv}^{(u,v)} = G_{st}^{(s,t)}.$$

Let $x_{uv} = G_{st}^{(u,v)}$ for $(u,v) \in \Omega$, $x_{ij} = 0$ for $(i,j) \in \partial\Omega$, and let $p_{uv} = -\Xi x_{uv}/x_{uv}$ for $(u,v) \in \Omega$. Then clearly,

$$\begin{aligned}\Xi x_{uv} + p_{uv}x_{uv} &= 0,\ (u,v) \in \Omega, \\ x_{uv} &= 0,\ (u,v) \in \partial\Omega,\end{aligned}$$

and

$$\sum_{(u,v)\in\Omega} p_{uv} = \sum_{(u,v)\in\Omega} -\frac{\Xi x_{uv}}{x_{uv}} = \sum_{(u,v)\in\Omega} \frac{\delta_{st}^{(u,v)}}{G_{st}^{(u,v)}} = \frac{1}{G_{st}^{(s,t)}},$$

as required.

As mentioned in Section 5.6, there are several cases where the Green's functions and their maxima can be found. For instance, when Ω is a chain with n lattice points,

$$\max_{(u,v)\in\Omega} G_{uv}^{(u,v)} = \begin{cases} X_{m-1}X_mX_n^{-1} & n = 2m \\ X_m^2 X_n^{-1} & n = 2m+1 \end{cases},$$

which yields

$$\sum_{(i,j)\in\Omega} |p_{ij}| \ge \begin{cases} X_{m-1}^{-1}X_m^{-1}X_n & n = 2m \\ X_m^{-2}X_n & n = 2m+1 \end{cases}.$$

8.6 Notes and Remarks

Averaging techniques have been employed quite often for showing nonexistence of positive solutions of partial difference equations. In particular, the results in Sections 8.1, 8.3.3, 8.4.2, 8.4.4 are obtained by such means. These methods reduce the original problem to showing the nonexistence of positive solutions of associated ordinary difference equations. Since there are a large number of such oscillation results for ordinary difference equations, explicit conditions can be stated for partial difference equations without too much trouble. For example, the reader may consult Cheng and Zhang [41], Cheng *et al.* [40, 29, 30] and [42].

The method of separation of variables is also a powerful method for showing nonexistence of positive solutions of partial difference equations. These methods also reduce our original problem to showing nonexistence of positive solutions of associated ordinary difference equations and the existence of positive eigensolutions. The examples in Sections 8.4.1, 8.4.3 and 8.4.5 make use of such methods and Theorem 130 and Theorem 135 are due to Cheng and Medina [28] and Cheng [16] respectively. Section 8.4.5 also makes use of Wirtinger's inequalities for showing nonexistence of positive solutions [43].

Theorems 126 and 127 are due to Lin and Cheng [108], while 129 improves a result in Zhang *et al.* [173]. A multivariate version of Theorem 129 can be found in Cheng *et al.* [38].

The frequently oscillation result in Section 8.3.5 is taken from Tian *et al.* [153]. Additional results along similar lines can be found in Tian [154].

The result for the ground state solution in the last section is in Cheng [14]. The final condition (8.79) is due to Cheng *et al.* [20]. This condition is usually called a Lyapunov type condition for existence of nontrivial solutions. A review article by Cheng [15] can be consulted for additional information.

Additional results on oscillation of partial difference equations can be found in Cheng [17], Domshlak and Cheng [57], Koehler and Braden [100], Liu and Wang [114], Liu and Cheng [112], Wong and Agarwal [163, 164], and Zhang and Liu [170, 171, 172].

Bibliography

[1] M. J. Ablowitz and J. F. Ladik, On the solution of a class of nonlinear partial difference equations, Studies Appl. Math., 57(1977), 1–12.

[2] V. Afraimovich and Y. Pesin, Travelling waves in lattice models of multi-dimensional and multi-component media: I. General hyperbolic properties, Nonlinearity, 6(1993), 429–455.

[3] A. C. Allen and B. H. Murdoch, A note on preharmonic function, Proc. Amer. Math. soc., 4(1953), 842–852.

[4] Y. Y. Azmy and V. Protopopescu, On the dynamics of a discrete reaction-diffusion system, Numer. Meth. Part. Diff. Eq., 7(1991), 385–405.

[5] D. Bainov and P. Simeonov, Integral Inequalities and Applications, Kluwer Academic Publishers, 1992.

[6] L. Berg, Operational Calculus, North-Holland, 1967.

[7] L. A. Bunimovich and Ya G. Sinai, Spacetime chaos in coupled map lattices, Nonlinearity, 1(1988), 491–516.

[8] J. W. Cahn, S. N. Chow and E. S. Van Vleck, Spatially discrete nonlinear diffusion equations, Rocky Mountain J. Math., 25(1995), 87–118.

[9] L. Carlitz and R. Scoville, Sequences of absolute differences, SIAM Review, 12(1970), 297–300.

[10] Ch. A. Charalambides, A new kind of numbers appearing in the n-fold convolution of truncated binomial and negative binomial distributions, SIAM J. Appl. Math., 33(1977), 279–288.

[11] S. S. Cheng, Maximum principles for solutions of second order partial difference inequalities, Symposium on Functional Analysis and Applications (held at the Tsing Hua University, Hsinchu, Taiwan, R. O. C., 1980), 395–401.

[12] S. S. Cheng, Sturmian comparison theorems for three term recurrence equations, J. Math. Anal. Appl., 111(1985), 464–474.

[13] S. S. Cheng, Discrete quadratic Wirtinger's inequalities, Linear Algebra and its Appl., 85(1987), 57–73.

[14] S. S. Cheng, A sharp condition for the ground state of difference equation, Applicable Analysis, 34(1989), 105–109.

[15] S. S. Cheng, Lyapunov inequality conditions for differential and difference equations, Fasiculi Mathematici, 23(1991), 25–41.

[16] S. S. Cheng, An oscillation criterion for a discrete elliptic equation, Annals Diff. Eq., 11(1995), 10–13.

[17] S. S. Cheng, An underdeveloped research area: Qualitative theory of functional partial difference equations, Proceedings of the International Mathematics Conference '94 at Sun Yat-Sen University, World Scientific, 1996, 65–75.

[18] S. S. Cheng, Sturmian theorems for hyperbolic type partial difference equations, J. Difference Eq. Appl., 2(1996), 375–387.

[19] S. S. Cheng and L. Y. Hsieh, Inverses of matrices arising from difference operators, Utilitas Mathematica, 38(1990), 65–77.

[20] S. S. Cheng, L. Y. Hsieh and Z. T. Chao, Discrete Lyapunov inequality conditions for partial difference equations, Hokkaido Math. J., 19(1990), 229–239.

[21] S. S. Cheng and S. S. Lin, Existence and uniqueness theorems for nonlinear difference equations, Utilitas Math., 39(1991), 167–186.

[22] S. S. Cheng and R. F. Lu, Discrete Wirtinger's inequalities and conditions for partial difference equations, Fasciculi Mathematici, 23(1991), 9–24.

[23] S. S. Cheng and R. F. Lu, A generalization of the discrete Hardy's inequality, Tamkang J. Math., 24(1993), 469–475.

[24] S. S. Cheng and T. T. Lu, The maximum of a bilinear form under rearrangement, Tamkang J. Math., 17(1986), 161–168.

[25] S. S. Cheng and R. Medina, Growth conditions for a discrete heat equation with delayed control, Dynamic Sys. Appl., 8(1999), 361–367.

[26] S. S. Cheng and R. Medina, Bounded and positive solutions of discrete steady state equations, Tamkang J. Math., 31(2)(2000), 131–135.

[27] S. S. Cheng and R. Medina, Positive and bounded solutions of discrete reaction-diffusion equations, preprint.

[28] S. S. Cheng and R. Medina, Necessary and sufficient oscillation criteria foe discrete reaction-diffusion equations, Proceedings of the Fifth International Conference on Difference Equations and Applications, to appear.

[29] S. S. Cheng, S. L. Xie and B. G. Zhang, Qualitative theory of partial difference equations (II): Oscillation criteria for direct control systems in several variables, Tamkang J. Math., 26(1995), 65–79.

[30] S. S. Cheng, S. L. Xie and B. G. Zhang, Qualitative theory of partial difference equations (III): Forced oscillations of parabolic type partial difference equations, Tamkang J. Math., 26(1995), 177–192.

[31] S. S. Cheng and G. H. Lin, Green's function and stability of a partial difference scheme, Comput. Math. Appl., 35(5)(1998), 27–41.

[32] S. S. Cheng and J. Y. Lin, Stability criteria for a discrete reaction-diffusion equation, Far East J. Math. Sci., 6(3)(1998), 425–435.

[33] S. S. Cheng and Y. Z. Lin, Bounds for solutions of a four-point partial difference equation, Far East J. Math. Sci., 5(2)(1997), 273–290.

[34] S. S. Cheng and Y. Z. Lin, General solutions of a four point partial difference equation, Far East J. Math. Sci., 1(4)(1999), 507–516.

[35] S. S. Cheng and Y. Z. Lin, Exponential stability of a partial difference equation with nonlinear pertubation, Acta Math. Appl. Sinica, 15(1), 98–108.

[36] S. S. Cheng, Y. Z. Lin and G. Zhang, Traveling waves of a discrete conservation law, PanAmerican J. Math., PanAmerican Math. J., 11(1)(2001), 45–52.

[37] S. S. Cheng, S. T. Liu and B. G. Zhang, Positive flows of an infinite network, Comm. Appl. Anal., 1(1997), 83–90.

[38] S. S. Cheng, S. T. Liu and G. Zhang, A multivariate oscillation theorem, Fasiculi Math, 30(1999), 15–22.

[39] S. S. Cheng and Y. F. Lu, General solutions of a three-level partial difference equations, Computers Math. Appl., 38(7–8)(1999), 65–79.

[40] S. S. Cheng and B. G. Zhang, Qualitative theory of partial difference equations (I): Oscillation of nonlinear partial difference equations, Tamkang J. Math., 25(1994), 279–288.

[41] S. S. Cheng and B. G. Zhang, Nonexistence criteria for positive solutions of a discrete elliptic equation, Fasiculi Math., 28(1998), 19–30.

[42] S. S. Cheng, B. G. Zhang and S. L. Xie, Qualitative theory of partial difference equations (IV): Forced oscillations of hyperbolic type nonlinear partial difference equations, Tamkang J. Math., 26(1995), 337–360.

[43] S. S. Cheng, B. G. Zhang and S. L. Xie, Qualitative theory of partial difference equations (V): Sturmian theorems for a class of partial difference equations, Tamkang J. Math., 27(1996), 89–97.

[44] S. S. Cheng and G. Zhang, Existence criteria for positive solutions of a nonlinear difference inequality, Ann. Polonici Math., 73(3)(2000), 197–220.

[45] S. N. Chow and J. Mallet-Paret, Pattern formation and spatial chaos in lattice dynamical systems: I, IEEE Trans. Circuits Systems-I: Fundamental Theory and Appl., 42(10)(1995), 1–6.

[46] S. N. Chow and W. Shen, Dynamics in a discrete Nagumo equation - spatial topological chaos, SIAM J. Appl. Math., 55(6)(1995), 1764–1781.

[47] L. O. Chua and L. Yang, Cellular neural network: theory, IEEE Trans. Circuits and Systems, CAS-35(1988), 1257–1272.

[48] L. O. Chua and L. Yang, Cellular neural networks: applications, IEEE Trans. Circuits and Systems, CAS-35(1988), 1273–1290.

[49] A. Corduneanu, Linear equations with differences in two variables, An. st. Univ. Iasi, 35(1989), 137–144.

[50] A. Corduneanu, The asymptotic behavior of the solution of some difference equations, Buletinul Institutului Politehnic Din Iasi, 39(1993), 45–52.

[51] A. F. Cornock, The numerical solution of Poisson's and the bi-harmonic equations by matrices, Proc. Cambridge Philos. Soc., 50(1954), 524–535.

[52] R. Courant, K. Friedrichs and H. Lewy, On the partial difference equations of mathematical physics, IBM J. 11(1967), 215–234.

[53] J. B. Crutchfield and K. Kaneko, Phenomenology and spatio-temporal chaos, Directions in Chaos (Ed. B. L. Hao), World Scientific, Singapore, 1987.

[54] H. Davis, Poisson's partial difference equation, Quart. J. Math. Oxford(2), 6(1955), 232–240.

[55] C. R. Deeter and J. M. Gray, The discrete Green's function and the discrete kernel function, Discrete Math., 10(1974), 29–42.

[56] J. B. Diaz and R. C. Roberts, Upper and lower bounds for the numerical solution of the Dirichlet difference boundary value problem, J. Math. Phys., 31(1952), 184–191.

[57] Y. Domshlak and S. S. Cheng, Sturmian theorems for a partial difference equation, Functional Differential Eq., 3(1996), 83–97.

[58] R. Dubisch, Introduction to Abstract Algebra, John Wiley & Sons, 1965.

[59] P. DuChateau and D. Zachmann, Applied Partial Differential Equations, Harper and Row, New York, 1989.

[60] R. J. Duffin, Discrete potential theory, Duke Math. J., 20(1953), 233–251.

[61] R. J. Duffin, Basic properties of discrete analytic functions, Duke Math. J., 23(1956), 335–363.

[62] R. J. Duffin and C. S. Duris, A convolution product for discrete function theory, duke Math. J., 31(1964), 199–220.

[63] R. J. Duffin and J. Rohrer, A convolution product for the solutions of partial difference equations, Duke Math. J., 35(1968), 683–698.

[64] R. J. Duffin and E. P. Shelly, Difference equations of polyharmonic type, Duke Math. J. 25(1958), 209–238.

[65] W. Feller, An Introduction to Probability and Its Applications, vol. 1, 3rd ed., John Wiley, New York, 1968.

[66] T. Fenyes and P. Kosik, The algebraic derivative and integral in the discrete operational calculus, Studia Sci. Math. Hungarica, 7(1972), 117–130.

[67] G. E. Forsythe and W. R. Wasow, Finite Difference Methods for Partial Difference Methods, 2nd edition, Clarendon Press, Oxford, 1978.

[68] T. Fort, Finite Differences and Difference Equations in the Real Domain, Oxford University Press, 1948.

[69] T. Fort, The loaded vibrating net and resulting boundary value problems for a partial difference equation of the second order, J. Math. Physics, 33(1954), 94–104.

[70] B. Freedman, The four number game, Scripta Math., 14(1948), 35–47.

[71] H. Glantz and E. Reissner, On finite sum equations for boundary value problems of partial difference equations, J. Math. Phys., 34(1956), 286–297.

[72] S. K. Godunov and V. S. Ryabenkii, Difference Schemes, Studies in Mathematics and its Applications, No. 19, Elsevier Science Publishers, 1987.

[73] D. Greenspan, Discrete Models, Addison-Wesley Publishing Co., 1973.

[74] J. Gregor, The multidimensional z-transform and its use in solution of partial difference equations, Supplement to the Journal Kybernetika, 24(1988), Number 1,2, 1–40.

[75] J. Gregor, Convolutional solutions of partial difference equations, Math. Control Signals Systems, 4(1991), 205–215.

[76] J. Gregor, Singular systems of partial difference equations, Multidim. Systems Signal Processing, 4(1993), 67–82.

[77] R. T. Gregory and D. L. Karney, A Collection of Matrices for Testing Computational Algorithms, Wiley Interscience, 1969.

[78] I. Győri and G. Ladas, Oscillation Theory of Delay Differential Equations with Applications, Oxford University Press, Oxford, 1991.

[79] S. Haruki, On the general solution of a nonsymmetric partial difference functional equation analogous to the wave equation, Aeq. Math., 36(1988), 20–31.

[80] H. A. Heilbronn, On discrete harmonic functions, Proc. Cambridge Philo. Soc., 45(1949), 194–206.

[81] F. B. Hildebrand, Finite Difference Equations and Simulations, Prentice Hall, 1968.

[82] R. Honsberger, Ingenuity in Mathematics, random House, New York, 1970.

[83] T. Roska and J. Vandewalle, Cellular Neural Networks, John Wiley & Sons, 1993.

[84] J. R. Hundhausen, A generating operator for solutions of certain partial difference and differential equations, SIAM J. Math. Anal., 4(1)(1973), 15–21.

[85] G. Jennings, Discrete Shocks, Comm. Pure Appl. Math., 27(1974), 25–37.

[86] W. Jentsch, On a partial difference equation of L. Carlitz, Fibonacci Quarterly, 4(3)(1964), 202–208.

[87] W. Jentsch, Operatorenrechnumg für Funktionen zweier diskreter Variabler, Wiss. Z. Univ. Halle XIV, M4(1965), 311–318.

[88] W. Jentsch, Charakterisierung der Quotienten in der zweidimensinalen diskreten Operatorenrechnung, Studia Math., 26(1965), 91–99.

[89] W. Jentsch, On an initial value problem for linear partial difference equations, Fibonacci Quart., 9(1971), 313–323.

[90] W. Jentsch, Über die Struktur der rationalen Operatoren in der zweidimensionalen diskreten Operatorenrechnung, Studia Math., 42(1972), 15–27.

[91] W. Jentsch, Über eine Erweiterung der zweidimensionalen diskreten Operatorenrechnung, Beiträge zur Analysis, 9(1976), 159–166.

[92] W. Jentsch, Ein Struktursatz über rationale Operatoren von Verschiebungsoperatoren, Beiträge zur Analysis, 10(1977), 41–45.

[93] W. Jentsch, Bemerkung zu den diskret-analytischen Funktionen, Wiss. Z. Univ. Halle-Witt., 27(1978), 43–45.

[94] J. Jeske, Linear recurrence relations - Part III, The Fibonacci Quarterly, 2(1964), 197–203.

[95] C. Jordan, Calculus of Finite Differences, 2nd edition, Chelsea, New York, 1947.

[96] K. Kaneko, Theory and Applications of Coupled Map Lattices, Chichester, 1993.

[97] E. M. Keberle and G. L. Montet, Explicit solutions of partial difference equations and random paths on plane nets, J. Math. Anal. Appl., 6(1963), 1–32.

[98] W. Kecs, The Convolution Product and Some Applications, D. Reidel Publishing Co., 1982.

[99] A. R. Khan, P. P. Choudhury, K. Dihidar and R. Verma, Text compression using two-dimensional cellular automata, Computers Math. Appl., 37(1999), 115–127.

[100] F. Koehler and C. M. Braden, An oscillation theorem for solutions of a class of partial difference equations, Proc. Amer. Math. soc., 10(1959), 762–766.

[101] H. J. Kuo and N. S. Trudinger, On the discrete maximum principle for parabolic difference operators, Math. Modelling and Numer. Anal., 27(1993), 719–737.

[102] S. A. Kuruklis, The asymptotic stability of $x_{n+1} - ax_n + bx_{n-k} = 0$, J. Math. Anal. Appl., 188(1994), 719–731.

[103] K. Kreith, Oscillation Theory, Lecture Notes in Mathematics, No. 324, Springer-Verlag, 1973.

[104] P. Laasonen, On the solution of Poisson's difference equation, J. Assoc. Comput. Mach., 5(1958), 370–382.

[105] M. Lees, Approximate solutions of parabolic equations, SIAM J. 7(1959), 167–183.

[106] T. Y. Li and J. A. Yorke, Period three implies chaos, Amer. Math. Monthly, 82(1974), 985–992.

[107] C. T. Long, On a problem in partial difference equations, Canad. Math. Bull., 13(3)(1970), 333–335.

[108] Y. Z. Lin and S. S. Cheng, Necessary and sufficient conditions for oscillations of linear partial difference equations with constant coefficients, PanAmerican Math. J., 6(1996), 61–67.

[109] Y. Z. Lin and S. S. Cheng, Stability criteria for two partial difference equations, Computers Math. Appl., 32(7)(1996), 87–103.

[110] Y. Z. Lin and S. S. Cheng, Subexponential solutions of two partial difference equations with delays, Taiwanese J. Math., 1(2)(1997), 181–194.

[111] Y. Z. Lin and S. S. Cheng, Bounds for solutions of a three-point partial difference equation, Acta Math. Scientia, 18(1)(1998), 107–112.

[112] S. T. Liu and S. S. Cheng, Nonexistence of positive solutions of a nonlinear partial difference equation, Tamkang J. Math., 28(1997), 51–58.

[113] S. T. Liu and S. S. Cheng, Existence of positive solutions of a partial difference equation, Far East J. Math. Sci., 5(1997), 387–402.

[114] S. T. Liu and H. Wang, Necessary and sufficient conditions for oscillations of a class of delay partial difference equations, Dynamic Sys. Appl., 7(1998), 495–500.

[115] M. Lotan, A problem in difference sets, Amer. Math. Monthly, 56(1949), 535–541.

[116] R. E. Lynch, J. R. Rice and D. H. Thomas, Direct solution of partial difference equations by tensor product methods, Numer. Math., 6(1964), 185–199.

[117] F. Y. Maeda, A. Murakami and M. Yamasaki, Discrete initial value problems and discrete parabolic potential theory, Hiroshima Math. J., 21(1991), 285–299.

[118] J. S. Maybee, Some structural theorems for partial difference operators, Numer. Math., 7(1965), 66–72.

[119] W. H. McCrea and F. J. W. Whipple, Random paths in two and three dimensions, Proc. Roy. Soc. Edinburgh 60(1939-40), 281–298.

[120] E. Merzrath, Direct solution of partial difference equations, Numer. Math., 9(1967), 431–436.

[121] J. Mikusinski, Operational Calculus, Vol. 1, 2nd ed. Pergamon Press, 1987.

[122] R. Miller, A Pascal triangle for the coefficients of a polynomial, Amer. Math. Monthly, 64(1957), 268–269.

[123] R. Miller, A game with n numbers, Amer. Math. Monthly, 85(3)(1978), 183–185.

[124] A. R. Mitchell and J. C. Bruch, Jr., A numerical study of chaos in a reaction-diffusion equation, Numer. Meth. Part. Diff. Eq., 1(1985), 13–23.

[125] A. R. Mitchell and D. F. Griffiths, The Finite Difference Method in Partial Differential Equations, John Wiley & Sons, 1980.

[126] D. H. Moore, Convolution products and quotients and algebraic derivatives of sequences, Amer. Math. Monthly, 69(1962), 132–138.

[127] D. H. Mugler, Green's functions for the finite difference heat, Laplace and wave equations, International Series of Numerical Mathematics, Vol. 65, 543–553, Birkhauser Verlag Basel, 1984.

[128] D. H. Mugler and R. A. Scott, Fast Fourier transform method for partial differential equations, case study: The 2-D diffusion equation, Computers Math. Applic., 16(3)(1988), 221–228.

[129] A. Murakami and M. Yamasaki, An introduction of Kuramochi boundary of an infinite network, Mem. Fac. Sci. Eng. Shimane Univ. Series B: Math. Sci., 30(1997), 57–89.

[130] A. Murakami, M. Yamasaki, Y. Yone-e, Some properties of reproducing kernels on an infinite network, Mem. Fac. Sci. Shimane Univ., 28(1994), 1–8.

[131] I. Niven, Formal power series, Amer. Math. Monthly, 76(1969), 871–889.

[132] S. Okamoto, A difference approach to Mikusinski's operational calculus, Proc. Japan Acad., 54, Ser. A (1978), 303–309.

[133] J. Ortega, Stability of difference equations and convergence of iterative processes, SIAM J. Numer. Anal., 10(1973), 268–282.

[134] C. V. Pao, Positive solutions and dynamics of a finite difference reaction-diffusion system, Numerical Methods for Partial Diff. Eq., 9(1993), 285–311.

[135] C. V. Pao, Block monotone iterative methods for numerical solutions of nonlinear elliptic equations, Numer. Math., 72(1995), 239–262.

[136] C. V. Pao, Finite difference reaction diffusion equations with nonlinear boundary conditions, Numer. Methods Partial Diff. Eq., 11(1995), 355–374.

[137] C. V. Pao, Monotone iterations for numerical solutions of reaction-diffusion-convection equations with time delay, 14(1998), 339–351.

[138] C. V. Pao, Dynamics of a finite difference system of reaction diffusion equations with time delay, J. Difference Eq. Appl., 4(1998), 1–11.

[139] H. B. Phillips and N. Wiener, Nets and the Dirichlet problem, J. Math. Phy., 2(1923), 105–124.

[140] R. B. Potts, Ordinary and partial difference equations, J. Austral. Math. Soc. Ser. B, 27(1986), 488–501.

[141] M. H. Protter and H. F. Weinberger, Maximum Principles in Differential Equations, Prentice Hall, 1967.

[142] M. Razpet, An application of the umbral calculus, J. Math. Anal. Appl., 149(1990), 1–16.

[143] J. Riordan, Generating Functions, Chapter 3 in Applied Combinatorial Mathematics, ed. by E. F. Beckenbach, Wiley, New York, 1964.

[144] J. Riordan, Combinatorial Identities, Wiley, New York, 1968.

[145] C. Rorres and H. Anton, Applications of Linear Algebra, John Wiley & Sons, 1984.

[146] A. Rosenfeld, Digital topology, Amer. Math. Monthly, 86(8)(1979), 621–630.

[147] Y. Shogenji and M. Yamasaki, Hardy's inequality on networks, preprint.

[148] Q. Sheng and R. P. Agarwal, Monotone methods for higher-order partial difference equations, Computers Math. Appl., 28(1-3)(1994), 291–307.

[149] H. J. Stetter, Maximum bounds for the solutions of initial value problems for partial difference equations, Numer. Math., 5(1963), 399–424.

[150] R. F. Streater, A bound for the difference Laplacian, Bull. London Math, Soc., 11(1979), 354-357.

[151] J. C. Strikwerda, Finite Difference Schemes and Partial Differential Equations, Wadsworth, 1989.

[152] V. Thomee, Stability theory for partial difference operators, SIAM Review, 11(1969), 152–195.

[153] C. J. Tian, S. L. Xie and S. S. Cheng, Measures for oscillating sequences, Computers Math. Appl., 36(10–12)(1998), 149–161.

[154] C. J. Tian and B. G. Zhang, Frequent oscillation of a class of partial difference equations, J. Anal. Appl., 18(1)(1999), 111–130.

[155] D. Ullman, More on the four numbers game, Math. Mag., 65(1992), 170–174.

[156] R. Varga, Matrix Iterative Analysis, Prentice Hall, Englewood cliffs, N.J. 1962.

[157] J. Veit, Subexponential solutions of multidimensional difference equations, Mutidimensional Systems and Signal Processing, 8(1997), 365–385.

[158] R. Vich, Z Transform Theory and Applications, Reidal Publishing company, 1987.

[159] Z. X. Wang, The solution of a nonlinear partial difference equation, Discrete Applied Mathematics, 28(1990), 177–181.

[160] J. J. Wachholz and J. C. Bruch, Jr., An investigation of chaos in reaction-diffusion equations, Numer. Math. Part. Diff. Eq., 3(1987), 139–168.

[161] D. V. Widder, The Heat Equation, Academic Press, New York, 1975.

[162] S. Wolfram, Theory and Applications of Cellular Automata, World Scientific, 1986.

[163] P. J. Wong and R. P. Agarwal, Oscillation criteria for nonlinear partial difference equations with delay, Computers Math. Applic., 32(6)(1996), 57–86.

[164] P. J. Wong and R. P. Agarwal, Nonexistence of unbounded nonoscillatory solutions of partial difference equations, J. Math. Anal. Appl., 214(1997), 503–523.

[165] S. L. Xie and S. S. Cheng, Stability for parabolic type partial difference equations, J. Comp. Appl. Math., 75(1996), 57–66.

[166] M. Yamasaki, The equation $\Delta u = qu$ on an infinite network, Mem. Fac. Sci. Shimane Univ., 21(1987), 31–46.

[167] M. Yamasaki, Nonlinear Poisson equations on an infinite network, Mem. Fac. Sci. Shimane Univ., 23(1989), 1–9.

[168] C. G. Yu, An explicit formula of solution for a kind of the homogeneous recurrence relation of constant coefficients with two indices, Acta Math. Appl. Sinica, 20(1)(1997), 119–127.

[169] B. G. Zhang and S. S. Cheng, A discrete boundary value problem with delay, Bull. Inst. Math. Acad. Sinica, 23(1995), 305–315.

[170] B. G. Zhang and S. T. Liu, Oscillation of partial difference equations, PanAmerican Math. J., 5(1995), 61–70.

[171] B. G. Zhang and S. T. Liu, Necessary and sufficient conditions for oscillations of delay difference equations, Discussiones Mathematicae-Differential Inclusions, 15(1995), 213–219.

[172] B. G. Zhang and S. T. Liu, On the oscillation of partial difference equations, J. Math. Anal. Appl., 206(1997), 480–492.

[173] B. G. Zhang, S. T. Liu and S. S. Cheng, Oscillation of a class of delay partial difference equations, J. Difference Eq. Appl., 1(1995), 215–226.

[174] B. G. Zhang and C. J. Tian, Stability criteria for a class of linear delay partial difference equations, Computers Math. Appl., 38(1997), 37–43.

[175] G. Zhang and S. S. Cheng, Elementary nonexistence criteria for a recurrence relation, Chinese J. Math., 24(1996), 229–235.

[176] Y. L. Zhou, Applications of Discrete Functional Analysis to the Finite Difference Method, International Academic Publishers, 1991.

[177] W. Zhuang, Y. Chen and S. S. Cheng, Monotone methods for a discrete boundary value problem, Comput. Math. Appl., 32(12)(1996), 41–49.

Index

Milton Keynes UK
Ingram Content Group UK Ltd.
UKHW040444071024
449327UK00020B/979

9 780367 395476